SPACECRAFT
SYSTEMS
ENGINEERING

SPACECRAFT SYSTEMS ENGINEERING

Edited by

Peter W. Fortescue

and

John P. W. Stark
Department of Aeronautics and Astronautics
University of Southampton, UK

JOHN WILEY AND SONS
Chichester • New York • Brisbane • Toronto • Singapore

Reprinted with corrections February 1992
Reprinted October 1992
Reprinted September 1994

Other Wiley Editorial Offices

John Wiley & Sons, Inc., 605 Third Avenue,
New York, NY 10158-0012, USA

Jacaranda Wiley Ltd, G.P.O. Box 859, Brisbane,
Queensland 4001, Australia

John Wiley & Sons (Canada) Ltd, 22 Worcester Road,
Rexdale, Ontario M9W 1L1, Canada

John Wiley & Sons (SEA) Pte Ltd, 37 Jalan Pemimpin 05-04,
Block B, Union Industrial Building, Singapore 2057

Library of Congress Cataloging-in-Publication Data:

Spacecraft systems engineering / edited by Peter W. Fortescue and John
P.W. Stark.
p. cm.
Includes bibliographical references and index.
ISBN 0 471 92794 5
1. Space vehicles—Design and construction. I. Fortescue, Peter
W. II. Stark, John P.W.
TL875.S68 1990
629.47′4—dc20
90-12599
CIP

British Library Cataloguing in Publication Data:

Spacecraft systems engineering.
1. Space vehicles. Systems engineering
I. Fortescue, Peter W. II. Stark, John P.W.
629.474
ISBN 0 471 92794 5
ISBN 0 471 93451 8 (pbk)

Typeset by Asco Trade Typesetting Ltd, Hong Kong
Printed in Great Britain by Redwood Books, Trowbridge, Wiltshire

CONTENTS

LIST OF CONTRIBUTORS

EDITORS

Peter W. Fortescue,
Department of Aeronautics and Astronautics,
University of Southampton, U.K.

Dr John P.W. Stark,
Department of Aeronautics and Astronautics,
University of Southampton, U.K.

AUTHORS

Professor Robin A. East,
Department of Aeronautics and Astronautics,
University of Southampton, U.K.

Dr Ross A. Henderson,
Department of Aeronautics and Astronautics,
University of Southampton, U.K.

Dr Richard Holdaway,
British National Space Centre,
Rutherford Appleton Laboratory, U.K.

Thomas A. Meaker,
European Space Agency.

Professor J. Barrie Moss.
Department of Propulsion. Power and Automative Engineering,
Cranfield Institute of Technology, U.K.

Julian A. Robinson,
Marconi Space Systems, U.K.

Dr Howard Smith,
Department of Electrical and Electronic Engineering,
Portsmouth Polytechnic, U.K.

Raymond F. Turner,
British National Space Centre,
Rutherford Appleton Laboratory, U.K.

Les J.C. Woolliscroft,
Department of Control Engineering,
University of Sheffield, U.K.

PREFACE

This book has grown out of a set of Course Notes which accompany a series of short courses given at Southampton University. These courses started in 1974 with a two-week 'space technology' course, and they are aimed at the recent science or engineering graduate who wishes to become a spacecraft engineer. The courses are still thriving, now serving much of European industry, with one-week versions for experienced engineers, sometimes senior ones, who are specialists in their own fields.

On the courses the attendees work in competing teams on a project which involves designing a spacecraft in response to an overall objective. Over the years mission designs have been directed at all application areas: science, astronomy, communications, and Earth observations. There is now a 'museum' of models which demonstrate vehicle layouts and support the attendees' presentations covering operation, subsystem specification, and launch constraints. These models demonstrate system viability rather than detailed design. The projects are designed at 'system level', and their supervision has provided a basis for deciding the level of detail which should be included in this book.

The coverage in this book is therefore aimed at giving the breadth which is needed by system engineers, with an emphasis on the bus aspect rather than on the payload. The specialist engineer is well served with textbooks which cover many of the subsystems in detail and in depth. He is unlikely to learn very much about his own specialist topic from this book. But he may well learn something about other specialists' disciplines, and, it is hoped, enough for him to appreciate the trade-offs which affect his own subsystem in relation to others.

Chapters 2–5 set the general scene for spacecraft, and particularly for satellites. They must operate in an environment which is generally hostile compared to that with which we are familiar on Earth, and the main features of this are described in Chapter 2. Chapters 3 and 4 address the dynamics of objects in space, where the vehicles will respond to forces and moments which are minute, and which would be discounted as of no significance if they occurred on Earth. Indeed, most of them do occur here, but we do not often operate in a fully free state, and our Earth-bound vehicles are subject to other, much larger forces. Chapter 5 relates the motion of the spacecraft to Earth rather than to the inertially based reference system of celestial mechanics.

Chapters 6–15 address the main subsystems. Chapters 7 and 8 cover the subjects of getting off the ground and return through the atmosphere. Chapters 6, 9–12, and 14 deal with the main subsystems on board the spacecraft, including the on-board end of the telemetry and control link (Chapter 14) with ground control (Chapter 15). The communication link is covered in Chapter 13, where the fundamentals of the subject are included together with their rather special application to spacecraft. This is relevant to the telemetry and control link, and to a communications payload.

Chapter 16 introduces electromagnetic compatibility (EMC), one of the subjects which must be addressed by the systems engineer if the various subsystems are to work in harmony.

Product assurance is of vital concern to spacecraft engineers. Their product(s) must survive a hostile launch environment and then must last many years without the luxury of any maintenance. It does great credit to the discipline they exercise, that so many of their products do so.

We editors would like to express our thanks to the authors who have contributed chapters in the book. Most of them have lectured on the courses mentioned above. Our task has been to whittle down the material they have provided since they have been very generous. We are grateful too for their patience. The conversion of course notes into a book was expected to be a short process. How wrong we were!!

We would also like to thank colleagues Graham Swinerd and Adrian Tatnall, who read some of the texts and gave advice. And finally our thanks to Sally Mulford, who has converted some much-abused text into typescript, with patience and good humour.

LIST OF ACRONYMS

ABM	Apogee boost motor
ACS	Attitude control system
AGC	Automatic gain control
AKM	Apogee kick motor
AM	Amplitude modulation
AMOOS	Aero-manoeuvring orbit-to-orbit shuttle
AOCS	Attitude and orbit control system
AOTV	Aero-assisted orbital transfer vehicle
ASK	Amplitude-shift keying
ASW	Address and synchronization word
BCDT	Binary code data transfer
BER	Bit error rate
BRTS	Bilateration ranging transponder system
BSF	Back-surface field
BSR	Back-surface reflector
C & DH	Control & data handling
CCB	Configuration control board
CCIR	Comité Consultatif International de Radiocommunication
CCITT	Comité Consultatif International de Téléphonie et de Télégraphie
CCSDS	Consultative Committee for Space Data Systems
CDMA	Code-division multiple access
CMG	Control moment gyroscope
CMOS	Complementary metal oxide semiconductor
COMSAT	Communications satellite
CR	Corrosion resistance
CSG	Centre Spatial Guyanais
CVCM	Collected volatile condensible materials
CW	Continuous Sine-wave
DPL	Declared parts list
DPSK	Differential phase-shift keying
DSBSC	Double side-band suppressed carrier modulation
EIRP	Equivalent isotropic radiated power
EMC	Electro-magnetic compatibility
EMI	Electro-magnetic interference
ER-MIL	Established reliability-MIL
ERS	Earth resources satellite
ESA	European Space Agency

FDMA	Frequency-division multiple access
FE	Finite element
FEA	Finite element analysis
FEEP	Field emission electric propulsion
FET	Field effect transistor
FFSK	Fast frequency-shift keying
FM	Frequency modulation
FMECA	Failure mode effects and criticality analysis
FRSI	Flexible reusable surface insulation
FSK	Frequency-shift keying
GEO	Geostationary Earth orbit
GPS	Ground positioning system
GSFC	Goddard Space Flight Center
GTO	Geostationary transfer orbit
HEO	Highly elliptical orbits
HGAS	High gain antenna system
HOTOL	Horizontal take-off and landing
HRSI	High-temperature reusable surface insulation
HST	High speed telemetry
IC	Integrated circuit
ICBM	Inter-continental ballistic missile
IF	Intermediate frequency
IFRB	International Frequency Registration Board
IRAS	Infra-red Astronomical Satellite
ITU	International Telecommunications Union
KSA	K-band steerable antenna
KSC	Kennedy Space Center
LAM	Liquid apogee motor
LEO	Low Earth orbit
LNA	Low noise amplifier
LO	Local oscillator
LRSI	Low-temperature reusable surface insulation
LST	Low speed telemetry
MA	Multiple access
MFR	Multi-function receiver
MMS	Multi-mission modular spacecraft
MOP	Multiple on-line peripherals
MOS	Metal oxide semiconductor
MOSFET	Metal oxide semiconductor field effect transistor
MPD	Magneto-plasma-dynamic
MTBF	Mean time between failures
MW	Momentum wheel
N-MOS	N-type metal oxide semiconductor
NASA	National Aeronautics and Space Administration

NBFM	Narrow-band frequency modulation
NDT	Non-destructive testing
NERVA	Nuclear engine for rocket vehicle applications
OBC	On-board computer
OBDH	On-board data handling
OCC	Operations control centre
OMS	Orbital manoeuvring subsystem
OSR	Optical solar reflector
OTA	Optical telescope assembly
OTV	Orbital transfer vehicle
P-MOS	P-type metal oxide semiconductor
PA	Product assurance
PAEHT	Power-augmented hydrazine thruster
PAM	Payload assist module
PAM-A	Payload assist module – Atlas-sized
PAM-D	Payload assist module – Delta-sized
PCM	Pulse code modulation
PDF	Probability density function
PID	Proportional, integral and differential
PIM	Passive intermodulation products
PM	Phase modulation
PN	Pseudo-random noise
PPL	Preferred parts list
PRK	Phase-reversal keying
PSK	Phase-shift keying
QPL	Qualified parts list
QPSK	Quadrature phase-shift keying
RAM	Random access memory, and also radio-frequency anechoic material
RCC	Reinforced carbon–carbon
RCE	Reaction control equipment
RF	Radio frequency
RIG	Rate-integrating gyroscope
RMS	Remote manipulator system
ROM	Read-only memory
RS	Reed-Solomon
RTG	Radioisotope generator
RW	Reaction wheel
S^3R	Sequential switching shunt regulation
SAS	Solar array system
SAW	Surface acoustic wave
SBE	S-band exciter
SCC	Stress corrosion cracking
SCPC	Single channel per carrier
SCRAMJET	Supersonic combustion ramjet
SGL	Space-to-ground link
SMM	Solar maximum mission

SNAP-19	System for nuclear auxiliary power
SNR	Signal-to-noise ratio
SOP	Spacecraft overhead pass
SPE	Solid polymer electrolyte
SPELDA	Structure Porteuse pour Lancement Double Ariane
SPS	Satellite power system
SRE	Spacecraft ranging equipment
SSA	S-band steerable antenna
SSB	Single side-band
SSM	Second surface mirror
SSMA	Spread-spectrum multiple access
SSME	Space shuttle main engine
SSUS	Solid spinning upper stage
STDN	Spaceflight Tracking and Data Network
STS	Space Transportation System
SYLDA	Systeme de Lancement Double Ariane
TC & R	Telemetry, command and ranging
TDMA	Time-division multiple access
TDPS	Tracking and data processing station
TDRS	Tracking and data relay satellite
TDRSS	Tracking and data relay satellite system
TIU	Time interval unit
TML	Total mass loss
TPS	Thermal protection subsystem
TT & C	Tracking, telemetry and control
TTL	Transistor–transistor logic
TWT	Travelling wave tube
TWTA	Travelling wave tube amplifier
UARS	Upper Atmosphere Research Spacecraft
USB	Upper side-band
USNO	US Naval Observatory
UT	Universal Time
VCDU	Virtual channel data unit
VCHP	Variable-conductance heat pipe
VCO	Voltage controlled oscillator
VCXO	Voltage controlled crystal oscillator
WARC	World Administrative Radio Conference
WBFM	Wide-band frequency modulation
WSGT	White Sands Ground Terminal
WTR	Western Test Range
XPD	Cross-polar discrimination

1 *INTRODUCTION*

Peter W. Fortescue and John P.W. Stark

Department of Aeronautics and Astronautics, University of Southampton

Man has only had the ability to operate spacecraft successfully since 1957, when the Russian Sputnik I was launched into orbit. In a few decades the technology has made great strides, to the extent that the Americans' manned expedition to the Moon and back is already history. In little more than 30 years unmanned explorer spacecraft have flown past all the major bodies of the solar system except for Pluto. Vehicles have landed on the Moon, Venus and Mars. A spacecraft is now on its way to 'land' on Jupiter and initial plans for a lander mission to Titan, one of Saturn's moons, are laid.

Many countries have the capability of putting spacecraft into orbit; satellites have now established a firm foothold as part of the infrastructure of society. There is every expectation that they have much more to offer in the future.

Before the twentieth century space travel was largely a flight of fancy. Most authors during that time failed to understand the nature of a spacecraft's motion, and this resulted in the idea of 'lighter-than-air' travel for most would-be space-farers [1, 2]. At the turn of the century however, a Russian teacher, K.E. Tsiolkovsky, laid the foundation stone for rocketry by providing insight into the nature of propulsive motion. In 1903 he published a paper in the *Moscow Technical Review* deriving what we now term the rocket equation, or Tsiolkovsky's equation (equation 3.17). Due to the small circulation of this journal the results of his work were largely unknown in the West prior to the work of Hermann Oberth which was published in 1923.

These analyses provided an understanding of propulsive requirements, but they did not provide the technology. This eventually came, following work by R.H. Goddard in America and Wernher von Braun in Germany. The Germans demonstrated their achievements with the V-2 rocket which they used towards the end of World War 2. Their rockets were the first reliable propulsive systems, and whilst they were not capable of placing a vehicle into orbit, they could deliver a warhead of approximately 1000 kg over a range of 300 km. It was largely the work of these same German engineers which led to the first successful flight of Sputnik I on 4 October 1957, closely followed by the first American satellite, Explorer I, on 31 January 1958.

Three decades have seen major advances in space technology. It has not always been

Spacecraft Systems Engineering. Edited by P.W. Fortescue and J.P.W. Stark
© 1991 John Wiley & Sons Ltd

smooth, as evidenced by the major impact that the Challenger disaster had on the American space programme. Technological advances in many areas have however been achieved. Particularly notable are the developments in energy-conversion technologies— especially solar photovoltaics, fuel cells and batteries. Developments in heat-pipe technology have also occurred in the space arena, with ground-based application to the oil pipelines of Alaska [3] as a spin-off. Perhaps the most notable developments in this period, however, have been in electronic computers and software. Although these have not necessarily been driven by space technology, the new capabilities which they afford have been rapidly assimilated, and they have revolutionized the flexibility of spacecraft. In some cases they have even turned a potential mission failure into a grand success, as evidenced by Voyager 2.

But the spacecraft has also presented a challenge to Man's ingenuity and understanding. Even something as fundamental as the unconstrained rotational motion of a body is now better understood as a consequence of placing a spacecraft's dynamics under close scrutiny. Man has had to devise designs for spacecraft which will withstand a hostile space environment; and he has come up with many solutions, not just one.

1.1 PAYLOADS AND MISSIONS

Payloads and missions for spacecraft are many and varied. Some have reached the stage of being economically viable, such as satellites for communications purposes. Others play an important part as a service to mankind, such as weather and navigation satellites. Yet others monitor Earth for its resources, the health of its crops, and pollution. Of vast importance to mankind has been the identification of ozone 'holes' over both poles of our planet. Other satellites serve the scientific community of today and perhaps the layman of tomorrow by adding to Man's knowledge of the Earth's environment, the solar system and the universe.

Each of these peaceful applications is paralleled by inevitable military ones. By means of global observations both superpowers acquire knowledge of military activities on the surface of the planet, and the deployment of aircraft. Communications satellites serve the military user, as do weather satellites. In the future, following the full deployment of the Navstar/GPS navigational satellite constellation, an infantryman, sailor or fighter pilot will know his location to an accuracy of about a metre.

Table 1.1 presents a list of payloads/missions with an attempt at placing them into categories based upon the types of trajectory they may follow. The satellites may be categorized in number of ways, such as by orbit altitude, eccentricity or inclination.

It is important to note that the specific orbit adopted for a mission will have a strong impact on the design of the vehicle, as illustrated in the following paragraphs.

Consider geostationary (GEO) missions; these are characterized by the vehicle having a fixed position relative to the features of the Earth. The propulsive requirement to achieve such an orbit is large, and thus the 'dry mass' (exclusive of propellant) is only a modest fraction of the all-up 'wet mass' of the vehicle. With the cost per kilogram-in-orbit being as high as it currently is—of the order $50 000 per kilogram in geostationary orbit—it usually becomes necessary to optimize the design to achieve minimum weight, and this leads to a large number of vehicle designs, each suitable only for a narrow range of payloads and missions.

Table 1.1 Payload/mission types

Mission	Trajectory type
Communications	Geostationary for low latitudes, Molniya for high latitudes (mainly Russian)
Earth resources	Polar LEO for global coverage
Weather	Polar LEO, or geostationary
Navigation	Polar LEO for global coverage
Astronomy	Various high altitude
Space environment	Various, including sounding rockets
Military	Polar LEO for global coverage, but various
Space stations	LEO
Technology proving	Various

Considering the communication between the vehicle and the ground it is evident that the large distance involved means that the received power is considerably less than the transmitted value. But the vehicle is continuously visible at its ground control station, and this enables its health to be monitored continuously, and reduces the need for it to be autonomous or to have a complex data handling/storage system.

Low Earth orbit (LEO) missions are altogether different. Communication with such craft is more complex due to the intermittent nature of ground station passes. This has resulted in the development of a new breed of spacecraft—the tracking and data relay satellite (TDRS)—operating in GEO to provide a link between craft in LEO and a ground centre.

The power subsystem is also notably different between LEO and GEO satellites. A dominant feature is the relative period spent in sunlight and eclipse in these orbits. LEO is characterized by a high fraction of the orbit being spent in eclipse, and hence a need for substantial oversizing of the solar array to meet battery-charging requirements. In GEO on the other hand, a long time (up to 70 minutes) spent in eclipse leads to deep discharge requirements on the battery, although the eclipse itself is only a small fraction of the total orbit period. Additional differences in the power system are also partly due to the changing solar aspect angle to the orbit plane during the course of the year. This may be off-set, however, in the case of the sun-synchronous orbit (see Section 5.3), which maintains a near-constant aspect angle. (This is not normally done for the benefit of the spacecraft bus designer, but rather because it leads to constant illumination of the ground at the subsatellite point, enabling the passive instruments to function effectively.)

It soon becomes clear that changes of mission parameters of almost any type have potentially large effects upon the specifications for the subsystems which comprise and support a spacecraft.

1.2 A SYSTEM VIEW OF SPACECRAFT

This book is concerned with spacecraft systems. The variety of types and shapes of these is extremely wide. When considering spacecraft it is convenient to subdivide them into functional elements or subsystems. But it is also important to recognize that the satellite itself is only an element within a larger system. There must be a supporting ground control

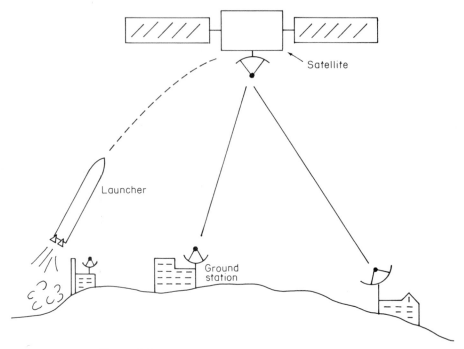

Figure 1.1   The spacecraft and attendant subsystems

system (Figure. 1.1), which enables commands to be sent up to the vehicle and status information to be returned to the ground. There must also be a launcher system which sets the vehicle on its way to its final orbit. Each of the elements of the overall system must interact with the other elements, and it is the job of the system designer to achieve an overall optimum in which the mission objectives are realized efficiently.

Turn now to the spacecraft system itself. This may be divided conveniently into two principal elements, the payload, and the bus. It is of course the payload which is the motivation for the mission itself. In order that this may function it requires certain resources which will be provided by the bus. In particular it is possible to identify the following functional requirements:

1. The payload must be pointed in the correct direction.

2. The payload must be operable.

3. The data from the payload must be communicated to the ground.

4. The desired orbit for the mission must be maintained.

5. The payload must be held together, and on to the platform on which it is mounted.

6. The payload must operate and be reliable over some specified period of time.

7. An energy source must be provided, to enable the above functions to be performed.

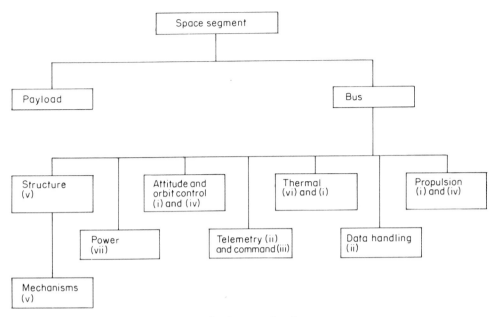

Figure 1.2 Spacecraft subsystems

These requirements lead on to the breakdown into subsystems which is shown in Figure 1.2. Inset in each of these is a number which relates it to the functions above.

One facet of these subsystems is that the design of any one has impacts and resource implications on the other. A most important feature of spacecraft system design is to identify what aspects of the mission and what elements of the design provide the major influences on the type of satellite which may meet the specific mission requirements. This process is the identification of the 'design drivers'. In some cases the drivers will affect major features of the spacecraft hardware. The varied mission requirements, coupled with the need to minimize mass and hence power, has thus led to a wide variety of individual design solutions being realized. However the spacecraft industry is now evolving towards greater standardization—in the shape of the specific buses which may be used to provide the resources for a variety of missions (for example the Eurostar bus of British Aerospace and Matra).

It is not simply the nature of its payload which determines the design which is selected for a given mission, although this will have a considerable influence. Commercial and political influences are strongly felt in spacecraft engineering. Individual companies have specialist expertise; system engineering is dependent on the individual experience within this expertise. This is perhaps most notably demonstrated by the Hughes Company, which has advanced the art of the spin-stabilized satellite. Spacecraft systems engineering is not all science—there is indeed an art to the discipline.

This leads to another major feature of spacecraft system design, namely the impact which reliability has. The majority of terrestrial systems may be maintained, and their reliability, whilst being important, is not generally critical to their survival. If a major component fails the maintenance team can be called in. In space this luxury is not afforded—the system must be fault-tolerant, and when this tolerance is exceeded the system is no longer operable and the mission has ended.

There are two principal methods used to obtain high reliability. The first is to use a design which is well proven. This is true for both system and component selection. The requirement for environmental compatibility validation of components (Chapter 2) leads to relatively old types being used in mature technology, especially in electronic components. This tends to lead to a greater demand for power than the terrestrial 'state-of-the-art' technology does. At system level a 'tried and tested' solution will minimize development risk, reducing system cost whilst also achieving high reliability.

The second method of achieving high reliability is via de-rating (Chapter 17). By reducing the power of the many electronic components, for example, a greater life expectancy can be obtained. This leads to an overall increase in mass.

The net effect of designing for high reliability is that spacecraft design is conservative—'if it has been done before then so much the better'. Much of satellite design is thus not state-of-the-art technology. Design teams evolve a particular design solution to meet varied missions—because it is a design they understand—and hence system design is an art as well as a science.

In making the selection of subsystems for his spacecraft the designer must have a good grasp of the way in which the subsystems work, and the complex interactions between them, and he must recognize how the craft fits into the larger system. He must be able to trade off advantages in one area with the disadvantages in another and achieve a balance in which the end result will work as a harmonious whole. Whilst each subsystem will have its own performance criterion, nevertheless its performance must be subordinated to that of the system as a whole.

1.3 THE FUTURE

We are at present approaching a new frontier in space. We have been able to demonstrate no more than a rudimentary access to the space environment—our regular utilisation of it is still limited. Beyond the frontier we shall have an established space infrastructure, including the prime elements of transportation and communications, with a permanent presence of man in space—initially on space stations, but perhaps also on the Moon. It is interesting to note that if US funding had continued at the same rate as it did during the Apollo missions then a manned base on the Moon would most probably already exist.

Clearly, to set up the infrastructure there must be a transportation system. The first stage, getting off the ground, already sees the vertical-launch reusable-vehicle (Shuttle) competing with expendable launchers, and it in turn is likely to be overtaken by horizontal-take-off-and-landing (HOTOL) versions. Already retrieval of spacecraft from LEO has been demonstrated and further developments in two-way traffic between ground and LEO may be expected, with reductions in the cost of this operation.

Staging posts, where 'a new team of horses' can be obtained, are perhaps not too far away. Surely staging posts will eventually become assembly and servicing posts too, so that spacecraft do not all have to be designed to withstand the full rigours of launch (Section 2.2) when their subsequent stages of travel can be relatively stress-free. Perhaps manufacturing in space will be a way of proceeding, not only for exotic materials such as are currently being considered, but for lightweight structural materials extruded in zero gravity, for use in zero gravity. A communications infrastructure is already in being (TDRSS). There will need to be accommodation units; and we see greater and greater knowledge of man's ability to live in space being gathered each year. There needs to be a power generating and supply system, ... and so on.

This new frontier will be achieved only if it is the will of the people of this planet, and the politicians commit the resources apolitically for durations far greater than the term of office of any one administration. Perhaps the energies of our engineers and the taxes of the people could be of more benefit in the furtherance of a space infrastructure—rather than in the weaponry of mass destruction and military prowess.

There is a whole new exciting arena waiting to be explored, occupied and used for the benefit of terrestrial mankind.

REFERENCES

[1]   de Bergerac, Cyrano (1649) *Voyage dans la lune.*
[2]   Winter, F.H. (1983) *Prelude to the Space Age*, Smithsonian
[3]   Briscoe, M. and Toussaint, J., (1989) *European Space Technology*, ESA BR-55.

THE SPACE ENVIRONMENT AND ITS EFFECTS ON SPACECRAFT DESIGN

2

John P.W. Stark

Department of Aeronautics and Astronautics, University of Southampton

2.1 INTRODUCTION

Spacecraft operation is characterized by its remoteness from the Earth and thus the loss of the Earth's protective shield, namely the atmosphere. This atmosphere evidently provides a suitable stable environment in which the human species has been able to evolve. Coupled with the gravitational force of the Earth, 'the one-g environment', it provides familiarity in design and its removal has significant and sometimes unexpected implications. The aim of this chapter is to introduce the reader to the nature of the spacecraft's environment and the implications that it has on spacecraft design by considering both the way materials behave and the way in which systems as a whole are influenced. The final section looks at the implications for manned spaceflight.

Before considering the environment in detail it should be noted that the different phases in the life of a space vehicle, namely manufacture, pre-launch, launch, and finally space operation, all have their own distinctive features. Although a space vehicle spends the majority of its life in space it is evident that it must survive the other environments for complete success. Whilst the manufacturing phase is not specifically identified in the following section it has an effect upon the reliability and the ability to meet design goals. Cleanliness, humidity and codes of practice are critical in the success of spacecraft missions, and these are dealt with separately, in Chapter 17.

Spacecraft Systems Engineering. Edited by P.W. Fortescue and J.P.W. Stark
© 1991 John Wiley & Sons Ltd

2.2 SPACECRAFT ENVIRONMENTS

2.2.1 Pre-launch environment

The design, manufacture and assembly of a spacecraft, and its final integration into a launch vehicle is a lengthy process, lasting typically 5–10 years. Components and sub-systems may be stored for months or even years prior to launch (for example the Ulysses mission). Careful environmental control during such periods of time is essential if degradation of the spacecraft system as a whole is to be avoided.

2.2.2 The launch phase

From an observer's viewpoint, the launch of a space vehicle is evidently associated with gross noise levels. This impinges itself on the structure of a satellite contained within the launcher's shroud. The launch sequence itself provides high levels of vibration, associated both with the noise field and structural vibration, modest-to-high levels of acceleration during ascent, mechanical shock due to pyrotechnique device operation, a

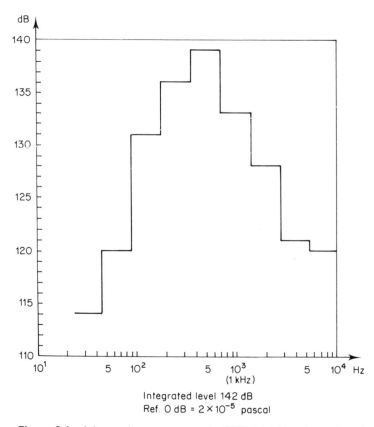

Integrated level 142 dB
Ref. 0 dB = 2×10^{-5} pascal

Figure 2.1   Ariane noise spectrum under SPELDA fairing (reproduced by permission of Arianespace [2])

thermal environment which differs from both laboratory and space environments, and for most launch vehicles, a rapidly declining ambient pressure. These features are described separately below.

The *severe acoustic/vibration environment* during launch is due to both the operation of the launch vehicle's main engines, and also the aerodynamic buffeting as the vehicle rises through the lower region of the earth's atmosphere. Two peak levels occur.

The first peak occurs at the moment of lift-off. The overall build-up of the rocket motor firing and the exhaust products reflected from the ground reaches a peak at launcher release. During ascent the contribution from ground reflection decreases, but a variety of mechanical components, such as liquid fuel turbopump operation, continue to excite the overall launch vehicle structure. The vibration is not only directly transmitted to the spacecraft through structural components but it also excites the launch shroud to generate a secondary acoustic field. For light, flexible components such as the solar array, the acoustic environment may be more severe than the mechanically induced vibration [1].

The second peak in the acoustic field occurs during transonic flight. The launch shroud is again excited, this time by the unsteady flow field around the vechicle.

Measurement of the acoustic field is generally made with reference to a pressure of 2×10^{-5} Pa. The sound pressure level I is then measured in decibels, given by

$$I = 20 \log_{10}(F/2 \times 10^{-5}) \, \mathrm{dB}$$

where F is the acoustic field intensity. The frequency spectrum of the noise field will be

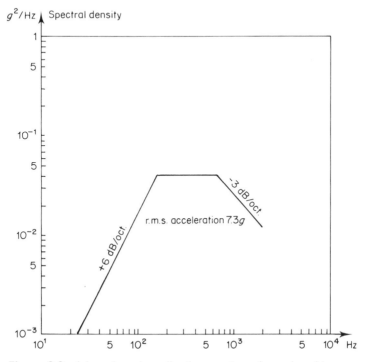

Figure 2.2   Ariane 4 random vibration spectrum (reproduced by permission of Arianespace [2])

dependent on the launch vehicle, and will change during the launch sequence. Data is provided for design purposes, thus enabling the designer to quantify vibrational inputs that individual components and the whole space vehicle will experience. The design noise spectrum for Ariane 4 is indicated in Figure 2.1 [2], and for comparison the random vibration is indicated in Figure 2.2.

The *steady component of launch acceleration* must achieve a speed increase of about 9.5 km/s. Its time history is dependent on the launch vehicle used. Low-mass payload

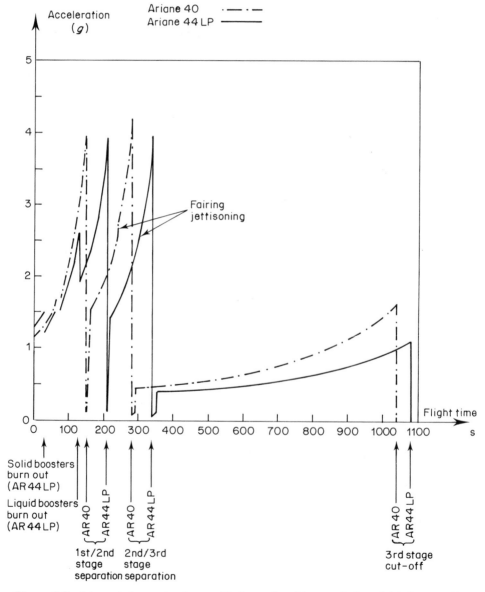

Figure 2.3   Ariane static acceleration profile (reproduced by permission of Arianespace [2])

vehicles, such as Scout, sounding rockets, and missiles generally exhibit high peak acceleration, whereas those of large payload/manned vehicles are much smaller. Thus the peak for the Skylark sounding rocket is $\sim 13.5g_0$ whereas for Shuttle it is less than $3g_0$ (g_0 is the suface gravitational acceleration, 9.8 m/s^2). There are significant deviations from this general rule, however; for example the peak acceleration test level required for the solar array design on Saturn/Apollo was $8-12g_0$. For multi-stage vehicles the acceleration increases during the burn of each stage, and peaks at burn-out/stage-separation. This is demonstrated for Ariane 4 in Figure 2.3.

For manned flight it is necessary to place the astronaut in a suitable position to withstand peak acceleration levels; Figure 2.4 shows the typical maximum levels which may be withstood. Greatest protection is provided when the astronaut is on his back during launch.

Mechanical shock is experienced when devices such as latches or explosive bolts are used, or at ignition of rocket motor stages and their subsequent separation, launch vehicle/payload separation, or when docking or landing.

These instantaneous events can provide extremely high acceleration levels lasting only a few milliseconds locally, and in some cases to the complete system. Their frequency spectrum is characterized by high-frequency components. In the case of Ariane during payload separation the peak excitation that the satellite must survive is some $2000g_0$ at frequencies above 1.5 kHz.

The *thermal environment* experienced during launch is determined generally by the temperature reached by the launch shroud. Its high temperature arises from the aerodynamic frictional forces of the vehicle moving at high velocity through the atmosphere. The temperature reached is determined by the specific heat of the shroud material and a balance between friction heating and radiative and convective heat losses. The subsequent temperature rise of the payload within the shroud is dominated by radiative and heat

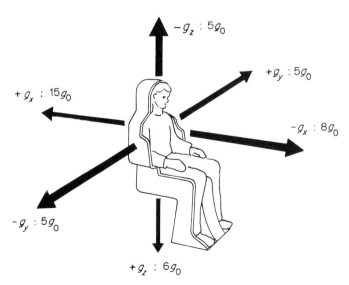

Figure 2.4 Typical tolerance to sustained linear acceleration as a function of direction of acceleration

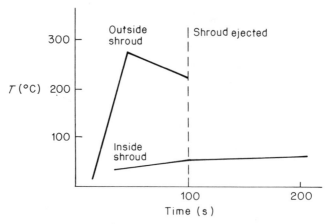

Figure 2.5 Heating profile inside and outside shroud for Scout launch vehicle

conduction paths between shroud and payload. Once the shroud has been jettisoned, payload heating arises directly as a result of frictional forces, but the low density of the atmosphere then results in only modest heat fluxes.

For Ariane 4, the peak heat flux experienced by the payload within the shroud is less than 500 W/m² at any point. Following shroud jettison this rises to a maximum of 1135 W/m² [2]. Figure 2.5 shows a representative heating profile for the Scout launch vehicle [3].

The *ambient atmospheric pressure* declines during launch. The rate at which depressurization occurs depends on the venting of the shroud volume. Generally this is fixed by the inclusion of venting ports; for example on Ariane the static pressure declines at a rate of 10 mbar/s, but for Shuttle the venting of the cargo bay can be controlled. For Shuttle, which has both pressurized and unpressurized elements, venting control is particularly important due to possible adverse static loads being placed on structural members. A detailed description of its venting system is given in [4]. It should be noted that venting of electronic boxes is also generally required within the vehicle.

2.2.3   The space environment

The Sun provides all the heat input to the solar system (excluding radioactive decay processes), and its mass is 99.9% of the total. It is not surprising that it dominates the space environment of the whole solar system, and it is therefore appropriate to outline its significant features as they affect the near-Earth environment. Earth is just one of the nine planets (see Table 2.1).

The Sun itself is not a particularly significant body in the universe. It is a star of mass $\sim 2 \times 10^{30}$ kg, modest by stellar standards, and is one of $\sim 10^{11}$ stars which form our galaxy. It is classified as a G2V star, having a yellowish appearance because its radiated light peaks at ~ 460 nm, and it is termed a yellow-dwarf star. Its radius is 7×10^8 m.

The nearest star is 3.5 light years away (1 light year $= 9.46 \times 10^{12}$ km) and between

Table 2.1 Solar system parameters

Object	Mass (10^{24} kg)	Radius* (10^4 km)	Density (g/cm^3)	Angular momentum (10^{46} g cm^2/s)	Average velocity (10^3 km/s)	r	e	i	τ
Sun	1.99×10^6	69.6	1.409	170†	—	—	—	—	—
Mercury	0.33	0.243	5.46	0.906	47.9	57.9	0.2056	7.004	0.241
Venus	4.87	0.605	5.23	18.5	35.1	108.2	0.0068	3.394	0.615
Earth	5.97	0.638	5.52	26.7	29.8	149.6	0.0167	0.0	1.000
Mars	0.642	0.340	3.92	3.52	24.2	227.9	0.0934	1.850	1.881
Jupiter	1899	7.16	1.31	19400	13.1	778.3	0.0484	1.305	11.862
Saturn	568	6.0	0.7	7840	9.64	1427	0.0557	2.490	29.46
Uranus	87.2	2.54	1.3	1700	6.81	2870	0.0472	0.773	84.01
Neptune	102	2.47	4.66	2500	5.44	4497	0.0086	1.774	164.79
Pluto	0.66	0.32	4.9	17.9	4.75	5900	0.253	17.14	248.43

* Values of equatorial radius
† Spin angular momentum of Sun
r = mean distance to Sun (10^6 km) ($r_e = 1$ AU)
e = eccentricity of orbit
i = inclination of orbit plane relative to ecliptic
τ = sidereal period in years

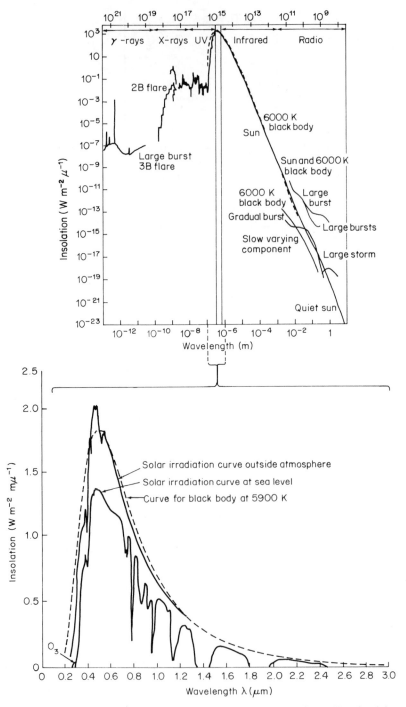

Figure 2.6 Solar spectrum (reproduced by permission from Hynek, J.A. (1951) *Astrophysics*, McGraw-Hill)

the stars the gas density is low, with hydrogen as the dominant species. The density amounts to only 3 atoms/cm³, in comparison to the nominal number density of our own atmosphere at sea level of $\sim 3 \times 10^{19}$ molecules/cm³.

The Sun is fundamentally a giant thermonuclear fusion reactor whose surface temperature is ~ 5800 K. The photosphere is optically thick, and its spectrum approximates to that of a black body.

The *solar spectrum* is shown in Figure 2.6, with a 5800 K black body for comparison. It is evident that it departs from the black-body spectrum at some wavelengths, these discrepancies arising in the solar atmosphere. There are two primary regions of this. The lower, or chromosphere, extends to a few thousand kilometres above the photosphere, and is a region of increasing temperature, peaking at $\sim 10\,000$ K; it is responsible for enhanced ultraviolet (u.v.) emission. The upper atmosphere, called the corona, becomes more tenuous and extends to several solar radii. Its nominal temperature is around 2×10^6 K and it emits substantial amounts of X-rays. The nominal release of energy from the Sun is at a rate of 3.85×10^{26} W. In order to gain an insight to this power level, if the Earth's fossil fuel resources were to be burnt at this rate then they would be exhausted in a mere 50 milliseconds.

The *solar wind* is another outward flux from the Sun. It is a flow of plasma expelled at high velocity. In reality it forms the outermost layer of the solar atmosphere, being continuously driven outward due to the Sun's radiation pressure. At Earth the speed of the wind is ~ 450 km/s, its density is ~ 9 protons/cm³, and its kinetic temperature is $\sim 100\,000$ K.

Sunspots are an indication that there are significant disturbances taking place on the Sun's surface and through its atmosphere. These, first observed by Galileo, are regions of its disc which are cooler than the surrounding surface. They emit less radiation and thus appear as dark spots. Periods of high solar activity occur when there are a large number of sun-spots and then enhanced emission of radiation occurs, most notably at radio wavelengths and at X-ray and γ-ray energies. This enhanced emission is generally associated with solar flares, which occur at sites near sunspots. They may last for periods of minutes to several hours, and occur as frequently as one every 2 hours [5] during high solar activity.

The Zurich sun-spot number R_z is used to quantify the overall number of sun-spots on the Sun at any time. It is defined as $R_z = K(10g + f)$, where f is the number of sun-spots which exhibit umbrae, and g is the number of groups into which these spots

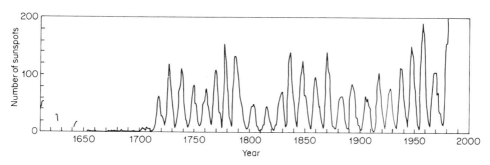

Figure 2.7   Historical observation of number of sun-spots (reproduced by permission from Kraus, J.D. (1986) *Radio Astronomy*, Cygnus-Quasar)

SPACECRAFT SYSTEMS ENGINEERING

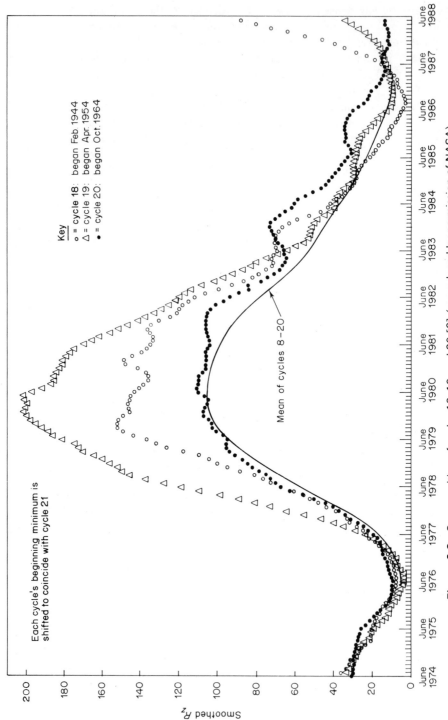

Figure 2.8 Superposition of cycles 18, 19, and 20 [6] (reproduced by permission of NASA)

Table 2.2 Solar variability [6] (reproduced by permission of NASA)

Spectral region	Wavelength	Flux $(\text{J m}^{-2}\,\text{s}^{-1}\,\mu\text{m}^{-1})$	Variability
Radio	$\lambda > 1$ mm	10^{-11}–10^{-17}	$\times 100$
Far infra-red	1 mm $\geq \lambda > 10\ \mu\text{m}$	10^{-5}	Uncertain
Infra-red	$10\ \mu\text{m} \geq \lambda > 0.75\ \mu\text{m}$	10^{-3}–10^{2}	Uncertain
Visible	$0.75\ \mu\text{m} \geq \lambda > 0.3\ \mu\text{m}$	10^{3}	$<1\%$
Ultraviolet	$0.3\ \mu\text{m} \geq \lambda > 0.12\ \mu\text{m}$	10^{-1}–10^{2}	1–200%
Extreme ultraviolet	$0.12\ \mu\text{m} \geq \lambda > 0.01\ \mu\text{m}$	10^{-1}	$\times 10$
Soft X-ray	$0.01\ \mu\text{m} \geq \lambda > 1\ \overset{\circ}{\text{A}}$	10^{-1}–10^{-7}	$\times 100$
Hard X-ray	$1\ \overset{\circ}{\text{A}} \geq \lambda$	10^{-7}–10^{-8}	$\times 10$–$\times 100$

fall. K is a factor which relates to the observing instrument, and is used as a normalization factor.

The detailed prediction of individual flares is not yet possible, but the general level of activity has a well-defined 11-year cycle as shown in Figure 2.7. However, due to magnetic pole reversal of the Sun at peak solar activity, the real period is 22 years. It may be seen

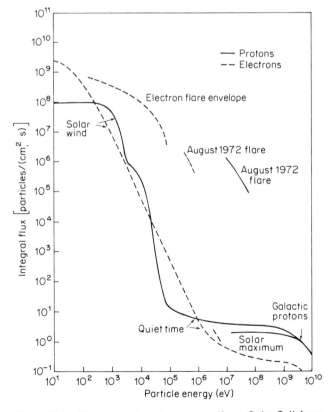

Figure 2.9 Electron and proton spectra (from *Solar Cell Array Design Handbook* by H.S. Rauschenback, copyright © 1980 by Van Nostrand Reinhold. All rights reserved)

that the magnitude of the peak varies from cycle to cycle. The present cycle, beginning June 1976, is significantly greater than the mean value of the previous 16, as shown in Figure 2.8 [6].

Table 2.2 shows the typical intensity variability of the Sun at particular wavelengths where it has been established. There is good correlation between sunspot number and radiated power at some specific wavelengths and this is used to provide an additional measure of the overall level of solar activity. The most frequently used monitor is the solar flux at 10.7 cm. One source for such data is from NOAA, in the US.

An enhanced flux of high-energy particles is also associated with major solar flares. These have energies up to approximately 30 MeV. Two components may be distinguished when it arrives at the Earth. The first occurs approximately 20 minutes following the flare (observed through its electromagnetic emissions), and then a more prolonged component arrives about a day after it. These latter particles appear to be an enhanced component of the solar wind, travelling at velocities of $\sim 10^3$ km/s. Figure 2.9 shows the variability of particle fluxes in the interplanetary medium, caused by solar activity.

2.2.4   The near-Earth environment

The Earth orbits the Sun at a mean distance of one Astronomical Unit (AU), equal to 1.496×10^8 km. It is a nearly spherical body having a mass of only 3×10^{-6} times that of the Sun, but having a gravitationally bound atmosphere and a significant magnetic field. Each of these features is important in determining the near-Earth environment, in which the majority of space vehicles operate.

The *atmosphere* at sea level is predominantly molecular nitrogen (78%) and molecular oxygen (21%), with a variety of trace elements, the most significant being argon. Standard sea-level pressure is accepted to be 1.013×10^5 Pa [7]. The lower atmosphere, up to ~ 86 km, is sufficiently turbulent to result in a homogeneous gas mixture, albeit with a pressure decreasing exponentially. Above this height photochemical processes disturb the homogeneity. Of particular importance is the absorption of u.v. radiation from the Sun, leading to the dissociation of oxygen in the upper atmosphere. The details of the processes are beyond the scope of this chapter and the interested reader is referred to [7] and [8]; however, the resultant atmosphere above ~ 120 km is one in which each atmospheric constituent is decoupled from all the others. For each species it is then possible to write down an equation of the diffusive equilibrium of the form

$$n_i v_i + D_i \left(\frac{\mathrm{d}n_i}{\mathrm{d}Z} + n_i \frac{(1+\alpha_i)}{T} \frac{\mathrm{d}T}{\mathrm{d}Z} + \frac{g n_i M_i}{R^* T} \right) = 0 \qquad (2.1)$$

where n_i is the number density of species i, having a molecular weight M_i, at altitude Z, v_i is the vertical transport velocity of the species and D_i and α_i are its molecular and thermal diffusion coefficients, T is the atmospheric temperature, R^* is the universal gas constant and g is the height-dependent acceleration due to gravity. If negligible vertical transport takes place, and for species where thermal diffusion is negligible, equation (2.1) reduces to a hydrostatic equilibrium equation wherein the number density profile is driven by the atmospheric temperature. Figure 2.10 shows the variation of number density with height for different species.

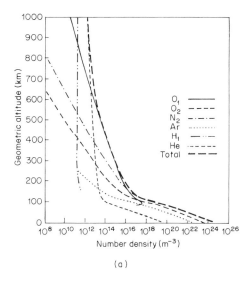

(a)

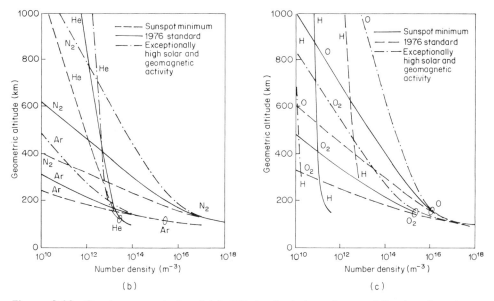

(b)   (c)

Figure 2.10 Species concentration (a) in US standard atmosphere and (b, c) under extreme conditions

For detailed calculation of the atmosphere a variety of models are available [e.g. 9, 10]. Each of these requires the specification of a temperature at some height, and then by using a predefined temperature profile the density may be determined through equation (2.1). Figure 2.11 shows the general form of such profiles. It can be seen from this that at extreme altitudes the temperature tends to a limiting value, the so-called exospheric temperature, T_∞. Within the exosphere the atmosphere is effectively isothermal. T_∞ rises

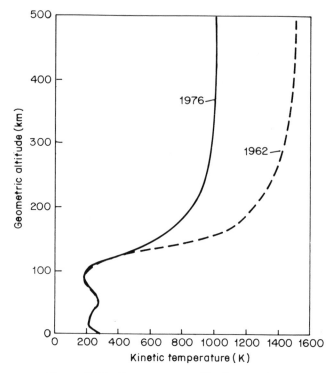

Figure 2.11   Temperature profile of atmosphere

through increased solar activity; most models relate it to the flux of solar radiation at 10.7 cm ($F_{10.7}$) through an algorithm, and also include the effects of geomagnetic activity, stimulated by the interaction of the solar wind and the Earth's magnetic environment or magnetosphere (see below). Since under conditions of hydrostatic equilibrium the density falls at a rate:

$$\rho = \rho_{SL} \exp\left(\frac{-gM_i}{R^*T} Z\right),$$

(2.2)

then for a given altitude the atmospheric density will increase with solar activity, and this will result in reduced lifetimes for low orbiting vehicles.

The US standard atmosphere 1976 [7] is shown in Figure 2.10. At altitudes typical of low Earth orbiting (LEO) vehicles (300–900 km) the density is low, but not insignificant. At geostationary altitude, $\sim 36\,000$ km, the density of the neutral atmosphere is approximately the same as in the interplanetary medium, namely 10^{-20} kg/m^3, and the pressure is $\sim 10^{-15}$ Pa [11]. The dominant species of the atmosphere in LEO is generally either atomic oxygen or helium. The effects of atomic oxygen on surfaces are considered in Section 2.3.

Few *atomic/molecular collisions* take place between components of the atmosphere in both LEO and GEO. Table 2.3 shows that above 200 km altitude the mean free path is significantly greater than the dimensions of most space vehicles. This has two consequences.

Table 2.3   Mean free path as a function of altitude

Altitude (km)	λ_0 (m)	Altitude (km)	λ_0 (m)
100	0.142	300	2.6×10^3
120	3.31	400	16×10^3
140	18	500	77×10^3
160	53	600	280×10^3
180	120	700	730×10^3
200	240	800	1400×10^3

Firstly the ability to exchange heat energy with the environment is solely as a result of radiation. Thus the dominant radiative heat input is due to solar radiation which in the near-Earth environment is 1371 ± 5 W/m^2. Secondary input occurs due to Earth albedo (the reflection of solar radiation from the top of the atmosphere) and Earth shine (the black-body radiation of the Earth), and has a magnitude of ~ 200 W/m^2. The neutral atmosphere at $\sim 10^3$ K and the solar wind at $\sim 2 \times 10^5$ K provide negligible heating. The temperature that a space vehicle reaches is thus dependent upon a balance of radiative heat input and output (see Chapter 12).

Secondly the aerodynamics of spacecraft at orbiting altitudes must be based upon free molecular flow as briefly described in Chapter 4. Since the density is low the frictional heating forces are negligible, even though the relative velocity approaches 8 km/s.

The *ionosphere*, above ~ 86 km, is a region of increasing plasma density caused by photoionization, due to incident u.v. photons. The plasma has a significantly lower density than the neutral density below an altitude of ~ 1000 km, even though its peak value occurs at 300–400 km. It has significant influences upon the propagation of radio

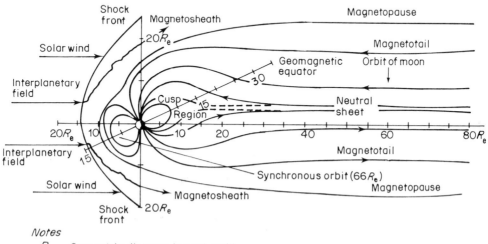

Notes

R_e Geocentric distance in earth radii

→   Direction of magnetic flux lines

Figure 2.12   The Earth's magnetosphere (reproduced by permission of Kluwer Academic Publishers)

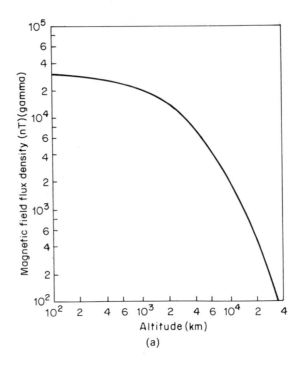

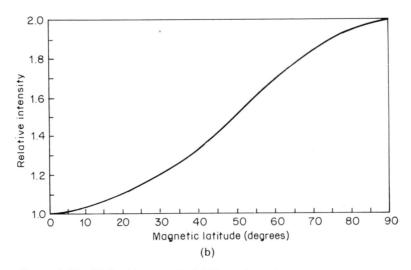

Figure 2.13 (a) Earth's magnetic field intensity at the magnetic equator as a function of altitude (adapted from [12]); (b) relative intensity of the Earth's magnetic field as a function of magnetic latitude (reproduced by permission of Kluwer Academic Publishers)

waves; for an electron density n_e electrons cm^{-3}, frequencies below the plasma frequency, given by $f_p \sim 9000\sqrt{n_e}$ Hz, cannot propagate, and thus radio waves of lower frequency incident upon such a plasma will be reflected. If there is additionally a magnetic field present, as is the case in the near-Earth environment, then the polarization of any electromagnetic radiation propagating through the plasma will be rotated due to Faraday rotation [5]; inefficiencies can then occur in communication systems if linearly polarized radio waves are used (see Chapter 13).

The Earth's magnetic field has two primary sources. The dominant one at its surface is due to currents circulating within its core, whilst at higher altitudes the currents caused by the differential motion of electrons and ions in the magnetosphere play a significant role. The solar wind plasma, carrying its own magnetic field, distorts the Earth's simple dipole field into the shape shown in Figure 2.12, with both open and closed magnetic field lines.

From Figure 2.12 it is apparent that at high altitude the magnetic field structure is complex; however, at lower altitude it is possible to make certain observations. Firstly the overall strength of the magnetic field is not constant, but is decreasing at $\sim 0.05\%$/year. This field is weakest on the equator, and Figure 2.13 shows its dependence on both latitude and altitude.

The Van Allen radiation belts are regions of high-flux, high-energy particles—protons and electrons of both solar and cosmic origin. Some of these become trapped in the magnetosphere and accelerate through a magnetic mirror effect [12]. Particles may reside in the belts for periods of hours to years. Two primary belts exist and the quiescent or mean radiation levels are shown schematically in Figure 2.14. Not surprisingly the radiation intensity in these belts (intensity, energy spectrum, etc.) is a function of solar

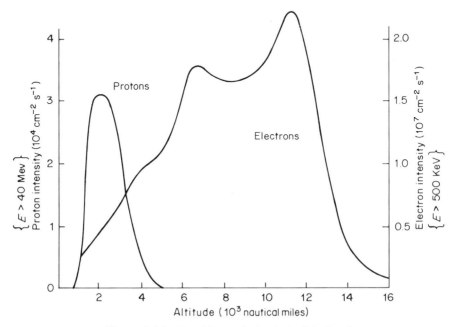

Figure 2.14 Van Allen radiation belts (idealized)

activity. Various models qualifying the fluence (time integrated flux) of such radiation are available to the designer [see e.g. 13, 14]. Whilst the belts provide a severe radiation environment for space vehicles travelling through them (components and man), they do to some extent protect low Earth orbiting vehicles from sources of external particulate radiation.

Electrostatic charging of a spacecraft travelling through the near-Earth space environment will occur, whether it is in or out of the radiation belts. As a consequence currents will occur between the space vehicle and the plasma, imbalance of which will cause spacecraft to develop a charge which may be returned to balance through arcing. The two major sources of currents are [12] the ambient plasma itself and photoelectron emission due to solar sunlight, and in particular the short wavelength component of this radiation. The latter phenomenon is particularly important when the spacecraft enters and leaves eclipses [15]. Severe problems arise if differential charging of spacecraft surfaces occurs. The simplest method of preventing this is to use conductive surfaces wherever possible. One primary area where this is not possible is on the solar array; an alternative solution is then to apply a near transparent coating of indium oxide to the cell cover glass material, which typically reduces the resistivity of the glass surface to less than 5000 Ω/cm^2 [16].

Meteoroids and micrometeoroids occur with a frequency which varies considerably with

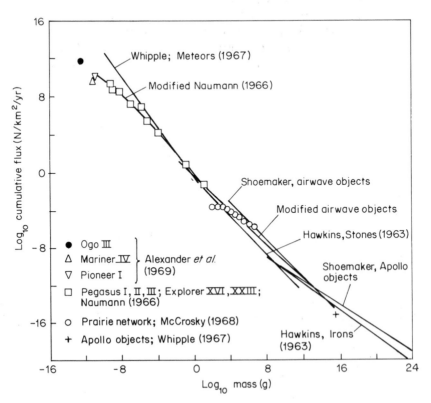

Figure 2.15 Terrestrial mass-influx rates of meteoroids. N is the flux of particles with mass greater than m [26] (reproduced by permission of NASA)

the type of space mission. These are solid objects whose mass and size vary over many orders of magnitude. Their mass spectrum is shown in Figure 2.15. Near large gravitational masses such as the Earth their fluxes tend to be enhanced. The asteroid belt is also a region of enhanced meteoroid density. Impact of micrometeorites generally causes a degradation of surface thermal properties, although the possibility of component failure clearly exists. The most dramatic evidence of particle impacts on a spacecraft is from the Giotto spacecraft and its passage near to Halley's comet during 1986. Particle impacts led to the failure of some experiments and a change in the attitude of the vehicle at closest encounter.

2.3 ENVIRONMENTAL EFFECTS ON MATERIALS

Outgassing or sublimation refers to the vaporization of surface atoms of a material when it is subjected to an ambient pressure which is comparable with its own vapour pressure. Such a pressure, 10^{-11}–10^{-15} Pa, occurs at spacecraft altitudes. This process occurs at an increasing rate as temperature rises. A summary of rates for different metals at a variety of temperatures is shown in Table 2.4 but they do not apply to alloy materials or those having a covering surface layer of a different material [3]. Since the precise surface conditions of orbiting vehicles are difficult to determine, particularly for vehicles in low Earth orbit, exact figures for mass loss are not at present available. Whilst structural problems arising due to outgassing are unlikely, the subsequent deposition of the material is hazardous both to optical and electrically sensitive surfaces. Thin plastic layers and oxide coatings are particularly sensitive to mass loss, especially if the material is used for its thermal properties; if mass loss is associated with specific constituents then modified emissive properties may result.

For plastic materials high-vapour-pressure components evaporate rapidly, although initially mass loss is usually associated with the loss of adsorbed gases and water vapour.

Traditional lubricants used on the ground are clearly not appropriate to spacecraft operation. Generally they have high vapour pressure and would thus outgas rapidly, but in many the lubricative action arises due to the presence of adsorbed gases and water vapour. Whilst low-volatility oils are used, solid lubricant coatings such as MoS_2 are frequently adopted. Reference [11] provides a detailed listing of NASA recommended lubricants.

Table 2.4 Temperature for given sublimation rate (°C)

Element	0.1 μm/yr	10 μm/yr	1 mm/yr
Cd	38	77	122
Zn	71	127	177
Mg	110	171	233
Au	660	800	950
Ti	920	1070	1250
Mo	1380	1630	1900
W	1870	2150	2480

2.3.1 Atomic oxygen erosion

The atmospheric composition shown in Figure 2.10 indicates that atomic oxygen forms the major atmosphere species in low Earth orbit. Following the severe erosion noted for Kapton on STS3 [17], it has become apparent that atomic oxygen provides an aggressive environment for materials used on space vehicles in LEO. This arises not only from its chemical activity, but most significantly from the fact that its atoms are travelling at ~8 km/s relative to the vehicle (due to the vehicle's orbital velocity). Silver is one of the few metals attacked by this environment, so due to its extensive use on solar arrays it is important to avoid bare silver exposure. This environment erodes many plastics to a varying extent. Simulation of the atomic oxygen environment has become particularly important and there is much active research due in part to the materials requirements for the international space station. Further information on this critical feature of the environment may be found in Refs. [18] and [19].

Material strength and fatigue life are also affected by a high-vacuum environment. Generally fatigue life is improved. For many materials it may be extended by more than an order of magnitude [20], although in some cases, for example, pure nickel and Inconel 550, the improvement is uncertain [20, 21]. The physical reasons for such changes in mechancial properties, and also changes in strength, are believed to arise due to one of two principal reasons. One of these is that gases absorbed into surface cracks either aid or hinder crack propagation. The second possible physical cause is that oxidation and gas diffusion absorbed into the material bulk influences material properties. A well-documented example of material whose strength is improved in vacuum is that of glass, wherein a threefold improvement in strength has been noted at a pressure which is one thousandth that of atmospheric pressure.

Embrittlement is a form of material damage which is caused by exposure to u.v. radiation. Many polymers are particularly sensitive to such photons, whose energy is high enough to modify their chemical bonding structure.

Ultraviolet exposure also causes *electrical changes* in the form of resistivity modification, and *optical changes* affecting both thermal characteristics and opacity. A particularly u.v. sensitive element is the solar array. More specifically the solar cell coverglass and its attendant adhesive are subject to darkening. This results in reduced cell illumination and an enhanced operating temperature, both being deleterious to cell operation (see Chapter 11).

Radiation damage affects all materials to some extent, and man. High-energy corpuscular radiation is experienced most severely in the Van Allen radiation belts, but is also at a significant level in any space operation. Under even modest radiation doses some metals such as cadmium and zinc may form metallic whiskers and cause electrical shorting to take place and these materials are generally avoided.

Semiconductor materials and biological tissue are particularly sensitive to damage caused by high-energy charged and neutral particles. This generally arises due to both the displacement of atoms from crystal lattice sites and the attendant local ionization, together with an ionization track caused by the penetrating radiation. Protection is a complex task, since screening material will itself generate secondary radiations due to the passage of a primary high-energy particle through it. It should be noted that 'heavy' particles such as protons and neutrons cause significantly more damage than light ones such as electrons. They cause a dense track of ionization through a material until they are stopped by an atomic collision. When this occurs a significant displacement of the atom arises, leading to further ionization which does not occur for electrons.

In semiconductor materials two damaging effects have been noted. Primarily radiation damage reduces the effectiveness of semiconductor operation. Specifically in solar cells it results in a reduction in the efficiency of conversion from sunlight to electrical energy. In order to quantify the effect of this the spacecraft designer usually resorts to tables of fluence for particular orbits. The upper, exposed surface of the cell may be protected to some extent by the use of a coverglass. The lower surface is generally protected by the substrate on which it is mounted. Manufacturers' calibration data is generally available to enable the expected degradation in solar cell performance to be assessed during the mission lifetime. This topic will be returned to in greater detail in Chapter 11.

The second damaging effect is the possibility of software errors caused by single-particle events. It is becoming more apparent as the processing power of on-board computers is increasing. This is a challenging area of present spacecraft design for which solutions are being sought.

2.4   ENVIRONMENTAL EFFECTS ON MAN

The removal of man from his natural environment or habitat necessarily introduces deleterious effects. A suitable life-support system will guard against the more obvious ones such as lack of atmosphere, his thermal environment and possible high-acceleration loads. Details applicable to the design of a suitable life-support environment may be found in Refs. [22] and [23]. Here we consider the environmental problems for which a life-support system offers no solution. It should be noted that the response of man to extended exposure to the space environment is neither well documented nor well under-stood, principally due to the modest experience of manned flight to date. A review may be found in [24].

In general terms the response of man to spaceflight may be classified under one of four categories: transient effects lasting for short periods on initial exposure to, or removal from, the space environment; flight duration adaptation; cumulative effects dependent upon the length of the flight which are reversible on return to the ground; cumulative effects which are irreversible.

The dominant effect of the space environment, which cannot be removed through life-support systems, is zero or microgravity. This causes a major disturbance to the human system, with effects which fall into each of the first three categories above. The most notable of these are the following:

1. *Blood volume redistribution.* On Earth the blood pressure of a person whilst standing decreases with height above the feet. Typically the pressure in the brain is only one third that in the feet. The immediate effect of a 'zero g' environment is to cause a redistribution of the blood volume, resulting in a 'puffy' face. Whilst on Earth, the body system is familiar with orientation changes (lying down, etc.), a complex system of hormonal secretion results in a control system which adjusts heart rate to these conditions. In zero gravity, however, this hormonal action appears to result in a loss of sodium with consequent fluid loss. Typically stabilization occurs after $\sim$4 days

when a loss of fluid of some 2–4 kg has occurred. Readaptation on return to Earth occurs over a short period of time, during which giddiness may be expected.

2. *Muscular atrophy.* Atrophy of all muscles occurs through long periods of inactivity. In zero *g* the heart itself is required to do less work since it is not pumping blood against gravity. Reduction in both heart muscle mass and heart rate occurs. It is believed that anaemia (a reduction in red blood cell count) is an additional side effect which has been noted in astronauts. To combat general atrophy of muscles astronauts spend much of their time in space undergoing physical exercise.

3. *Vestibular problems.* The human vestibular system is dependent upon both visual and inner ear sensors. In the inner ear the sensor has two orthogonal components, horizontal and vertical, which rely on gravity for their operation. Removal of gravity results in enhanced sensitivity to acceleration (including rotation) because of fluid motion in these sensors. Conflict between visual and inner ear sensors therefore arises during motion of the astronaut. The most notable effect is motion sickness, but astronauts also experience enhanced clumsiness. Adaptation generally takes 3–4 days.

4. *Locomotor system.* The major effect of zero gravity on the locomotor system is cumulative bone decalcification. This directly results in bone fragility and indirectly leads to problems of recalcification, external to the bones in, for example the kidneys, forming kidney stones. High calcium diets do not appear to stop this increased calcium mobility and on long duration flights such as Salyut 6, a 2–8% loss of calcium was noted. Restabilization occurs on return to Earth.

The corpuscular radiation environment noted in preceding sections evidently provides a hazardous environment for man. Shielding materials must be chosen with care, as poor design can lead to a secondary radiation which is more hazardous than the primary [22]. Typical unscreened radiation dose rates are indicated in Table 2.5. From this it is evident that solar flares provide a particularly hazardous environment for man, and thus permanently manned space stations will require some type of flare shield for personnel protec-

Table 2.5 Radiation dose in space-laboratory-type orbit (free space)

Source	Dose rate
Galactic radiation	0.01–0.05 rad/day ($\sim$0.3 rem/day)
Radiation belts	protons: 1–10 rad/h [behind 1 cm Al] electrons: 10^2–10^3 rad/h [at surface of S/C] ($\sim$1–30 rem/h)
Solar flares	12–350 rad/event ($\sim$10 500 rem)

mean dose on Earth 250–300 millirem/year

tion. For a geostationary orbit base, it has been estimated [25] that an equivalent shield protection of 21 g/cm^2 of aluminium would be required. This would result in a mass of ∼ 3.6 tons for a 6-man habitat. For comparison purposes it should be noted that for the Apollo command module's lunar flights the shield, made of aluminum, stainless steel and phenolic epoxy, provided a shield thickness equivalent to 7.5 g/cm^2 of aluminium. The maximum radiation dose experienced was ∼ 1.0 rad. Radiation dose in man is a cumulative hazard which cannot be reversed on return to the ground.

REFERENCES

[1]   Rauschenback, H.S. (1980) *Solar Cell Array Design Handbook*, Van Nostrand, New York.
[2]   Ariane 4 Users Manual Issue 1 (1983) Arianspace.
[3]   NASA SP-3051 (1969).
[4]   Lufti, M.S. and Neider, R.L. (1983) NASA CP 2283, 231.
[5]   Kraus, J.D. (1966) *Radio Astronomy*, McGraw-Hill, New York.
[6]   NASA TM 82478, Volume 1 (1982).
[7]   U.S. Standard Atmosphere 1976 (1976) NOAA, Washington, D.C.
[8]   Akasofu, S. and Chapman, S. (1972) *Solar Terrestrial Physics* Oxford University Press, Oxford.
[9]   CIRA; COSPAR International Reference Atmosphere 1972 (1972) Compiled by members of the Cospar Working Group IV, Pergamon Press, Oxford.
[10] Jacchia, L.G. (1977) SAO Special Report 375.
[11] NASA SP-8021 (1973).
[12] Garrett, H.B. (1979) *Rev. Geophys. and Space Phys.* 17, 397.
[13] Sawyer, D.J. and Vette, J.I. (1976) INDC-A-RDS-76-06, National Space Science Data Centre.
[14] Teague, M.J., Chou, J.W., and Vette, J.I. National Space Science Data Centre 76-04.
[15] Garrett, H.B. and Gaunt, D.M. (1980) *Prog. Astron. and Aeron.* 71, 227.
[16] Pilkington Space Technology (1984) Coverglass Specification PS 292.
[17] Leger, L.J. (1983) AIAA Paper No. AIAA-83-0073.
[18] Fourth European Symposium on Spacecraft Materials in the Space Environment (1988) CERT.
[19] Brinza, D.E. (ed.) (1987) *Proc. NASA Workshp on Atomic Oxygen Effects* JPL Publication 87-14.
[20] NASA TN-D 2563 (1965).
[21] NASA TN-D 2898 (1965).
[22] NASA SP-3006 (1973).
[23] Sharpe, M.R. (1969) *Living in Space* Aldus Books 1, London.
[24] ESA BR-17 (1984).
[25] NASA SP-413 (1977).
[26] Gault, D.E. (1970) *Radio Science*, **5**, 273.

3 DYNAMICS OF SPACECRAFT

Peter W. Fortescue

Department of Aeronautics and Astronautics, University of Southampton

3.1 INTRODUCTION

This chapter serves as a general introduction to the subject of the dynamics of bodies, and sets a framework for the subjects of celestial mechanics and for attitude control (Chapters 4 and 10). For both of these, Newtonian dynamics will provide a sufficient means of forecasting and of understanding a spacecraft's behaviour. The summary presented here is chosen with a view to its relevance to spacecraft.

The approach adopted is to develop an understanding of dynamics in two stages. The first is to express the dynamics of both translation and rotation in terms of the appropriate form of momentum—linear or angular. Momentum becomes an important concept, in terms of which it is relatively easy to determine the consequences of forces or moments.

The second stage is to interpret the momenta in terms of the physical movement—the velocities, linear and angular. This is straightforward for linear momentum since momentum and velocity are in the same direction. More difficult is the relationship between rotational movement and angular momentum.

3.1.1 Translation/rotation separation

One feature which is peculiar to spacecraft is that their translational (trajectory) motion is virtually independent of their rotational motion. This is due to the fact that the moments or torques which cause their rotation are not dependent upon their direction of travel, and the forces which determine their trajectory are not dependent on their attitude. Whilst this is not entirely true it is approximately so, and spacecraft designers will normally aim to preserve this independence.

At a fundamental level it is convenient to think of a spacecraft as being a collection of

Spacecraft Systems Engineering. Edited by P.W. Fortescue and J.P.W. Stark
© 1991 John Wiley & Sons Ltd

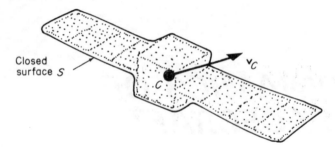

Closed
surface S

Figure 3.1 The closed surface S

those particles and bodies which lie within a closed surface S. This concept allows the surface to be chosen at one's convenience, to embrace the complete craft or just part of it or, when considering a docking manoeuvre or a tethered configuration, it may be chosen to embrace more than just the one spacecraft. But for the following assume that it envelops just one spacecraft, unless stated otherwise.

The dynamics of the craft (or whatever is within the surface S) may now be described in terms of its momenta—its linear momentum **L**, and its angular momentum **H**—the former leading to equations which describe the trajectory, and the latter to those which describe the attitude motion.

In particular it is the centre of mass C whose trajectory will be described. It may seem to be pedantic to pick on one specific point for this purpose, but the centre of mass has special properties which lead to its being chosen for the development of Newtonian dynamics, properties which make it sensible to separate the motion into:

1. the motion of the centre of mass, C; and

2. the motion relative to the centre of mass.

For highly specialized purposes, such as when using the orbit to improve man's knowledge of the Geoid, the accuracy with which the orbit is determined may be commensurate with the dimensions of the spacecraft, and then it is no longer pedantic to consider to which point of the vehicle its orbit refers.

3.1.2 The centre of mass, *C*

The centre of mass of the particles in S, relative to an arbitrary point O, is the point whose position vector $\mathbf{r}_{OC}$ obeys:

$$M\mathbf{r}_{OC} = \Sigma(m\mathbf{r}_{OP}) \tag{3.1}$$

where $\mathbf{r}_{OP}$ is the position vector of a general particle P,
             m is the mass of the general particle,
and M is the total mass within S.

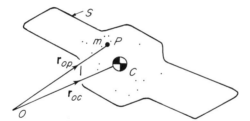

Figure 3.2   The centre of mass, C

When the location of C is being determined for a spacecraft then it is useful to note that equation (3.1) can be applied to objects such as items of equipment rather than to particles. In that case $\mathbf{r}_{OP}$ refers to the centre of mass of the item and m is its mass.

When an object has a continuous mass distribution then the integral equivalent of equation (3.1) should be used, this being:

$$M\mathbf{r}_{OC} = \int \mathbf{r}_{OP}\, dm \qquad (3.2)$$

The centre of mass is an important reference point. In the development of the theory of dynamics it is useful as an origin or reference point for the particles in S. Putting the origin at C leads to $\mathbf{r}_{OC} = \mathbf{0}$, and so equation (3.1) becomes

$$\Sigma(m\mathbf{r}_{CP}) = \mathbf{0} \qquad (3.3)$$

This is true at all times and so all its derivatives are zero; for example

$$\Sigma(m\mathbf{v}_{CP}) = \mathbf{0} \qquad (3.4)$$

where $\mathbf{v}_{CP}$ is the velocity of a particle relative to C.

3.2 TRAJECTORY DYNAMICS

For trajectory purposes it is convenient to treat a spacecraft as if it were a particle, as is done in Chapter 4. This will be called the *equivalent particle* (e.p.), having a mass M equal to that of the spacecraft, and situated at its centre of mass C. The e.p. therefore moves with the velocity $\mathbf{v}_C$ of C, and its momentum $\mathbf{L}$ is the aggregate of all the particles and bodies comprising the spacecraft (Figure 3.3).

$$\mathbf{L} = M\mathbf{v}_C \qquad (3.5)$$

This interpretation admits the possibility that the spacecraft may have moving parts or appendages. The surface S (Figure 3.1) may in fact be chosen to embrace any objects whose joint orbit is of interest.

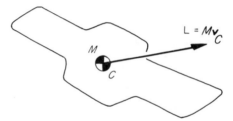

Figure 3.3   The equivalent particle at C

According to Newtonian mechanics there are two ways in which the momentum **L** of the e.p. may be changed:

1. *Application of external forces.* These include gravitational attraction towards the heavenly bodies, solar radiation pressure, aerodynamic forces, etc.

    Internal forces, acting between any two particles or bodies, will not produce any change in the total momentum. It follows that objects moving within a spacecraft, such as fuel, astronauts, mechanisms, or flexing structures, etc., will not cause the total linear momentum to change. Likewise the total momentum of two docking spacecraft will not be changed by any forces between them.

    This mechanism applies to a spacecraft which has no propulsion; its characteristic is that the mass M remains constant.

2. *Ejection of some of the particles from within S.* This occurs during rocket propulsion. The ejected particles take away both their mass and their momentum, leading to a change in the residual mass and momentum of the e.p.

In the above it is important to recognize that the momentum must in all cases involve the velocity relative to inertial space rather than to the spacecraft.

It is convenient to deal separately with the two cases:

1. when there is no propulsion,

and the more general case.

2. when there is propulsion.

3.2.1   Translational motion with no propulsion

The effect of external forces $\mathbf{F}_{ext}$ on the momentum can be forecast from a diagram such as is shown in Figure 3.4. Their effect will depend upon not only the magnitude and

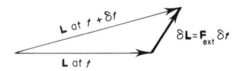

Figure 3.4   The incremental increase in momentum caused by an external force

(a)   (b)

Figure 3.5 The effect of forces along (a) the tangent, and (b) the normal to the trajectory

direction of the forces, but also upon the magnitude of the momentum **L** when the forces act. The additional momentum $\delta\mathbf{L}$ produced during an interval of time δt is equal to

$$\delta\mathbf{L} = \mathbf{F}_{ext}\delta t \qquad (3.6)$$

This must be added vectorially as shown in Figure 3.4 in order to see the effect upon **L** during the time interval.

Equation (3.6) may be written as the Newtonian equation:

$$d(\mathbf{L})/dt = d(M\mathbf{v}_C)/dt = \mathbf{F}_{ext} \qquad (3.7)$$

Equations (3.6) and (3.7) are vector relationships, and it is useful to note the consequences under special circumstances:

1. When there are no external forces, or when they add up to zero, then the total momentum **L** is conserved. The centre of mass moves in a straight line in inertial space, with constant velocity.

2. When the external forces $\mathbf{F}_{ext}$ are in the same direction as the momentum **L**, then they will increase its magnitude without any change of direction (Figure 3.5a).

3. When the external forces are at right angles to the momentum **L**, then they will change its direction but not its magnitude. This is illustrated in the vector diagram in Figure 3.5b, which shows that in the time interval δt the change of direction is $\delta\psi = (F_{ext}/Mv_C)\delta t$. The direction of travel therefore changes at a rate equal to

$$\frac{d\psi}{dt} = \frac{F_{ext}}{Mv_C} \qquad (3.8)$$

If a thruster is to be used in order to change the direction of travel of a spacecraft, then it is most effective if it is used when the speed is least.

It will later be seen that angular momentum **H** responds to a torque **T** in the same way as linear momentum responds to a force, and so does moment of momentum $M\mathbf{h}_I$ in response to a moment $\mathbf{M}_I$.

3.2.2 Moment of momentum mh, and angular momentum

Moment of momentum is a useful concept in celestial mechanics (Section 4.2), and it is also a useful stepping stone towards angular momentum and the study of attitude motion.

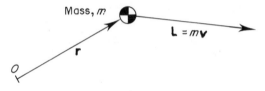

Figure 3.6

The *moment of momentum* $m\mathbf{h}_O$ referred to a point O may be defined as

$$m\mathbf{h}_O = \mathbf{r} \times m\mathbf{v} \tag{3.9}$$

$\mathbf{r}$ is the position vector from O to any point on the line of action of the momentum vector $m\mathbf{v}$ (Figure 3.6), and $\mathbf{v}$ is the velocity relative to the reference point O. This definition is equivalent to the product of the momentum with the perpendicular distance from O to the momentum vector.

For the trajectory of (the e.p. of) a spacecraft whose mass is M it is useful to take the centre of gravitational attraction as the reference point, I, say. Then the moment of momentum vector $M\mathbf{h}_I$ will be normal to the orbit plane (Figure 3.7).

In the study of attitude dynamics the aggregate of all the moments of the momenta of the particles comprising the spacecraft is a useful concept and it is defined as the *angular momentum*, $\mathbf{H}$. In this case the most useful reference point is the centre of mass, C.

Note that 'moment of momentum' is often loosely referred to as 'angular momentum'. This can lead to some confusion. Only in special cases are the two identical, such as when there is only one particle. In the more general case of a spacecraft its angular momentum referred to I is the sum of the moment of momentum of its e.p. referred to I, and its angular momentum referred to C:

$$\mathbf{H}_I = M\mathbf{h}_I + \mathbf{H}_C \tag{3.10}$$

In practice $M\mathbf{h}_I$ will be much larger than $\mathbf{H}_C$.

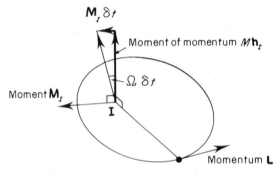

Figure 3.7   Precession of an orbit's plane due to a moment lying in the plane

3.2.3 The rate of change of moment of momentum, Mh_I

For the translational motion of a particle the moment of momentum $M\mathbf{h}_I$ referred to an inertially fixed point I is only changed if the forces on it have a moment $\mathbf{M}_I$ about I. The relationship may be expressed in the form of the Newtonian equation:

$$d(M\mathbf{h}_I)/dt = \mathbf{M}_I \qquad (3.11)$$

This is also true for the e.p. of a spacecraft. The forces on it are conventionally separated into:

(a) the dominant central gravitational force, and

(b) additional small perturbative forces.

The former acts towards a point I which may be taken to be inertially fixed, such as the centre of the Earth for Earth-orbiters, for example, and so it contributes nothing towards the moment $\mathbf{M}_I$.

The similarity between equations (3.11) and (3.7) indicates that the moment of momentum $M\mathbf{h}_I$ responds to the moment $\mathbf{M}_I$ in the same way as momentum $\mathbf{L}$ responds to a force $\mathbf{F}_{\text{ext}}$. So when the moment $\mathbf{M}_I$ is zero, such as when there are no perturbative forces, then $M\mathbf{h}_I$ will remain constant, by analogy with (1) above. (This is equivalent to Kepler's second law, see Section 4.1.) The constant direction of $M\mathbf{h}_I$, normal to the orbit plane, means that the plane is fixed in inertial space.

Case (3) above leads by analogy to the observation that perturbative forces which produce a moment vector $\mathbf{M}_I$ which lies in the orbit plane will cause the plane to change, i.e. precess (see Section 4.5). Its normal, in the direction of $\mathbf{h}_I$, will rotate at a rate Ω given by

$$\Omega = M_I/Mh_I \qquad (3.12)$$

see Figure 3.7.

3.2.4 Translational motion under propulsion

When a rocket motor is fired there is an expulsion of particles out of the spacecraft—out of the surface S referred to above. Each particle of the exhaust gases takes away its contributions to two of the properties of the e.p., its mass M, and its momentum $M\mathbf{v}_C$—the total momentum within S.

If the rocket's mass flow is σ, then this is the rate at which the mass of the e.p. decreases, i.e.

$$dM/dt = -\sigma \qquad (3.13)$$

Also the absolute velocity of the exhaust is $(\mathbf{v}_C + \mathbf{v}_{\text{ex}})$, and so the rate at which the momentum of the e.p. changes is

$$d(M\mathbf{v}_C)/dt = -\sigma(\mathbf{v}_C + \mathbf{v}_{ex}) \tag{3.14}$$

Here $\mathbf{v}_{ex}$ is the exhaust velocity relative to the centre of mass.

In the general case there will be an external force $\mathbf{F}_{ext}$ acting in addition, part of this being due to the back-pressure if the rocket operates in an atmosphere. Then the total rate of increase of momentum obeys the Newtonian equation:

$$d(M\mathbf{v}_C)/dt = \mathbf{F}_{ext} - \sigma(\mathbf{v}_C + \mathbf{v}_{ex}) \tag{3.15}$$

From equations (3.13) and (3.15) it follows that the absolute acceleration $\mathbf{a}_C$ of the e.p. obeys

$$M\mathbf{a}_C = \mathbf{F}_{ext} - \sigma\mathbf{v}_{ex} \tag{3.16}$$

Clearly when the rocket is restrained, as on a test-bed, there will be no acceleration, and the restraining force $\mathbf{F}_{ext}$ will be equal to the thrust, $\sigma\mathbf{v}_{ex}$.

The *rocket equation* is the integral of equation (3.16) for the case of no external forces, and the thruster points in a constant direction. The velocity increment $\Delta\mathbf{v}$ due to the burn is then

$$\Delta\mathbf{v} = -\mathbf{v}_{ex} \ln(M_0/M_1) \tag{3.17}$$

M_0/M_1 is the *mass ratio*, the ratio between the masses M_0 before and M_1 after the burn. In the form above $\mathbf{v}_{ex}$ is taken to be positive in the direction of travel; the thrust vector is in the opposite direction.

Equation (3.17) is frequently used as an approximation when the burn is short enough that it may be considered to be impulsive. Then by pointing the rocket in different directions it may be used to achieve an increase or decrease in speed, or a change of direction, or any combination which is consistent with the vector diagram shown in Figure 3.8.

A *change of direction* $\Delta\psi$, for example, without any change of speed, will occur if the initial and final velocity vectors in Figure 3.8 form an isosceles triangle with Δv as base. The angle turned through will be

$$\Delta\psi = 2 \arcsin(\tfrac{1}{2}\Delta v/v) \tag{3.18}$$

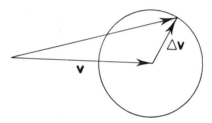

Figure 3.8   Possible changes of velocity due to a rocket's $\Delta\mathbf{v}$

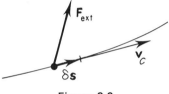

Figure 3.9

and the thrust needs to be directed at an angle $(90 + \psi/2)$ degs to the initial direction of travel.

3.2.5 Translational kinetic energy

An energy equation may be obtained as the first integral of equation (3.7), and this relationship can be very useful in the subject of celestial mechanics.

The translational kinetic energy may be expressed as

$$\text{K.E.} = \tfrac{1}{2}Mv_C^2 \quad \text{or} \quad \tfrac{1}{2}M\mathbf{v}_C \cdot \mathbf{v}_C, \quad \text{or} \quad \tfrac{1}{2}\mathbf{L} \cdot \mathbf{v}_C \tag{3.19}$$

The change in this is equal to the work done by the external forces $\mathbf{F}_{\text{ext}}$, and it may be expressed as

$$\Delta\text{K.E.} = \Delta\tfrac{1}{2}M\mathbf{v}_C \cdot \mathbf{v}_C = \int \mathbf{F}_{\text{ext}} \cdot \text{d}\mathbf{s} \tag{3.20}$$

where ds is the incremental change in position (see Figure 3.9).

The conservative forces may be separated from the nonconservative ones since their contribution to the integral may be expressed in terms of potential energy (P.E.). The resulting energy equation becomes

$$\Delta\text{K.E.} + \Delta\text{P.E.} = \int \mathbf{F}_{\text{nc}} \cdot \text{d}\mathbf{s} \tag{3.21}$$

where $\mathbf{F}_{\text{nc}}$ is the nonconservative contribution to the total external force $\mathbf{F}_{\text{ext}}$.

When the only force on the spacecraft is that due to gravitational attraction, a conservative force whose potential energy is $-\mu M/r$, then the orbital energy equation becomes

$$\tfrac{1}{2}M\mathbf{v}_C \cdot \mathbf{v}_C - \mu M/r = \text{constant.} \tag{3.22}$$

3.3 GENERAL ATTITUDE DYNAMICS

Trajectory dynamics supplies rules governing the motion of the centre of mass C relative to some inertially fixed reference point I. Attitude dynamics on the other hand uses the centre of mass as a reference point.

In terms of momentum, attitude dynamics is mathematically identical to trajectory dynamics. That is to say angular momentum **H** responds to a torque **T** in exactly the same way as linear momentum **L** does to a force **F**. But the physical motions associated with the two types of momentum are quite different.

In order to establish the fundamental principles in terms of momentum it is convenient to refer once more to the closed surface S (Figure 3.1), the boundary which separates particles which are of interest from those which are not. These results are then quite general, covering spacecraft with fluids and moving parts as well as the important cases of a rigid body and multiple bodies; the rules for the physical motions of the latter are dealt with later.

3.3.1 Angular momentum H

The angular momentum $\mathbf{H}_O$ referred to a point O is defined as the aggregate of the moments of the momenta of all the particles within S. In mathematical terms angular momentum referred to O is defined as

$$\mathbf{H}_O = \Sigma(\mathbf{r} \times m\mathbf{v}) \tag{3.23}$$

where both **r** and **v** are relative to O. It is a product of rotation in much the same way as linear momentum is a product of translation.

The *reference point* which is most useful for attitude dynamics is the centre of mass C. For example the rotational motion of a spacecraft is normally derived from the equation (3.25), using its momentum $\mathbf{H}_C$ referred to its centre of mass.

The rule governing the transfer of reference point from C to some other point O is

$$\mathbf{H}_O = \mathbf{H}_C + M\mathbf{h}_O = \mathbf{H}_C + (\mathbf{r} \times M\dot{\mathbf{r}}) \tag{3.24}$$

Figure 3.10 illustrates the terms used. The transfer involves adding on the moment of the momentum of the e.p., a process similar to the transfer of inertia from C to O (Appendix, equation (3.A9)). Equation (3.24) could be used, for example, for expressing the contribution to a spacecraft's total angular momentum referred to *its* centre of mass, which arises from a momentum wheel. But see also equation (3.32).

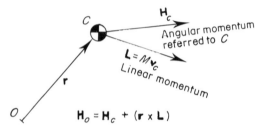

Figure 3.10 Transfer of reference point from C to O

3.3.2 Rate of change of angular momentum, H

The following explanation applies only when the reference point is the centre of mass C or an inertially fixed point I. It does not apply for other points.

The angular momentum may be changed in two ways, the first being

- by applying an external couple, or a force which has a moment about the reference point. The term 'external torque' (**T**) will be used as a general term to cover both of these. Its consequence is a rate of change of **H**, referred to C or I, described by the Newtonian equation:

$$d(\mathbf{H})/dt = \mathbf{T} \qquad (3.25)$$

Internal torques, acting between particles or bodies, will not change the total momentum. Thus mechanisms, fuel movement, etc., will not change the total angular momentum of a spacecraft. Similarly the forces between two docking spacecraft will not affect their combined angular momentum.

There is an important consequence for spacecraft designers since there will always be naturally occurring external disturbance torques (See Section 10.4). Their mean level will therefore cause a progressive build-up of the angular momentum over the lifetime of the craft. The rotational motion associated with this would be quite unacceptable. It follows that spacecraft must be fitted with means of controlling this build-up, and only external torquers are capable of doing so.

The second means of changing **H** is

- by the ejection from S of some particles whose momenta have moments about the reference point. This will occur during the firing of rockets, when their thrust vector does not pass precisely through the centre of mass. An analysis of this process will not be covered in this book. (A procedure similar to that used above in the development of equations (3.13), (3.14), and (3.17) can be followed if required.)

The consequences of equation (3.25) may be seen to be mathematically the same as those of equation (3.7). The deductions (1)–(3) on page 37 can therefore be rephrased so as to apply to angular momentum in response to torque. Thus

1. When there is no external torque **T** then the total angular momentum **H** is constant, in magnitude and direction.

   For example under a central gravitational force the force on each particle of the spacecraft has zero moment about the gravitational centre I. It follows that $\mathbf{H}_I$ is constant. In view of equation (3.10) it follows that fluctuations in $\mathbf{H}_C$ due to gravity–gradient torque will be accompanied by equal and opposite fluctuations in the moment of the momentum of the e.p., $M\mathbf{h}_I$. But it is a very small effect.

2. An external torque in the same direction as the angular momentum **H** will increase its magnitude without changing its direction, the equivalent of Figure 3.5(a).

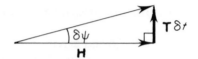

Figure 3.11 Precession due to
an external torque

3. An external torque which is at all times at right angles to the angular momentum
    will change its direction but not its magnitude. This is the equivalent of the situation
    shown in Figure 3.5(b) and reproduced as Figure 3.11. The best-known physical
    illustration of this is the precession of a gyroscope.

From Figure 3.11 it may be seen that as the angular momentum becomes large then
the change in direction due to a given torque impulse $T\ \delta t$ becomes small, leading to the
phrase *gyroscopic rigidity*. The rate of turn or precession rate may be seen to be

$$d\psi/dt = T/H \qquad (3.26)$$

This characteristic is made use of by spacecraft designers when they give their craft
momentum bias, as a means of making the bias direction insensitive to disturbance
torques. Making H large causes the precession rate to become small in response to a given
torque.

The *rate of change of the components of* $\mathbf{H}$ depends upon the rotation of the axis system
chosen, in addition to the change described by the Newtonian equation (3.25). When the
axis system has an angular velocity Ω relative to inertial space, then equation (3.25) must
be interpreted using the Coriolis theorem, as

$$d(\mathbf{H}_C)/dt = d/dt_{\text{compts}}(\mathbf{H}_C) + (\Omega \times \mathbf{H}_C) = \mathbf{T} \qquad (3.27)$$

where $d/dt_{\text{compts}}(\ )$ means the rate of change of the components of $(\ )$.

3.3.3 Angular momentum of rigid bodies

The angular momentum of a rigid body may be expressed in terms of its angular velocity
ω. The angular momentum equations above can then be used in order to develop
equations in ω which describe the attitude and rotational motion of such a body, and of
systems made up of rigid bodies.

The angular momentum $\mathbf{H}_C$ of a single rigid body referred to its centre of mass C may
be expressed as

$$\mathbf{H}_C = [I_C]\omega \qquad (3.28)$$

where ω is its angular velocity relative to an inertial (non-rotating) frame of reference
        $[I_C]$ is the inertia matrix based upon the centre of mass C.

In general $[I_C]$ may be expressed as

$$[I_C] = \begin{bmatrix} I_{xx} & -I_{xy} & -I_{zx} \\ -I_{xy} & I_{yy} & -I_{yz} \\ -I_{zx} & -I_{yz} & I_{zz} \end{bmatrix} \tag{3.29}$$

where I_{xx}, I_{yy}, I_{zz} are the moments of inertia.

I_{xy}, I_{yz}, I_{zx} are the products of inertia, these broadly representing a measure of the lack of mass symmetry, leading to cross-coupled behaviour, as will be seen later.

The Appendix to this chapter contains a full treatment of the subject of inertia, including the dependency of the elements upon the choice of axes. It is sufficient to record here that every body has a set of orthogonal axes at each point, for which the products of inertia are zero. These are called *principal axes*, and there may well be more than one set of them, depending upon the mass symmetry of the body. Principal axes are eigenvectors of the inertia matrix.

It is evident from equation (3.28) that the components of the angular momentum may in general be expressed as

$$\mathbf{H}_C = \begin{bmatrix} (I_{xx}\omega_x - I_{xy}\omega_y - I_{zx}\omega_z) \\ (I_{yy}\omega_y - I_{yz}\omega_z - I_{xy}\omega_x) \\ (I_{zz}\omega_z - I_{zx}\omega_x - I_{yz}\omega_y) \end{bmatrix} \tag{3.30}$$

When principal axes are used, then

$$\mathbf{H}_C = \{I_{xx}\omega_x, I_{yy}\omega_y, I_{zz}\omega_z\}^T \tag{3.31}$$

It is usual to develop the rotational equations for a body by choosing axes in which the inertias are constant. Axes fixed in the body will always achieve this, but there are other options if there is a mass symmetry. For example, a spinning object which has cylindrical mass symmetry about its spin axis, such as a typical wheel, will have constant inertias in non-spinning coordinate axes.

The *angular momentum of a rigid body with spinning wheels*, such as a spacecraft fitted with momentum or reaction wheels, can be expressed as the sum of the angular momentum of the rigid body containing the wheels in their non-spinning state (equation (3.30) or (3.31)), together with the extra momentum due to the angular velocities of the wheels *relative to the body*.

For example, suppose a spacecraft is fitted with a wheel which is spinning with an angular velocity $\boldsymbol{\omega}_{wh}$ relative to the craft. The vector $\boldsymbol{\omega}_{wh}$ will be along the wheel's axis of symmetry, which will be in a fixed direction in the spacecraft's axes. If its moment of inertia about its own axis is I_{wh}, then an additional angular momentum equal to $I_{wh}\boldsymbol{\omega}_{wh}$ must be added to that of the spacecraft's body. If there are several wheels and as a whole they contribute the additional momentum components $\{H_x, H_y, H_z\}^T$, say, then the total angular momentum of the body plus wheels becomes

$$\mathbf{H}_C = \begin{bmatrix} I_{xx}\omega_x - I_{xy}\omega_y - I_{zx}\omega_z + H_x \\ I_{yy}\omega_y - I_{yz}\omega_z - I_{xy}\omega_x + H_y \\ I_{zz}\omega_z - I_{zx}\omega_x - I_{yz}\omega_y + H_z \end{bmatrix} \tag{3.32}$$

If the body's principal axes are used, then this becomes

$$\mathbf{H}_C = \{(I_{xx}\omega_x + H_x), (I_{yy}\omega_y + H_y), (I_{zz}\omega_z + H_z)\}^{\mathrm{T}} \tag{3.33}$$

Multiple rigid bodies of a more general nature need to be addressed in order to deal with the docking manoeuvre, ejecting payloads from a launcher's cargo bay, yo-yo despin, astronauts repairing spacecraft, tethered spacecraft etc. For these multi-body situations each separate body will obey the dynamic equations which are stated elsewhere in this chapter. But it is worth noting that their combined momenta will not be changed by any forces or moments of interaction between them, and so will not be affected by a collision, separation or tethering. This applies to their absolute linear momentum, and also to their angular momentum about their combined centre of mass, C.

For example, consider two bodies with masses M_1, M_2, as in Figure 3.12, and total mass M. Assume that their centres of mass at C_1, C_2 have absolute velocities $\mathbf{v}_1$, $\mathbf{v}_2$, and that referred to these they have angular momenta $\mathbf{H}_1$, $\mathbf{H}_2$.

Then during a collision or separation, or due to a tether:

- their total absolute linear momentum remains constant, so

$$M\mathbf{v}_G = M_1\mathbf{v}_1 + M_2\mathbf{v}_2 \text{ remains constant} \tag{3.34}$$

- their angular momentum referred to C remains constant, so

$$\mathbf{H}_G = (M_1 M_2/M)(\mathbf{r}_{12} \times \mathbf{v}_{12}) + \mathbf{H}_1 + \mathbf{H}_2 \text{ remains constant} \tag{3.35}$$

where $\mathbf{r}_{12}$, $\mathbf{v}_{12}$ are the position and velocity vectors of C_2 relative to C_1.

These conservation of momentum laws also apply to the erection of solar arrays, etc., and when using pointing mechanisms.

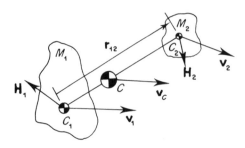

Figure 3.12 The combined momenta of two rigid bodies

3.3.4 Rotational kinetic energy

The rotational energy of a rigid body may be expressed as

$$E = \tfrac{1}{2}(\mathbf{H}_C \cdot \boldsymbol{\omega}) \quad \text{or} \quad \tfrac{1}{2}([I_C]\boldsymbol{\omega} \cdot \boldsymbol{\omega}) \tag{3.36}$$

This is additional to the translational energy of the e.p. quoted in equation (3.19).

Work is done by a torque $\mathbf{T}$ at a rate $\mathbf{T} \cdot \boldsymbol{\omega}$, and the rate at which the rotational energy is increased by a torque $\mathbf{T}_C$ about C is $\mathbf{T}_C \cdot \boldsymbol{\omega}$. It follows that in the absence of any external torque $\mathbf{T}_C$ the rotational energy will remain constant.

When the body is not rigid, however, or when there are moving parts, then the energy level can change without there being any external torque. In a spacecraft there will be internal dissipative mechanisms such as flexure, passive nutation dampers, etc., which lead to the loss of kinetic energy but at the same time the angular momentum $\mathbf{H}_C$ remains constant. A notable consequence of this is that the long-term tendency of a spinning body is towards spinning about its axis of maximum inertia, this being its minimum-energy state.

3.4 ATTITUDE MOTION OF SPECIFIC TYPES OF SPACECRAFT

The general theory in Section 3.3 may now be applied to specific types of spacecraft. For this purpose it will be assumed that all craft have rigid bodies, and have rigid moving parts unless otherwise stated. Whilst the response of their angular momentum to a torque is well ordered and straightforward, as explained above, the associated rotational motion is by no means necessarily so well behaved.

Spacecraft may be classified for convenience, as shown in Figure 3.13. The main subdivision depends upon whether the spacecraft has momentum bias or not. Bias means that the craft has a significant amount of angular momentum due to the spin of part or all of it. It then behaves like a gyroscope, with the associated characteristics of gyroscopic rigidity and a precessional type of response to a torque, as described in equation (3.26).

Some spacecraft may be spun-up for only a short time. It is common practice to do so prior to the firing of a high thrust rocket for example. This is primarily to prevent any thrust offset from causing the craft to veer off course, the spin causing the mean path to be straight. Subsequently the craft will be spun-down to the level of bias which is required for normal operation.

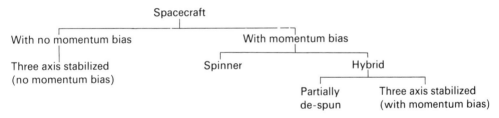

Figure 3.13 Categories of spacecraft

Cross-coupling is an essential feature of precessional response. The precession axis is at right angles to the axis of the torque which caused it. It is nevertheless a systematic and orderly effect. Normally, large cross-couplings are to be avoided so that control about each axis can proceed without interfering with the other axes, and vice versa. The rigid body dynamic equations developed in this section enable the sources of cross-couplings to be identified.

3.4.1 Three-axis-stabilized spacecraft with no momentum bias

Spacecraft in this class are usually large, with extensive solar arrays. They have a variety of different shapes such as the notional one shown in Figure 3.14. Their angular velocity is normally small, perhaps one revolution per orbit in order to maintain one face pointing towards Earth. The solar arrays may have even less angular movement, being required to point towards the Sun, and they may be mounted on a bearing to permit this.

The craft will be treated as a single rigid body. This is a reasonable approximation for the main structure, but the solar arrays are very flexible. Nevertheless the presentation brings out some important points which are relevant to the 'rigid body' responses of these craft.

Using principal axes which are fixed in the body, their angular velocity Ω will be that of the body, ω. Inserting $\mathbf{H}_C$ from equation (3.31) into equation (3.27) leads to

$$
\left.
\begin{aligned}
I_{xx}\dot{\omega}_x - (I_{yy} - I_{zz})\omega_y\omega_z &= T_x \\
I_{yy}\dot{\omega}_y - (I_{zz} - I_{xx})\omega_z\omega_x &= T_y \\
I_{zz}\dot{\omega}_z - (I_{xx} - I_{yy})\omega_x\omega_y &= T_z
\end{aligned}
\right\} \qquad (3.37)
$$

The alignment of the torquer axes with principal axes has the advantage that from an initially stationary condition the use of a single torquer, T_x, say, will produce a response about its own axis, without any cross-coupling into the other axes. The response is an

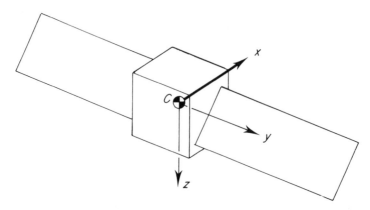

Figure 3.14   The three-axis stabilizd spacecraft

angular acceleration given by

$$\dot{\omega}_x = T_x/I_{xx} \tag{3.38}$$

Since torquers produce couples this condition is met when their axes are parallel to the principal axes through C. They do not need to coincide with them.

In the more general case when there are products of inertia, the initial acceleration from rest will be

$$\dot{\omega} = [I_C]^{-1}\mathbf{T} \tag{3.39}$$

Cross-coupling will occur, and if a pure response $\dot{\omega}_x$ is required, then potentially all three torquers must be used.

Cross-coupled motion will also result if the x-torquer is used when there is already an angular rate about the y- or z-axes. It must be assessed when a re-pointing manoeuvre is needed. It may be that the manoeuvre will best be carried out as a sequence of separate rotations about the principal axes, but there are a number of possible routes which may be taken in changing from one attitude to another.

In the absence of any external torque a stationary spacecraft will remain in that condition indefinitely. Once it is rotating then its motion may be divergent. The stability of rotating motion is dealt with in Section 3.4.2.

3.4.2 Spinning spacecraft

When the whole of the spacecraft structure spins then it has angular momentum. At very low values this leads to minor cross-couplings such as are mentioned above. But when the momentum becomes large and deliberate, then the cross-couplings become so great that the behaviour needs to be looked at afresh. The behaviour of the spacecraft becomes like that of a gyroscope rotor, or usually its designer would like it to be so. This will only

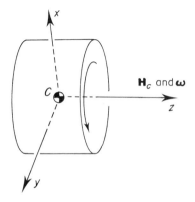

Figure 3.15 The spinning space-craft

be achieved if he observes constraints upon the choice of spin axis and upon the mass distribution, as indicated below.

The *choice of spin axis* is important. If a body is spinning with angular velocity ω then the spin axis will not in general remain in a fixed direction in space. It is the angular momentum $\mathbf{H}_C$ which has the fixed direction. If it is required that ω is also fixed, then it follows that it must be aligned with $\mathbf{H}_C$, i.e. $[I_C]\omega$. This can only happen if the spin axis is along an eigenvector of the inertia matrix, which is to say that the spin must take place about a principal axis.

If rotation is initiated about some axis other than a principal axis then the physical behaviour depends upon a number of factors. In general the angular velocity vector varies in direction and magnitude. The motion may be anything from an untidy tumbling type of motion to an oscillatory coning motion, depending upon the magnitudes of the initial rotation, the departure of its axis from a principal axis, and the disparity between the inertias. When faced with this type of behaviour, and also the instabilities referred to in the following paragraphs, it is difficult to believe that the angular momentum vector $\mathbf{H}_C$ is fixed in magnitude and direction.

The *stability of a spinning body* may be explored by using equation (3.37), with no torque components present. It will be assumed that the spin is about the z-axis and has a rate $\omega_z = S$, consistent with Figure 3.15. Two of equations (3.37) then become

$$I_{xx}\dot{\omega}_x - (I_{yy} - I_{zz})S\omega_y = 0$$

and   (3.40)

$$I_{yy}\dot{\omega}_y - (I_{zz} - I_{xx})S\omega_x = 0$$

These are linear equations whose characteristic equation in terms of the Laplace operator, s, is:

$$s^2 + (1 - I_{zz}/I_{xx})(1 - I_{zz}/I_{yy})S^2 = 0$$ (3.41)

It may be seen that for the spinning motion to be stable I_{zz} must be the maximum or the least of the moments of inertia.

Bodies which are designed to spin will normally have a mass distribution which is axisymmetric, meaning that $I_{xx} = I_{yy}$. The moments of inertia about all radii through C then have this same value.

There are many examples of such axially symmetric spinning bodies. Gyroscope rotors, vehicle wheels, propellers, etc. all spin about their axes of maximum inertia. On the other hand rifle bullets, expendable launchers, guided missiles, etc., spin about their axes of least inertia.

Long-term stability is important for spacecraft which must remain spinning for their lifetime. If they spin about their axis of least inertia, they will be stable in the short term only. Their long-term behaviour will be unstable if there is a loss of rotational energy brought about by internal dissipation rather than by an external torque. They will eventually adopt a cartwheeling type of motion—a spin motion about the axis of maximum inertia. During this process the angular momentum $\mathbf{H}_C$ remains constant whilst the energy $\frac{1}{2}\mathbf{H}_C \cdot \omega$ decreases, and it follows that the motion must move towards the minimum-energy state in which ω is least. With $[I_C]\omega$ remaining constant this means that it moves towards rotation about an axis of maximum inertia.

Thus it follows that spacecraft which are total spinners will spin about their axis of maximum moment of inertia.

The *effect of a torque* **T** is to change the angular momentum $\mathbf{H}_C$, as explained in Section 3.3. For the physical response its effect upon the angular velocity $\boldsymbol{\omega}$ is required, and the appropriate equations may be obtained by inserting $\mathbf{H}_C$ in the form $[I_C]\boldsymbol{\omega}$ into equation (3.27).

In the following analysis it will be assumed that the spacecraft is initially spinning about its z-axis, a principal axis of maximum or least inertia. When this is the only angular motion the craft is in a state of stable equilibrium in the short term, as shown above.

Repointing of the spin axis will normally call for a rotation about an axis which is at right angles to the spin axis, and which is fixed in space rather than fixed in the rotating structure. Coordinate axes which are convenient for the analysis are therefore non-spinning ones with the z-axis aligned with the spin axis. These axes have angular velocity $\boldsymbol{\Omega} = \{\Omega_x, \Omega_y, 0\}^T$, say.

The spacecraft spins at a rate $\{0, 0, S\}^T$ relative to the coordinate axes, and so its angular velocity $\boldsymbol{\omega}$ is $\{\Omega_x, \Omega_y, S\}^T$.

The *inertia matrix* $[I_C]$ referred to the coordinate axes has elements which are in general changing. If the moments of inertia referred to the spacecraft's principal axes are $\{I_{xx}, I_{yy}, I_{zz}\}$, and these axes are at an angle ψ to the coordinate axes, where $\dot{\psi} = S$, then the inertia matrix $[I_C]$ is obtainable from Appendix equation (3.A10):

$$[I_C] = \begin{bmatrix} (I_+ - I_- c) & I_- s & 0 \\ I_- s & (I_+ + I_- c) & 0 \\ 0 & 0 & I_{zz} \end{bmatrix} \tag{3.42}$$

where $I_+ = \frac{1}{2}(I_{yy} + I_{xx})$, the mean inertia orthogonal to the spin axis,
$I_- = \frac{1}{2}(I_{yy} - I_{xx})$, a measure of lack of axial symmetry,
and $c, s = \cos 2\psi$ and $\sin 2\psi$ respectively.

The angular momentum $\mathbf{H}_C$ is then

$$\mathbf{H}_C = [I_C]\boldsymbol{\omega} = \begin{bmatrix} \Omega_x(I_+ - I_- c) + \Omega_y I_- s \\ \Omega_x I_- s + \Omega_y(I_+ + I_- c) \\ S I_{zz} \end{bmatrix} \tag{3.43}$$

Noting that $dc/dt = -2Ss$ and $ds/dt = 2Sc$, then equation (3.27) leads to

$$\left. \begin{array}{l} I_+ \dot{\Omega}_x + I_{zz} S\Omega_y + I_- \{-(c\Omega_x - s\Omega_y) + 2S(s\Omega_x + c\Omega_y)\} = T_x \\ I_+ \dot{\Omega}_y - I_{zz} S\Omega_x + I_- \{(s\Omega_x + c\Omega_y) + 2S(c\Omega_x - s\Omega_y)\} = T_y \\ I_{zz}\dot{S} + \qquad\qquad I_- \{s\Omega_x^2 + 2c\Omega_x\Omega_y - s\Omega_y^2\} \qquad\qquad = T_z \end{array} \right\} \tag{3.44}$$

A number of conclusions can be drawn:

1. A torque T_z about the z-axis causes a simple acceleration $\dot{S}$ about that axis when the spacecraft is in its equilibrium state, equal to

$$\dot{S} = T_z/I_{zz} \tag{3.45}$$

2. Repointing the spin axis by means of a constant precession rate Ω_y about the y-axis,

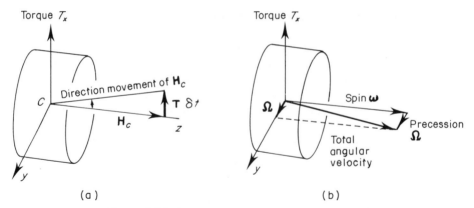

Figure 3.16 Precession of a spinner due to a torque

say, will require a torque T_x about the x-axis. Only if the moments of inertia I_{xx} and I_{yy} are equal, i.e. ($I_- = 0$), will the torque be constant. Under this condition,

$$\Omega_y = T_x/I_{zz}S \tag{3.46}$$

This represents complete cross-coupling between x- and y-axes, which is a characteristic of a gyroscope's precessional behaviour. But in addition it should be noted that a torque produces an angular velocity rather than the acceleration produced by T_z. Figure 3.16(a) shows how the angular momentum $\mathbf{H}_C$ is rotated towards the torque vector $\mathbf{T}$, and Figure 3.16(b) shows the total angular velocity, with contributions from $\boldsymbol{\omega}$ and $\boldsymbol{\Omega}$.

If I_{xx}, I_{yy} differ, then additional oscillating torques will be needed, namely

$$T_x = 2cSI_-\Omega_y \qquad \text{and} \qquad T_y = 2sSI_-\Omega_y \tag{3.47}$$

Furthermore, if a constant rate of spin S is to be maintained, then a torque will be needed, equal to

$$T_z = -sI_-\Omega_y^2 \tag{3.48}$$

Failure to provide these oscillating torques will lead to an unsteady precession in response to a constant torque T_x. Most objects which are designed to spin will be given axial mass symmetry, i.e. $I_- = 0$, or $I_{xx} = I_{yy}$, and this applies to spacecraft too.

It should be noted that the components T_x and T_y are in non-spinning axes and that torquers fixed in the spacecraft will need to apply resolved versions of T_x, T_y, namely:

$$\begin{array}{c} \text{Torque components} \\ \text{in spacecraft axes} \end{array} = \begin{bmatrix} \cos\psi & \sin\psi & 0 \\ -\sin\psi & \cos\psi & 0 \\ 0 & 0 & 1 \end{bmatrix} \begin{bmatrix} T_x \\ T_y \\ T_z \end{bmatrix} \tag{3.49}$$

3. There is an oscillatory 'nutation' mode. This appears as a coning type of motion when the spacecraft is disturbed from its equilibrium state. Whenever precessional torques are applied the mode will be activated, and some engineered form of damping will be needed.

It may already be seen from the characteristic equation (3.40) that an oscillatory mode exists, whose frequency is

$$\omega_{\text{nut}} = S\sqrt{[(1 - I_{zz}/I_{xx})(1 - I_{zz}/I_{yy})]} \tag{3.50}$$

This is the frequency which may be observed on traces from rate gyroscopes which are mounted on the structure, i.e. when it is observed in the spinning axes.

In the axisymmetric case the nutation frequency becomes

$$\omega_{\text{nut}} = S|1 - I_{zz}/I_{xx}| \quad \text{when viewed from spinning axes,} \tag{3.51}$$

$$\text{or } S(I_{zz}/I_{xx}) \quad \text{when viewed from non-spinning axes.} \tag{3.52}$$

3.4.3 Hybrid spacecraft

Two classes of spacecraft bring together the advantages of having momentum bias and yet provide a non-spinning base for their payload and torquers. These are the partially despun or dual-spin craft, and the three-axis-stabilized craft with momentum bias. In each case momentum bias is provided by mounting a rotating body on the non-spinning part; it is part of the structure in the case of the dual-spin craft, but is a high-speed purpose-built wheel in the three-axis stabilized case.

The spinning parts of these hybrids have dynamic properties which are similar to those of the 'spinner' covered above. They will have axially symmetric mass properties for the reasons given. So their centres of mass lie on the spin axis, and about axes normal to this their moments of inertia will be equal. Their contribution to the total inertia $[I_C]$ of the spacecraft will then be independent of their rotation. Their rate of spin will remain constant unless there is a torque about their axis.

The non-spinning structure will behave like the three-axis-stabilized craft described above. Some of the torques on it will be applied via the bearing of the rotating part, and so it is desirable (but not essential) that the bearing and its torquer axis is parallel to a principal axis of the non-spinning part.

The angular momentum of the hybrid spacecraft is then given by the expressions in equation (3.32) or (3.33). These may be substituted into equation (3.27) in order to obtain equations obeyed by the components of the angular velocity of the non-spinning structure, in any coordinate axes.

If the momentum bias direction is along the z-axis, a principal axis, then

$$\mathbf{H}_C = \{I_{xx}\omega_x, I_{yy}\omega_y, (I_{zz}\omega_z + H_z)\}^{\text{T}} \tag{3.53}$$

The dynamic equations for the components of angular velocity along principal axes are

$$\left. \begin{array}{l} I_{xx}\dot{\omega}_x + \omega_y\omega_z(I_{zz} - I_{yy}) + \omega_y H_z = T_x \\ I_{yy}\dot{\omega}_y + \omega_z\omega_x(I_{xx} - I_{zz}) - \omega_x H_z = T_y \\ I_{zz}\dot{\omega}_z + \dot{H}_z + \omega_x\omega_y(I_{yy} - I_{xx}) = T_z \end{array} \right\} \tag{3.54}$$

A number of observations may be made:

1. *Stability.*   The momentum bias axis may be the axis of intermediate inertia without causing instability. The constraint imposed on the total spinner is not necessary here.

The long-term stability cannot be forecast from the above equations. The hybrid will however be stable provided that energy dissipation in the non-spinning part exceeds that in the spinning part. The bias direction may then be along the axis of least inertia. Passive nutation dampers for example will be placed in the non-spinning part of the spacecraft.

2. There is a *nutation mode* whose frequency, when observed in non-spinning axes, is

$$\omega_{nut} = H_z/\sqrt{(I_{xx}I_{yy})} \tag{3.55}$$

3. The *response to a torque.* The inclusion of momentum bias makes the whole structure behave as a gyrostat, with a response which is very similar to that of a spinner. A constant torque about the *x*-axis will cause a constant rate of precession about the *y*-axis, for example, whose magnitude is

$$\omega_y = T_x/H_z \tag{3.56}$$

On the other hand a torque about the bias direction will produce an angular acceleration about that direction if it is parallel to a principal axis. In other cases there will be cross-coupling.

3.5 IN CONCLUSION

The material in this chapter has been aimed primarily at the dynamics of the main categories of spacecraft. The benefits of adopting certain mass distributions and torquer axes have been shown, together with the consequence of doing otherwise.

There are likely to be increasingly many occasions in which the designer has to deal with non-optimum mass distributions, as, for example, when reusable launch vehicles part with their payload, and when large space structures are assembled or constructed in orbit. The fundamental laws still apply but the control systems will have to adjust to the changes in the mass distribution.

The methods contained in the chapter are not confined to the conventional configurations. Any configurations which may be treated as being made up of rigid bodies may be dealt with in the manner shown. Their linear momenta, moments of momentum and angular momenta will obey the Newtonian equations (3.7) or (3.15), (3.11), and (3.25) respectively. The components of their angular momentum $\mathbf{H}_C$ may be found by using equation (3.28) or (3.32), having first obtained their inertia matrix using the Appendix. All the Newtonian equations should be developed by using the Coriolis theorem as illustrated in equation (3.27), enabling the motion equations to be expressed in terms of the components of their velocities, linear and angular, along any desired axes.

APPENDIX: INERTIA

A1 Introduction

The inertia matrix $[I_0]$ referred to a point O is a property of the mass distribution about O. Its elements depend upon the directions of a right-handed orthogonal set of axes x, y, *z* through *O*, and in particular it contains the moments and products of inertia associated with these axes. The matrix at the centre of mass *C* plays an important part in the rotational behaviour of a spacecraft, and it must be evaluated and controlled during its design.

This appendix defines terms which are associated with mass distributions and presents formulae which are useful for the evaluation of the inertia matrix.

A2 Definitions

- *Moments of inertia I_{xx}, I_{yy}, I_{zz}.*
  A moment of inertia is the second moment of mass about an axis. The contribution of an increment of mass δm at a distance *d* from the axis is $d^2 \delta m$ (see Fig. 3.A1). The moment of inertia about the *x*-axis is, for example,

$$I_{xx} = \int (y^2 + z^2)\, dm \qquad (3.\text{A}1)$$

  where the integral extends over the whole mass distribution.

- *Products of inertia I_{xy}, I_{yz}, I_{zx}.*
  The product of inertia associated which the *x*-axis is

$$I_{yz} = \int yz\, dm \qquad (3.\text{A}2)$$

Products of inertia are measures of the lack of symmetry in a mass distribution.

If there is a plane of symmetry, then the product of inertia associated with all axes in that plane will be zero. For example an aircraft whose *xz*-plane is a plane of symmetry will have I_{xy} and I_{yz} equal to zero.

If two of the coordinate planes are planes of symmetry, then all three of the products of inertia will be zero. This applies to axially symmetric bodies such as many expendable launchers.

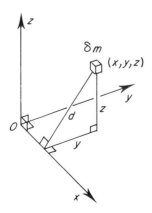

Figure 3.A1 An incremental mass δm

- *Principal axes* are sets of orthogonal axes for which all three products of inertia are zero. There is alway one such set at each point.

- *The inertia matrix* $[I_o]$ referred to $O(x, y, z)$ is defined as

$$[I_o] = \begin{bmatrix} I_{xx} & -I_{xy} & -I_{zx} \\ -I_{xy} & I_{yy} & -I_{yz} \\ -I_{zx} & -I_{yz} & I_{zz} \end{bmatrix} \tag{3.A3}$$

For a single particle with mass m at (x, y, z) this becomes $[I_{Om}]$, say, where

$$[I_{Om}] = \begin{bmatrix} m(y^2 + z^2) & -mxy & -mzx \\ -mxy & m(z^2 + x^2) & -myz \\ -mzx & -myz & m(x^2 + y^2) \end{bmatrix} \tag{3.A4}$$

- *Rotation matrix,* $[R]$

If the components of a vector $\mathbf{V}$ in one set of axes are expressed as the terms in a (3×1) column matrix $\mathbf{V}_1$, say, and $\mathbf{V}_2$ consists of its components in a second set which is rotated relative to the first, then $\mathbf{V}_2$ may be expressed as $\mathbf{V}_2 = [R]\mathbf{V}_1$. Then $[R]$ is known as a rotation matrix.

If Euler angles are used to describe the rotation as a sequence of separate rotations about the coordinate axes, then it is convenient to use a notation $[X(\phi)]$, $[Y(\theta)]$ and $[Z(\psi)]$ for the separate rotations. $[X(\phi)]$ is the rotation matrix for a clockwise

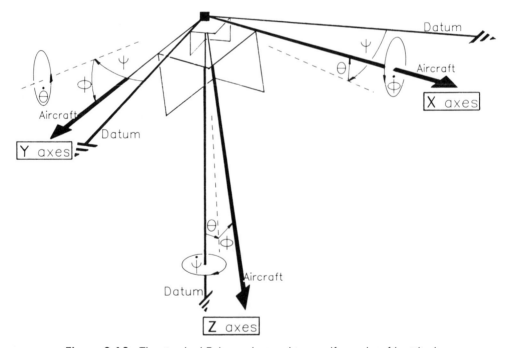

Figure 3.A2 The standard Euler angles used to specify an aircraft's attitude

rotation through an angle ϕ about the x-axis, and similarly for $[Y(\theta)]$ and $[Z(\psi)]$. These matrices are

$$[X(\phi)] = \begin{bmatrix} 1 & 0 & 0 \\ 0 & \cos\phi & -\sin\phi \\ 0 & \sin\phi & \cos\phi \end{bmatrix} \quad [Y(\theta)] = \begin{bmatrix} \cos\theta & 0 & \sin\theta \\ 0 & 1 & 0 \\ -\sin\theta & 0 & \cos\theta \end{bmatrix}$$

$$[Z(\psi)] = \begin{bmatrix} \cos\psi & -\sin\psi & 0 \\ \sin\psi & \cos\psi & 0 \\ 0 & 0 & 1 \end{bmatrix}$$

(3.A5)

For an aircraft whose attitude is defined in the conventional manner as the sequence of rotations yaw ψ, followed by pitch θ, followed by roll ϕ, as in Figure 3.A2, the matrix $[R]$ for converting the components of its velocity $\mathbf{V}$ say, from its own axes $\mathbf{V}_{a/c}$, to the datum axes $\mathbf{V}_{datum}$, is

$$\mathbf{V}_{datum} = [R]\mathbf{V}_{a/c}, \text{ where } [R] = [Z(\psi)][Y(\theta)][X(\phi)]$$

(3.A6)

A3 Useful formulae

- *Inertia invariant*

$$I_{xx} + I_{yy} + I_{zz} = 2\int r^2\, dm = 2I_O, \quad \text{say}$$

(3.A7)

where I_O is the second moment of mass about the origin O, and is independent of the direction of axes.

- *Perpendicular axis theorem*—for laminas only.
  For a lamina lying entirely in the yz-plane

$$I_{yy} + I_{zz} = I_{xx}$$

(3.A8)

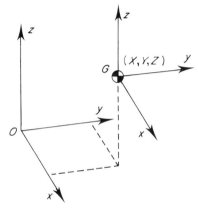

Figure 3.A3

- *Transfer of reference point* (parallel axis theorem)
  If an object whose centre of mass G is at (X, Y, Z) has an inertia matrix $[I_G]$ referred to G, then add on the inertia matrix of its equivalent particle referred to O, in order to obtain the inertia matrix $[I_O]$ referred parallel axes at O, i.e.,

$$[I_O] = [I_{OM}] + [I_G] \tag{3.A9}$$

where equation (3.A4) may be used for $[I_{OM}]$.

- *Rotated axes theorem*
  If the rotation matrix for transforming a vector's components from axes number 1 to axes number 2 is $[R]$ (see rotation matrices above), then the inertia matrix $[I]$ may be transformed between the same sets of axes by using

$$[I_2] = [R][I_1][R]^{-1} \tag{3.A10}$$

A4   Contribution of a piece of equipment to $[I_C]$

Suppose that a piece of equipment with mass M is installed with its centre of mass G at (x, y, z) relative to the spacecraft's axes at C. Suppose that $[I_{eq}]$ is its inertia matrix referred its own natural axes, and that these axes are rotated through an angle ϕ about the spacecraft's x-axis when it is installed.

Then its contribution to the inertia matrix $[I_C]$ of the spacecraft will be $[X(\phi)][I_{eq}]$ $[X(-\phi)]$ plus the inertia matrix of its e.p. referred to C (equation (3.A4)). In general,

$$[I_C] = [R][I_{eq}][R]^{-1} + [I_{CM}] \tag{3.A11}$$

4 CELESTIAL MECHANICS

John P.W. Stark

Department of Aeronautics and Astronautics, University of Southampton

4.1 INTRODUCTION

The theory of celestial mechanics underlies all the dynamical aspects of the orbital motion of spacecraft. The central feature is the mutual gravitational force of attraction which acts between any two bodies. This was first described by Newton, and together with his laws of motion (see Chapter 3), it provides us with the theoretical framework for celestial mechanics. The orbits which it forcasts will be relative to the stars, and the consequential motion relative to the ground will be covered in Chapter 5.

The simplified case in which the gravitational force acts between two point-like objects gives a good approximation to orbital motion for most spacecraft situations. It may easily be shown that if a body has a uniform mass distribution within a spherical surface, then outside it the gravitational force from the body does indeed appear to emanate from a point-like source. This so-called two-body problem has a solution which is a Keplerian orbit.

Kepler, whose major works were published during the first twenty years of the seventeenth century, consolidated the observations of planetary motion into three simple laws which are illustrated in Figure 4.1. These are:

1. The orbit of each planet is an ellipse with the Sun occupying one focus.

2. The line joining the Sun to a planet sweeps out equal areas in equal intervals of time.

3. A planet's orbital period is proportional to the mean distance between Sun and planet, raised to the power 3/2.

Newton's theory of gravity predicting the above results was not developed until more than 100 years later.

Various perturbing forces must be included in order to produce a better approximation to a spacecraft's orbit. At low altitude (< 1000 km) the departure from spherical symmetry of the Earth's mass and shape is important, together with aerodynamic forces. At high

Spacecraft Systems Engineering. Edited by P.W. Fortescue and J.P.W. Stark
© 1991 John Wiley & Sons Ltd

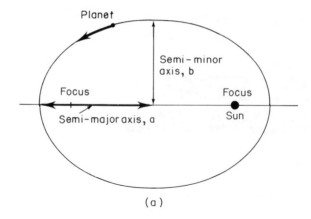

(a)

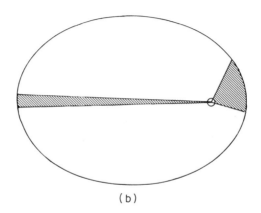

(b)

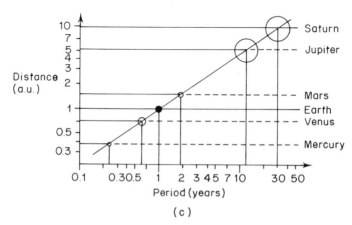

(c)

Figure 4.1 (a) Kepler's first law: orbit shape; (b) Kepler's second law: equal areas in equal time; (c) Kepler's third law: period $\tau \propto a^{1.5}$

altitudes such as in geostationary orbits ($\sim 36\,000$ km) the gravitational pull of the other celestial bodies such as the Moon and Sun become significant, together with radiation pressure from the Sun.

The theory of celestial mechanics is required in order that the motion of a spacecraft may be predicted. In most situations the accuracy required for the vehicle's position is relatively modest—typically a few kilometres. However, there are significant deviations from this, the most important being when the space vehicle must perform a manoeuvre to change its orbit. This topic is left to the next chapter. The problems of spacecraft tracking are left until Chapter 5. It should be noted that for some vehicles, particularly those which employ active remote sensing instrumentation, precise orbit prediction and subsequent determination are required. Accuracy of order 1 m was required for Seasat and values of ~ 10 cm have been quoted as a requirement for ERS1. Such precision in prediction is difficult at present because of our limited knowledge of the Earth's gravitational field and uncertainties surrounding aerodynamic drag. The precise orbit can only be determined retrospectively.

4.2 THE TWO-BODY PROBLEM—PARTICLE DYNAMICS

The simplest problem in celestial mechanics concerns the predicted motion of two co-rotating point masses. It is assumed in this analysis that the sole force acting between them is their mutual gravitational attraction. Newton demonstrated that the gravitational field from any point-like body is solely dependent upon its mass. The field's potential U at a distance R from a particle of mass M is simply

$$U = -\frac{GM}{R} \tag{4.1}$$

where G is the universal constant of gravitation, having a numerical value in SI units of $6.670 \times 10^{-11}\ \mathrm{N\,m^2\,kg^{-2}}$.

If a second particle, of mass m, is brought to a distance R from the first one as shown in Figure 4.2, then their joint centre of mass, the barycentre B, will lie on the line joining them. The forces $\mathbf{F}_1$, $\mathbf{F}_2$ on the particles may be expressed in terms of their distance from B, as

$$\mathbf{F}_1 = \frac{-GMm^3}{(M+m)^2 r_1^2}\frac{\mathbf{r}_1}{r_1} \quad \text{and} \quad \mathbf{F}_2 = \frac{-GM^3 m}{(M+m)^2 r_2^2}\frac{\mathbf{r}_2}{r_2} \tag{4.2}$$

These forces are both equal in magnitude to the force of attraction GMm/R^2, since $Mr_1 = mr_2$ and $r_1 + r_2 = R$.

The 'restricted two-body problem' is appropriate to spacecraft since it assumes that the

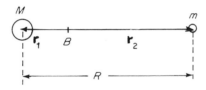

Figure 4.2

Table 4.1   Gravitational
parameters of selected bodies

Body	$\mu(\text{m}^3/\text{s}^2)$
Sun	1.327×10^{20}
Earth	3.986×10^{14}
Mercury	2.224×10^{13}
Venus	3.249×10^{14}
Mars	4.281×10^{13}
Jupiter	1.267×10^{17}
Saturn	3.794×10^{16}
Uranus	5.831×10^{15}
Neptune	6.865×10^{15}
Pluto	4.424×10^{13}
Moon	4.902×10^{12}

mass of one of the particles far exceeds that of the other, $m \ll M$ say; Earth (6×10^{24} kg) dominates the spacecraft ($\sim 10^3$–10^4 kg) for example. Then the barycentre is approximately at Earth's centre and for convenience $\mathbf{r}_2$ may be expressed as $\mathbf{r}$. Equation (4.2) then becomes

$$\mathbf{F} = -\frac{(GM)m}{r^3}\mathbf{r} \tag{4.3}$$

Using Newton's law in the form $\mathbf{F} = m\ddot{\mathbf{r}}$ leads to the orbital differential equation:

$$\ddot{\mathbf{r}} + \frac{(GM)\mathbf{r}}{r^2}\frac{}{r} = 0 \tag{4.4}$$

Equation (4.4) is also a good approximation for the planets and the Sun, since even the largest one, Jupiter ($\sim 2 \times 10^{27}$ kg), is dominated by the Sun ($\sim 2 \times 10^{30}$ kg).

The parameter GM, here referred to as μ, is the gravitational parameter for the body about which the motion is taking place. Table 4.1 lists its values for a selection of solar system bodies.

The orbit constants

Equation (4.4) may be solved and will lead to the orbit equation if the initial conditions are known. The constants associated with a particular orbit may be thought of as being constants of integration of equation (4.4) or they may be approached in physical terms via conservation laws, as in Chapter 3.

Moment of momentum conservation follows from the fact that the only force acting has no moment about the barycentre. The moment of the momentum vector $m\mathbf{h}$ is therefore constant, both in magnitude and direction.

By definition $\mathbf{h} = \mathbf{r} \times \dot{\mathbf{r}}$, and so its direction, perpendicular to both the position vector and the velocity vector, is normal to the orbit plane and is constant. The fact that the magnitude of $\mathbf{h}$ is constant is consistent with Kepler's second law. The value of this is $r^2\dot{\theta}$ (see Figure 4.3), which is twice the areal velocity $\dot{A}$.

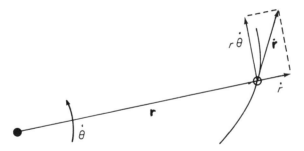

Figure 4.3

Energy conservation follows from the fact that the only external force on the spacecraft is due to the gravitational field, a conservative force whose potential energy per unit mass is $-\mu/r$.

The total energy per unit mass ε (the sum of the kinetic and gravitational energies), remains constant, and so

$$\frac{1}{2}V^2 - \frac{\mu}{r} = \varepsilon \qquad (4.5)$$

This is known as the energy equation or *vis-viva* integral.

The vector solution to equation (4.4) may be obtained by first taking the cross product with the constant $\mathbf{h}$ and integrating once with respect to time. This yields

$$\dot{\mathbf{r}} \times \mathbf{h} = \mu\left(\frac{\mathbf{r}}{r} + \mathbf{e}\right) \qquad (4.6)$$

where $\mathbf{e}$ is the vector constant of integration called the eccentricity vector, and lies in the plane of the orbit, as shown in Figure 4.4.

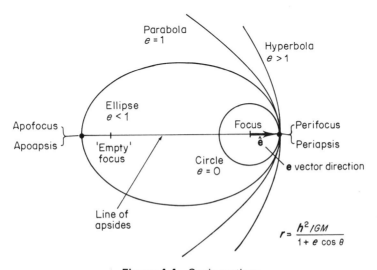

Figure 4.4   Conic sections

The final solution to equation (4.4), obtained by taking the dot product of (4.6) with r, is

$$r = \frac{h^2/\mu}{1 + e \cos \theta} \tag{4.7}$$

where θ is the angle between **r** and **e**. This is the equation of a conic section and demonstrates the first of Kepler's laws.

The *eccentricity* e determines the type of conic. It is a circle when $e = 0$, an ellipse when $0 < e < 1$, a parabola when $e = 1$, and a hyperbola when $e > 1$. These shapes are shown in Figure 4.4. The dominant mass is at one of the two foci.

The *ellipse* is the general form of a closed orbit and is shown in Figure 4.5. The major axis is called the *line of apsides*, recalling that half an ellipse is shaped like the apse of a church. The angle θ is termed *the true anomaly*. The dimensions are related and may be expressed in terms of the physical constants h and ε. For example:

The semi-latus rectum p and semi-major axis a obey:

$$p = a(1 - e^2) = h^2/\mu \tag{4.8}$$

The point of closest approach to the main body is termed the periapsis or perifocus and its distance is

$$r_p = p/(1 + e) = a(1 - e) \tag{4.9}$$

The most distant point, the apoapsis or apofocus, is at a distance

$$r_a = p/(1 - e) = a(1 + e) \tag{4.10}$$

Clearly $r_p + r_a = 2a$. The apses normally carry the name of the main body; so periapsis is called the perigee if the orbit is round Earth, perihelion if round the Sun etc. Equating the energy and the moments of momenta at the apses leads to

$$\varepsilon = -\mu/2a \tag{4.11}$$

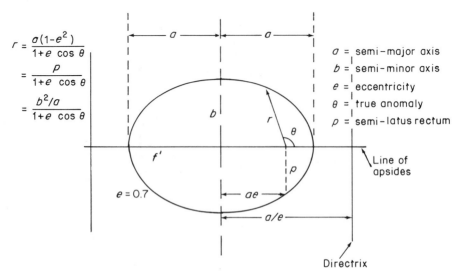

Figure 4.5 Ellipses

Kepler's third law is simply obtained, since the orbit period is the time taken for the complete area of the ellipse to be swept out by the radius vector. Since $\dot{A} = h/2$ and the area of an ellipse is $\pi a^2 \sqrt{(1 - e^2)}$, then by making use of the relationships above it follows that the orbit period τ is given by:

$$\tau = 2\pi \sqrt{(a^3/\mu)} \tag{4.12}$$

The *position versus time relationship* will be required for ground station passes, manoeuvres and other mission activities. One approach to providing this is to use $h = r^2\dot{\theta}$ together with equation (4.7). These yield

$$\dot{\theta} = \sqrt{\left(\frac{\mu}{p^3}\right)}(1 + e \cos \theta)^2 \tag{4.13}$$

This equation is valid for all types of conic. Whilst direct integration is possible, it is analytically more tractable to consider solutions separately for the three forms: ellipse (circle), parabola and hyperbola.

For *elliptical motion* $(0 \leq e \leq 1)$ a solution is obtained by constructing a circumscribing circle about the ellipse, as shown in Figure 4.6. This circle, of radius a, is called the *auxiliary circle*, and it is used to transform from the variable angle θ to a new variable E, called the *eccentric anomaly*. The position on the ellipse may be written in terms of E as

$$r = a(1 - e \cos E) \tag{4.14}$$

From geometrical considerations it can be shown that

$$\tan \theta/2 = \tan E/2 \sqrt{\left[\frac{1 + e}{1 - e}\right]} \tag{4.15}$$

It should also be noted that E is related to the shaded area indicated in Figure 4.6 by

$$E = \frac{2 \times \text{Area}}{a^2}$$

After some manipulation it is possible to show that

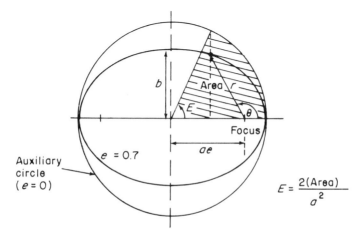

Figure 4.6 Eccentric anomaly definition

$$a(1 - e \cos E) \, dE = dt \sqrt{\frac{\mu}{a}} \tag{4.16}$$

which on integration yields:

$$t_p \sqrt{\frac{\mu}{a^3}} = E - e \sin E \tag{4.17}$$

where t_p is the time since perifocal passage. Equation (4.17) is known as Kepler's equation. It provides a simple relationship between time and position angle E which can be used in conjunction with (4.14) to provide a complete specification of position with time. It should be noted that $\sqrt{(\mu/a^3)}$ is the mean angular velocity or mean motion, about the focus. The *mean anomaly* M is defined as

$$M = t_p \sqrt{(\mu/a^3)} \tag{4.18}$$

and so Kepler's equation may be written

$$M = E - e \sin E \tag{4.19}$$

For a *parabolic trajectory* $e = 1$, and direct integration of equation (4.13) is straightforward. The solution, Barker's equation is

$$2\sqrt{\left(\frac{\mu}{p^3}\right)} t_p = 2M = \tan\frac{\theta}{2} + \frac{1}{3}\tan^3\frac{\theta}{2} \tag{4.20}$$

For a *hyperbolic trajectory* $(e > 1)$ a substitution analogous to the elliptic eccentric anomaly may be used. In this case the physical realization of the hyperbolic anomaly, F, is only in terms of the area noted in Figure 4.8. For the hyperbola the focal distance r is given by

$$r = a(1 - e \cosh F) \tag{4.21}$$

where

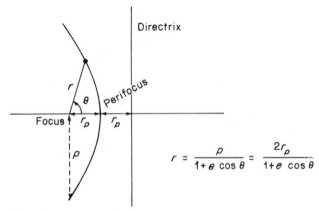

r_p = perifocal length

Figure 4.7   Parabola $(e = 1)$

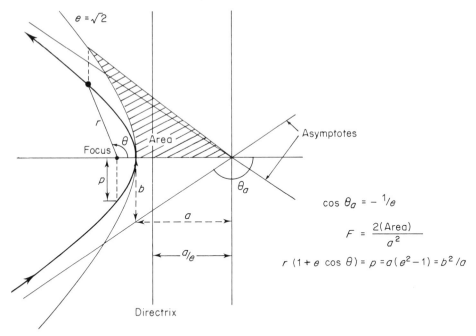

Figure 4.8 Hyperbola

$$\tan \theta/2 = \tanh(F/2)\sqrt{\left[\frac{e+1}{e-1}\right]} \qquad (4.22)$$

The resultant relationship between position and time is then found to be analogous to Kepler's equation:

$$M = e \sinh F - F \qquad (4.23)$$

The *velocity relationships* for Keplerian orbits are worth summarizing. They may be obtained primarily from equation (4.7) and the vis-viva integral (equations (4.5) and (4.11)), namely

$$\frac{1}{2}V^2 - \frac{\mu}{r} = \varepsilon = -\frac{\mu}{2a} \qquad (4.24)$$

For a *circular orbit* ($e = 0$) then the circular velocity is given by

$$V_{\text{circ}} = \sqrt{\frac{\mu}{r}} \qquad (4.25)$$

Using this, one can note that for Earth-orbiting space vehicles having a 24 hour orbit the radius will be 4.2×10^4 km, so the velocity is 3.1 km/s.

For a spacecraft to orbit the Earth at its surface (radius 6371 km), the velocity needed is 7.9 km/s. This is substantially larger than the Earth's equatorial velocity ~ 0.46 km/s. Thus this increment in velocity, 7.44 km/s, will be required just to orbit the Earth at zero altitude. This feature of orbit attainment will be developed further in Chapter 7 dealing with launch vehicles.

For an *elliptical orbit* $(0 < e < 1)$, the magnitude of the potential energy per unit mass is greater than the kinetic energy. The total energy ε per unit mass is negative. This is a necessary condition for a closed orbit, and may be used to determine planetary capture requirements for interplanetary vehicles.

The *parabolic 'orbit'* $e = 1$, and $\varepsilon = 0$, represents the boundary case between a captive, orbiting satellite and an escaping spacecraft. For equation (4.5), with $\varepsilon = 0$, the velocity required for escape is given by:

$$V_{esc} = \sqrt{\left(\frac{2\mu}{r}\right)} \tag{4.26}$$

This is a factor $\sqrt{2}$ greater than the local circular velocity.

For *hyperbolic trajectories* $e > 1$ the energy ε per unit mass is positive and a is negative. The energy of the motion is dominated by the kinetic energy.

As r becomes large $(r \to \infty)$ the velocity is given by

$$V_{\infty} = \sqrt{\left(\frac{\mu}{|a|}\right)} \tag{4.27}$$

V_{∞} is the *hyperbolic excess velocity*. The asymptotic direction along which the spacecraft escapes is obtained from equation (4.7) and Figure 4.8 as

$$\theta_{\infty} = \cos^{-1}\left\{-\frac{1}{e}\right\} \tag{4.28}$$

It may be shown from these equations [1] that the eccentricity, hyperbolic excess velocity and periapsis distance are simply related:

$$e = 1 + \frac{r_p V_{\infty}^2}{\mu} \tag{4.29}$$

4.3 KEPLERIAN ORBIT TRANSFERS

The situation frequently arises that a space vehicle must be transferred from one orbit to another. The detailed optimization of these manoeuvres is beyond the scope of this text. (For detailed analysis see [2] and [3].) However, it is appropriate to consider the nature of simple impulsive manoeuvres here, applied to Keplerian orbits.

If a single impulsive manoeuvre is performed, then the initial and final orbits intersect at the location of the manoeuvre; thus a single manoeuvre may only transfer a vehicle between intersecting orbits. *At least* two manoeuvres will be required to transfer a vehicle between two non-intersecting orbits.

A transfer from orbit A to an intersecting orbit B will occur if at some point, say r_1, the velocity vector is instantaneously changed from its value on orbit A to that which it would be on orbit B. The simplest case is that for a coplanar transfer from a circular orbit to an elliptical orbit. If on the circular orbit the velocity is increased, then the semi-major axis will be expanded.

A *Hohmann transfer* makes use of two such manoeuvres in order to transfer from one circular orbit to a larger coplanar one, as shown in Figure 4.9. It is the minimum energy, two-manoeuvre transfer, and is optimal (minimum velocity increment ΔV) if $r_2/r_1 < 11.8$.

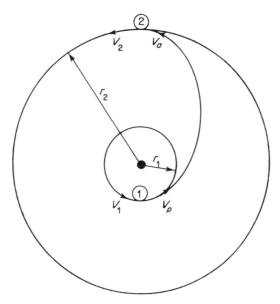

Figure 4.9 Transfer velocity definitions

The velocity increments needed for the two manoeuvres may be assessed by using the vis-viva integral equations (4.5) and (4.11). The first one requires an increase from the circular orbit speed $\sqrt{\mu/r_1}$ to the speed V_p at the periapsis of the transfer ellipse, where

$$V_p^2 = 2\mu \left\{ \frac{1}{r_1} - \frac{1}{r_1 + r_2} \right\}$$

This represents a ΔV given by:

$$\Delta V = \sqrt{\left(\frac{\mu}{r_1}\right)} \left\{ \sqrt{\frac{2r_2}{r_1 + r_2}} - 1 \right\} \tag{4.30}$$

A second, similar manoeuvre will transfer the vehicle into a circular orbit at apoapsis,

Table 4.2 Hohmann transfer data from the Earth to the planets

	a(AU) Transfer	T(years)	ΔV(km/s)
Mercury	0.847	0.289	5.6
Venus	0.931	0.400	3.5
Mars	1.131	0.709	3.6
Jupiter	2.051	2.731	6.3
Saturn	3.137	6.056	7.3
Uranus	5.534	15.972	8.0
Neptune	8.253	30.529	8.3
Pluto	10.572	45.208	8.4

with a radius of r_2. The ΔV required is

$$\Delta V = \sqrt{\left(\frac{\mu}{r_2}\right)}\left\{1 - \sqrt{\frac{2r_1}{r_1 + r_2}}\right\}$$

Table 4.2 provides details of Hohmann transfers within the solar system. Within this table, the transfer time is simply taken as half of the elliptical orbit period. The velocity increments indicated are from an assumed 185 km circular orbit about the Earth.

Plane rotation manoeuvres and rotation of the line of apsides may be treated similarly to the above analysis. They involve a change of direction for which the ΔV requirement is given in Chapter 3, equation (3.18). This can involve appreciable use of fuel. If no change of speed is entailed then the velocity increment required is proportional to the velocity at the time of the manoeuvre, and these transfers should, if possible, be performed when the velocity is at a minimum. This point will be developed further in Chapter 5, where aspects of mission planning and analysis are considered.

4.4 SPECIFYING THE ORBIT

The orbit equation (equation (4.4)) is equivalent to a sixth-order scalar differential equation which requires six initial conditions in order to determine the six constants of integration. At orbit injection these will be the three components of position and of velocity. A standard way of specifying an orbit is to use *orbital elements*, which refer the orbit to a frame of reference which is fixed relative to the stars. This is used both for astronomy and for satellites.

The *frame of reference* which is commonly used can be defined in terms of X-, Y-, and Z-axes (see Figure 4.10). The X- and Y-axes lie in Earth's equatorial plane and the Earth spins about the Z-axis.

The X-axis is in a direction from Earth to the Sun at the vernal equinox. The direction thus indicated is termed the *first point of Aries*. Some 2000 years ago this direction did

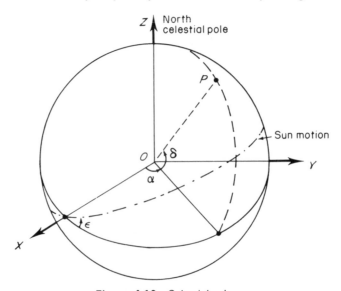

Figure 4.10   Celestial sphere

point towards the constellation of Aries, but at present it points to Aquarius, and is moving along the equator at about 0.8' per year.

The Z-axis is along the Earth's spin axis, in the northerly direction. It is at an angle of 23°27'8" to the normal of the ecliptic plane, changing slowly with respect to the stars, with a period of about 25 725 years. The overall precessional rate is $\sim 0.12''$ per year, and may be neglected for most spacecraft applications.

The Y-axis makes up a 'right-handed' orthogonal set with the X-, Z-axes.

The *celestial sphere* is the name given to a sphere with infinite radius centred on the Earth. The Z-axis meets it at a point known as the North celestial pole. The motion of the Sun on the celestial sphere is indicated in Figure 4.10. It moves in the ecliptic plane, tilted about the X-axis through 23°27'8", known as the obliquity of the ecliptic, ε.

The location of a point P on the celestial sphere, strictly a direction OP, can be expressed in terms of the two angles α and δ. The great circle through P and the North celestial pole is called the hour circle of P. The angle α between OX and the equatorial radius to the hour circle is called the right ascension of P. Its declination δ is the angle between OP and the equatorial plane (see Figure 4.10).

The *six orbital elements* of a spacecraft's orbit round the Earth are chosen to represent different features. The orbital plane in Figure 4.11 intersects the equatorial plane in a line called the line of nodes; the ascending node is the point on the equator at which the spacecraft moves from the southern into the northern hemisphere. The right ascension of this node, together with the inclination angle i, define the plane of the orbit. The orientation within the plane is defined by the angle ω, known as the argument of periapsis, which is the angle between the line of nodes and the vector $\mathbf{e}$ to the periapsis. The shape of the orbit is defined by its eccentricity $\mathbf{e}$, and its size by its semi-major axis a. The sixth element defines the position of the spacecraft in its orbit. The time (or epoch) of last passage through the periapsis may be used, or the mean anomaly M (equation (4.18)).

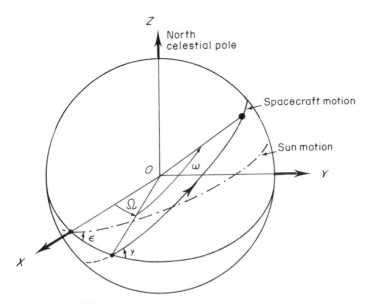

Figure 4.11   Spacecraft orbital elements

4.5 ORBIT PERTURBATIONS

At the distances of orbiting spacecraft from the Earth its asphericity and non-uniform mass distribution result in its gravitational potential departing from the simple $1/r$ function which was assumed in Section 4.2. Equation (4.1) is not valid in this situation and the equation of motion (equation (4.4)) must be modified to take account of the Earth's gravitational field.

There are additional forces which act on space vehicles, which were not included in the Keplerian formulation. Some are from additional masses which provide secondary gravitational fields; for Earth orbit, the Moon and the Sun provide such forces. Also at low altitudes (typically at less than 1000 km altitude) the Earth's atmosphere imposes a drag force. Table 4.3 lists the major perturbing forces, and their relative importance, for space vehicle orbital analysis.

The equation of motion for a space vehicle about a body taking into account perturbative influences may be written in the form:

$$\ddot{\mathbf{r}} = -\nabla U + \mathbf{b} \tag{4.31}$$

where U is the gravitational potential field and $\mathbf{b}$ is the force vector per unit mass, to which the vehicle is subject. A general closed solution is not possible, but there are a variety of solution methods which are appropriate for spacecraft dynamics. The method described here is the method of variation of orbital elements. Other methods such as those first proposed by Cowell and Crommelin [4] and Encke [5] are summarized by Cornelisse [6].

The method of the variation of orbital elements may be considered in the following way. The elements referred to in the preceding section are constants for a Keplerian orbit, as derived in Section 4.2. When perturbative forces exist they are no longer constant but for small forces they will change slowly.

A simple, physical model will serve to demonstrate this: consider a spacecraft in circular orbit about a spherically symmetrical planet containing an atmosphere of density ρ kg/m^3. If it is assumed that the perturbative drag force is small, then it is to be expected that the orbit will remain near circular. Now the velocity in a circular Keplerian orbit is given by $\sqrt{\mu/r}$. If the spacecraft's projected area in the direction of flight is S, then the work performed by the atmosphere as the vehicle moves round the orbit is given by $\sim \pi r \rho S C_D \mu/r$, where C_D is an appropriate drag coefficient for the vehicle. Since this reduces

Table 4.3 Magnitude of disturbing accelerations acting on a space vehicle whose area-to-mass ratio is A/M

Source	Acceleration (m/s^2)	
	500 km	Geostationary orbit
Air drag	$6 \times 10^{-5} A/M$	$1.8 \times 10^{-13} A/M$
Radiation pressure	$4.7 \times 10^{-6} A/M$	$4.7 \times 10^{-6} A/M$
Sun (mean)	5.6×10^{-7}	3.5×10^{-6}
Moon (mean)	1.2×10^{-6}	7.3×10^{-6}
Jupiter (max.)	8.5×10^{-12}	5.2×10^{-11}

the energy of the system it is apparent that the energy constant ε in equation (4.24) will decrease and the orbital element a must do so too.

The variation of orbital elements method assumes that the actual orbit of a body, at any given instant, may be considered to have instantaneous values of Keplerian orbital elements, defined so that if the perturbing forces are removed at that instant, thus leaving only a central gravitational field whose potential is proportional to $1/r$, then the orbit will follow the Keplerian orbit which has the instantaneous orbital elements. These are called the osculating elements. It must be emphasised that the method is only appropriate for perturbing forces having a magnitude significantly smaller than μ/r^2.

The normal method for so describing an orbit is with recourse to Lagrange's planetary equations (see for example [7]). One form of these is the following [8]:

$$
\left.
\begin{aligned}
\frac{da}{d\theta} &= \frac{2pr^2}{\mu(1-e^2)^2}\left\{ e\sin\theta\, S + \frac{p}{r}T \right\} \\[2mm]
\frac{de}{d\theta} &= \frac{r^2}{\mu}\left\{ \sin\theta\, S + \left(1+\frac{r}{p}\right)\cos\theta\, T + e\frac{r}{p}T \right\} \\[2mm]
\frac{di}{d\theta} &= \frac{r^3}{\mu p}\cos(\theta+\omega)\,W \qquad \frac{d\Omega}{d\theta} = \frac{r^3\sin(\theta+\omega)}{\mu p \sin i}\,W \\[2mm]
\frac{dw}{d\theta} &= \frac{r^2}{\mu e}\left\{ -\cos\theta\, S + \left(1+\frac{r}{p}\right)\sin\theta\, T \right\} - \cos i\,\frac{d\Omega}{d\theta} \\[2mm]
\frac{dt}{d\theta} &= \frac{r^2}{\mu p}\left\{ 1 - \frac{r^2}{\mu e}\left[\cos\theta\, S - \left(1+\frac{r}{p}\right)\sin\theta\, T \right] \right\}
\end{aligned}
\right\}
\tag{4.32}
$$

where S, T, W form a triad of forces in a spacecraft-centred coordinate reference frame, S acting radially, T transverse to S in the plane of motion and W normal to the orbit plane giving a right-handed system of forces.

It should be noted that the particular form of Lagrange's equations adopted depends on the type of orbit under investigation. For example the set defined in equation (4.32) fails for circular ($e = 0$) and equatorial orbits ($i = 0$). These may be dealt with by introducing parameter transformations of the type given for example by Roy [9].

Solutions of equations (4.32) must in general be performed using numerical techniques. This requires the formulation of functions for U and $\mathbf{b}$, and these are given below for certain cases of particular interest.

4.5.1 Gravitational potential of the Earth

The most convenient method for describing Earth's gravitational field outside its surface is to use a spherical harmonic expansion [10], given by

$$
\begin{aligned}
U(r, \Phi, \Lambda) = \frac{\mu}{r}\Bigg\{ 1 + &\sum_{n=2}^{\infty}\left[\left(\frac{R_E}{r}\right)^n J_n P_{n0}(\cos\Phi)\right. \\[2mm]
&\left. + \sum_{m=1}^{n}\left(\frac{R_E}{r}\right)^n (C_{nm}\cos m\Lambda + S_{nm}\sin m\Lambda)\, P_{nm}(\cos\Phi)\right]\Bigg\}
\end{aligned}
\tag{4.33}
$$

Table 4.4 Magnitude of low order J, C and S values for Earth

J_2 1082.7×10^{-6}	C_{21} 0	S_{21} 0
J_3 -2.56×10^{-6}	C_{22} 1.57×10^{-6}	S_{22} -0.897×10^{-6}
J_4 -1.58×10^{-6}	C_{31} 2.10×10^{-6}	S_{31} 0.16×10^{-6}
J_5 -0.15×10^{-6}	C_{32} 0.25×10^{-6}	S_{32} -0.27×10^{-6}
J_6 0.59×10^{-6}	C_{33} 0.077×10^{-6}	S_{33} 0.173×10^{-6}

where $U(r, \Phi, \Lambda)$ is the gravitational potential at a distance r from the centre of the Earth and Φ, Λ are the latitude and longitude. P_{nm} are Legendre polynomials. J_n, C_{nm}, and S_{nm} are functions of the mass distribution of the body, in this case the Earth. Terms of the form J_n are called *zonal harmonic coefficients*; they reflect the mass distribution of the Earth independently of longitude. C_{nm} and S_{nm} are the Earth's *tesseral harmonic coefficients* for $n \neq m$ and the *sectoral harmonic coefficients* for $n = m$. These coefficients have mainly been determined from the motion of Earth-orbiting spacecraft. Whilst the lower-order terms were determined during the early 1960s, determination of the Earth's gravitational field continues to be an area of active research [11]. One of the major problems in determination of the higher-order terms is due to their rapid decrease with altitude; from equation (4.33) terms decrease with $(R_E/r)^n$, and at low altitudes their effects are difficult to separate from those of variable air drag during space-vehicle orbit analysis. This situation should, however, be improved with missions such as GRAVSAT and ARISTOTLES at present planned for the 1990s.

Table 4.4 gives the magnitude of some of the lower-order coefficients. From this it is apparent that the term J_2 is some three orders of magnitude larger than the others, and to a first approximation it dominates the gravitational perturbative influences of the Earth. This term represents the polar flattening of the Earth (or equatorial bulge), and its effect is two-fold.

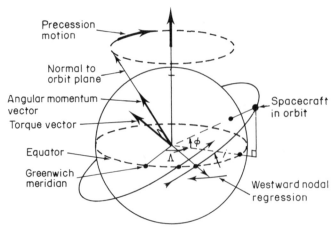

Figure 4.12 Nodal regression

Regression of the line of nodes

The equatorial bulge produces a torque which rotates the angular momentum vector. For prograde orbits ($i < 90°$) the orbit rotates in a westerly direction, leading to a regression of the line of nodes as shown pictorially in Figure 4.12. Neglecting all harmonic coefficients other than J_2 the rate of nodal regression may be written [9] to the first order in J_2, as

$$\bar{\Omega} = \Omega_0 - \frac{3}{2}\frac{J_2 R_E^2}{p^2}\bar{n}\cos i\, t + O[J_2^2] \qquad (4.34)$$

where $\bar{n}$ is the mean angular velocity, $\sqrt{\mu/a^3}$. Thus the secular rate of nodal regression per orbit is

$$\Delta\Omega = -\frac{3\pi J_2 R_E^2}{p^2}\cos i\ \text{rads/rev} \qquad (4.35)$$

Precession of the line of apsides

The second influence of the equatorial bulge may be considered in the following way. Since the mass 'seen' by the spacecraft crossing the equator is greater than the mean mass, the orbit tends to curve more rapidly. Since the gravitational field of the Earth is conservative, however, this leads to an overall rotation of the orbit, *within* the orbit plane, as demonstrated in Figure 4.13. This motion implies rotation of the semi-major axis and is termed precession of the line of apsides. The secular effect is given by

$$\bar{\omega} = \omega_0 + \frac{3}{2}\frac{J_2 R_E^2}{p^2}\bar{n}\left(2 - \frac{5}{2}\sin^2 i\right)t + O[J_2^2] \qquad (4.36)$$

or per orbit

$$\Delta\omega = 3\pi\frac{J_2 R_E^2}{p^2}\left(2 - \frac{5}{2}\sin^2 i\right)\text{rad/rev} \qquad (4.37)$$

At an inclination of $\sim 63.4°$ the precession is zero; this type of orbit is of particular interest as indicated in Chapter 5, and is frequently called a 'Molniya' orbit.

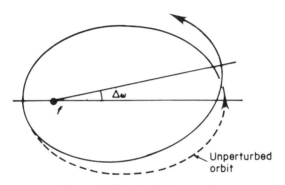

Figure 4.13 Apsidal precession

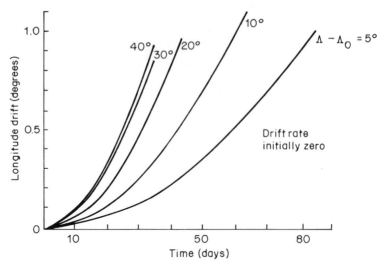

Figure 4.14   Longitudinal drift rate in geostationary orbit

Triaxiality perturbation

The terms representing the longitudinal variation of the Earth's gravitational field have their most significant influence on geostationary satellites, since non-synchronous orbits will average them out. The term J_{22} $((C_{22}^2 + S_{22}^2)^{1/2})$ has the greatest influence due to the $(R_{E/r})^n$ weighting of coefficients. This term represents the ellipticity or triaxiality of the equator, having its major axis aligned approximately along $15°\text{W}-165°\text{E}$. A satellite will be stable if it is at a minimum of the earth's potential field, that is, on the minor axis of the equator. Departure from these two longitude positions provides an increasing perturbation. Flury [12] gives the following expression for the rate of change of longitude:

$$\dot{\Lambda} = k^2 \sin^2(\Lambda - \Lambda_0) \tag{4.38}$$

where

$$k^2 = -18\omega_e \frac{a_e^2}{a_s^2} = -1.7 \times 10^{-13} \text{ deg/day}$$

and $\Lambda_0 = -14.79°$. ω_e is the sidereal rotation rate of Earth. a_e is the mean equatorial radius. a_s is the semi-major axis of the synchronous orbit.

This is shown graphically in Figure 4.14. Triaxiality can be seen to provide an East/West station-keeping problem for a geostationary satellite. This is referred to in Chapter 5.

4.5.2   Atmospheric drag

For low Earth orbiting spacecraft the perturbation due to the atmosphere cannot be neglected. Atmospheric effects lead to a force which may be expressed as two orthogonal components, drag along the direction of travel, and lift at right angles to it. The simplest form in which to write the drag force is

$$\mathbf{F}_D = \tfrac{1}{2}\rho S C_D V_r^2 \frac{\mathbf{V}_r}{|V_r|} \tag{4.39}$$

where $\mathbf{V}_r$ is the velocity vector of the spacecraft relative to the atmosphere, ρ is the atmospheric density, S a reference area for the vehicle and C_D the vehicle's coefficient of drag referred to the reference area. The lifting force is obtained by replacing C_D with the coefficient of lift C_L, in equation (4.39).

For most spacecraft the effects of lift are negligible compared with drag, but when precise orbit determination is required ($\sim$ few metres accuracy) its influence must be included [13].

The drag coefficient for space vehicles is somewhat complex due to the type of flow regime within which the motion takes place. Due to the large mean free path the flow cannot be treated in a continuum manner. Indeed the most suitable description is that of free molecular flow wherein the molecular description is such that molecules reflected from the spacecraft surfaces do not interact further with the flow field; no shock wave is therefore formed about a body moving through the atmosphere at orbit altitudes. The drag force experienced is dependent on the gas–surface interaction, for which there is only sparse experimental data at typical incident velocities. Stalder and Zurick [14] and Schaff and Chambre [15] and Schamberg [16] provide the basic theory for rarefied flow, and Cook [17] provides height-related C_D data adopting the Schamberg model. Typically values of $C_D \sim 2.5$ are predicted by these theories.

With reference to equation (4.39), drag is most significant at perigee where both the velocity and density are greatest. King-Hele [7] provides detailed analysis of the secular changes caused by drag. To first order, these approximate to an impulsive, negative velocity increment occurring at perigee. Through equation (4.30) this will lead to a reduction in the semi-major axis of an elliptical orbit. For a circular orbit drag will occur continuously around the orbit. Assuming that the change in the radius of such an orbit is small, the $\Delta\tau$ in orbit period for a circular orbit of radius r will be given by

$$\frac{\Delta\tau}{\tau} \approx -3\pi\rho r/(M/SC_D) \tag{4.40}$$

where ρ is the density at r (measured to the centre of the earth) and M/SC_D is called the vehicle ballistic parameter: M is the vehicle mass, and S the projected area.

The dominant influence of drag is thus to cause orbit contraction, with eventual re-entry. Since the Earth's atmosphere rotates approximately synchronously with the Earth, in general the drag force has a component perpendicular to the orbit plane. This results in a change in inclination for the orbit. King-Hele [7] gives expressions for the radial, transverse and normal force components for a vehicle moving through an atmosphere rotating at an angular rate α rad/s. These are

$$S = -\tfrac{1}{2}\rho v \delta \left(\frac{\mu}{PF}\right)^n \left\{ 1 + e \cos\theta - r\alpha \left(\frac{\rho}{\mu}\right)^{1/2} \cos i \right\}$$

$$T = -\tfrac{1}{2}\rho v \delta \left(\frac{\mu}{PF}\right)^{1/2} e \sin\theta \tag{4.41}$$

$$W = -\tfrac{1}{2}\rho v \delta \frac{r\alpha}{\sqrt{F}} \sin i \cos(\theta + \omega)$$

where

$$F = \left(1 - \frac{r_p \alpha}{V_p} \cos i\right)$$

and

$$P = a(1 - e^2)$$

and r_p, V_p represent conditions at perigee, V is the absolute velocity of the vehicle; δ is a modified ballistic parameter given by $\delta = FSC_D/M$,

4.5.3 Additional gravitational fields

Luni-solar perturbations

Other bodies in the solar system impose additional gravitational forces on spacecraft orbiting the Earth. The proximity and mass of the Moon provides the most significant influence but solar influences cannot be neglected. These perturbations are collectively termed *luni-solar perturbations*. Since in general these bodies will not lie in the same plane as the vehicle orbit, their most significant influence will be to change the inclination of the orbit with respect to the equator. The formulation of this three-body interaction does not admit a general closed form solution, and numerical techniques must be employed.

The disturbing acceleration a_d of a satellite due to a disturbing body having a mass M_d and gravitational parameter μ_d is given by

$$a_d = \mu_d \sqrt{(\mathbf{R} \cdot \mathbf{R})} \tag{4.42}$$

where

$$\mathbf{R} = \frac{\mathbf{r}_{sd}}{r_{sd}^3} - \frac{\mathbf{r}_d}{r_d^3}$$

and $\mathbf{r}_{sd}$ and $\mathbf{r}_d$ are defined in Figure 4.15. It can then be shown [6] that the maximum value of the ratio of the disturbing acceleration a_d to the central acceleration a_c is given by

$$\frac{a_d}{a_c} = \frac{M_d}{M_c}\left(\frac{r_s}{r_d}\right)^3 \sqrt{(1 + 3\cos^2 \beta)} \tag{4.43}$$

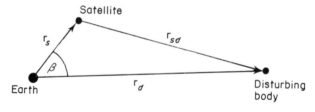

Figure 4.15 Disturbing body and satellite positions

or

$$\frac{a_d}{a_c} \leq 2\frac{M_d}{M_c}\left(\frac{r_s}{r_d}\right)^3$$

The angle β is shown in Figure 4.15; M_c is the mass of the central body about which the vehicle is in orbit.

At geostationary orbit this ratio provides values of 3.3×10^{-5} and 1.6×10^{-5} for the Moon and Sun respectively. The effect of Jupiter is some five orders of magnitude lower than these values.

Expressions which provide approximate average rates of change of orbital elements for a single disturbing body, given by Cook [18], are

$$\frac{da}{dt} \approx 0$$

$$\frac{de}{dt} \approx \frac{-15}{2}\frac{K}{n}e(1-e^2)^{1/2}[AB\cos 2\omega - \tfrac{1}{2}(A^2 - B^2)\sin 2\omega]$$

$$\frac{dr}{dt} \approx -ae$$

$$\frac{d\Omega}{dt} \approx \frac{3KC}{4n(1-e^2)^{1/2}\sin i}[5Ae^2\sin 2\omega + B(2 + 3e^2 - 5e^2\cos 2\omega]$$

$$\frac{d\omega}{dt} + \frac{d\Omega}{dt}\cos i = \frac{3K(1-e^2)^{1/2}}{2}\frac{}{n}[5\{AB\sin 2\omega + \tfrac{1}{2}(A^2 - B^2)\cos 2\omega\} - 1 + \tfrac{3}{2}(A^2 + B^2)]$$

$$\frac{di}{dt} = \frac{3KC}{4n(1-e^2)^{1/2}}[A(2 + 3e^2 + 5e^2\cos 2\omega) + 5Be^2\sin 2\omega]$$

$$(4.44)$$

where

$$K = \frac{GM_d}{r_d^3}$$

and

$$A = \cos(\Omega - \Omega_d)\cos u_d + \cos i_d \sin u_d \sin(\Omega - \Omega_d)$$

$$B = \cos i[-\sin(\Omega - \Omega_d)\cos u_d + \cos i_d \sin u_d \cos(\Omega - \Omega_d)] + \sin i \sin i_d \sin u_d$$

$$C = \sin i(\cos u_d \sin(\Omega - \Omega_d) - \cos i_d \sin u_d \cos(\Omega - \Omega_d)] + \cos i \sin i_d \sin u_d$$

Angles Ω_d, u_d, $(\theta_d + \omega_d)$ and i_d are the orbital elements of the disturbing body referred to the equatorial based system of Section 4.3.

4.6 INTERPLANETARY ORBITS

It is apparent that a spacecraft which moves between the planets must take due account of a variety of bodies within the solar system. One simplified method for examining interplanetary trajectories is the *patched conic method*. This assumes, in its simplest form, that only one body, a central body, is acting on the spacecraft at any one time. For this it is useful to define a sphere of influence as the region of space in which the influence of a planetary body is dominant. A spacecraft within this region is assumed to execute a Keplerian orbit. As the spacecraft leaves this region of space it enters the central force field of a new planet or the Sun, during which a new orbit or conic section is defined. Thus the entire trajectory requires the patching together of several conic sections.

Within the sphere of influence of the Sun the spacecraft will be in heliocentric orbit, the Sun occupying the focus of the conic. The sphere of influence thus indicates whether the motion of a body should be referred to, for example, either the Sun or the Earth. Recalling equation (4.42), if the spacecraft is in orbit about the disturbing body rather than the central body as indicated, then the perturbing acceleration due to the original central body may be written $a_{d'}$ where

$$a_{d'} = \mu_{\oplus} \sqrt{(\mathbf{R} \cdot \mathbf{R})} \qquad (4.45)$$

where $\mu_{\oplus}$ is the gravitational parameter of the Earth

and
$$\mathbf{R} = \frac{\mathbf{r}_s}{r_s^3} - \frac{\mathbf{r}_d}{r_d^3} \qquad (4.46)$$

$$\text{If} \quad \frac{r_s}{r_d} \ll 1 \quad \text{then} \quad a_{d'} = \frac{\mu_{\oplus}}{r_s^2}$$

Thus the new ratio of disturbing to central accelerations is given by

$$\frac{a_{d'}}{a_{m'}} = \frac{M_{\oplus}}{M_d} \frac{r_d^2}{r_s^2} \qquad (4.47)$$

Equating (4.47) and (4.43) we find

$$\left(\frac{r_s}{r_d}\right)^5 = \left(\frac{M_{\oplus}}{M_d}\right)^2 \frac{1}{\sqrt{(1 + 3\cos^2 \beta)}} \qquad (4.48)$$

Since the square root function differs little from unity, then

$$r_s \sim r_d \left(\frac{M_{\oplus}}{M_d}\right)^{2/5} \qquad (4.49)$$

and thus this locus for r_s defines approximately a sphere.

References [4] and [6] provide more detailed analysis.

REFERENCES

[1] Kaplan, M.H. (1976) *Modern Spacecraft Dynamics*, Wiley, London.
[2] De Veubeke, B.F. (ed.) (1969) *Advanced Problems and Methods for Spaceflight Optimization*, Pergamon Press, Oxford.
[3] Lawden, D.F. (1963) *Optimal Trajectories for Space Navigation*, Butterworth, London.
[4] Cowell, P.H. and Crommelin, A.C.D. (1910) *Investigation of the motion of Halley's comet from 1759–1910*, Appendix to Greenwich Observatory.
[5] Encke, J.F. (1852) *Astron. Nachr.* **33**, 377–98.
[6] Cornelisse, J.W., Schoyer, H.F.R., and Wakker K. F. (1979) *Rocket Propulsion and Spaceflight Dynamics*, Pitman, London.

[7]   King-Hele, D.G. (1987) *Satellite Orbits in an Atmosphere: Theory and Applications*, Blackie, London.

[8]   Brouwer, D. and Clemence, G.M. (1961) *Methods of Celestial Mechanics*, Academic Press, London.

[9]   Roy, A.E. (1982) *Orbital Motion*, Adam Hilger, Bristol.

[10] Wertz, J.R. (1978) *Spacecraft Attitude Determination and Control*, Reidel, Dordrecht.

[11] King-Hele, D.G. (1978) RAE Tech Report 78142.

[12] Flury *Eldo/ESRO Sci + Tech Review* **5** (1973).

[13] Stark, J.P.W. (1986) ESA SP 255, 239–246.

[14] Stalder, J.R. and Zurick, V.J. (1951) NACA Tech Note 2423.

[15] Schaff, S.A. and Chambre; P.L. (1958) H. Emmons (ed.) *Fundamentals of Gas Dynamics*, Oxford University Press, Oxford.

[16] Schamberg, R. (1959) Rand Corp. R.M. 2313.

[17] Cook G.E. (1965) *Planet Space Sci.* **13**, 929.

[18] Cook, G.E. (1962) *Geophys. J.*, **6**, 271.

5 *MISSION ANALYSIS*

John P.W. Stark

Department of Aeronautics and Astronautics, University of Southampton

5.1 INTRODUCTION

Whereas celestial mechanics deals with a spacecraft's orbit relative to the stars, mission analysis relates both orbit and attitude to the ground, the Sun etc. For study purposes it is convenient to place missions into categories based upon their orbits: geostationary Earth orbit (GEO), low Earth orbit (LEO), highly elliptical orbits (HEO), and non-geocentric orbits (lunar and interplanetary) cover most applications. There is also a potentially increasing miscellaneous category embracing space stations servicing/rescue, etc. This chapter will be restricted to GEO and LEO since they illustrate most of the aspects which link orbits to missions.

A history of each mission may be viewed as a sequence of events starting at pre-launch and extending to end-of-life. The *pre-launch phase* includes all those operations which are required in order to effect launch vehicle ignition, and separation from the umbilical cable between space system (vehicle plus launcher) and the ground launch facilities.

The *launch phase* involves a sequence of events, many of which are preprogrammed and automatic. Since control is effected through launch tapes which will be prepared for each mission, careful planning is required for placing the spacecraft into an appropriate intermediate orbit from which the operational one may be attained. One evident constraint during the launch phase is the requirement for continuous communications and tracking. These facilities may be required for a ground centre override to abort the mission, but also there is generally a requirement to monitor the performance of a launch vehicle. Figure 5.1 shows the ground centre network that was used for the Ariane launch site in French Guiana for the launch of Giotto in 1985. The times at which certain mission events took place are indicated in this figure.

Orbit transfer is the next major event. This involves transferring the spacecraft from the orbit into which the launch vehicle *actually* places it, to the one from which the operational phase of the mission can commence. There is *a priori* uncertainty in the orbit at launcher burn-out as shown by the performance envelope of the launch vehicle (see Chapter 7); but there is *a posteriori* uncertainty due to the orbit determination process which inevitably includes some errors in the range and range-rate measurements of the

Spacecraft Systems Engineering. Edited by P.W. Fortescue and J.P.W. Stark
© 1991 John Wiley & Sons Ltd

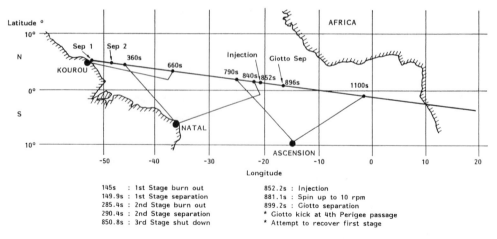

145s   : 1st Stage burn out
149.9s : 1st Stage separation
285.4s : 2nd Stage burn out
290.4s : 2nd Stage separation
850.8s : 3rd Stage shut down

852.2s : Injection
881.1s : Spin up to 10 rpm
899.2s : Giotto separation
* Giotto kick at 4th Perigee passage
* Attempt to recover first stage

Figure 5.1 Giotto injection; ground station visibility (elevation = 1°)

spacecraft trajectory (see Chapter 4). Transfer between these orbits requires propellant, and it is the task of the mission planners to determine how much is required in order to attain the desired orbit from any given initial one, with a given level of confidence. This is particularly important for geostationary spacecraft, since the final operations needed to obtain their orbit call for the use of a secondary propulsion system which may subsequently be used for attitude and orbit control. Excessive use or under-budgeting of fuel will therefore reduce the operational life of the space system as a whole.

On-station operations include various mission events. Fuel budgeting for station-keeping and attitude control will be necessary.

For surveyor/observatory missions the scientific goals may require continuous changing of the orientation of the spacecraft, in such a way that a wide variety of directions or even complete coverage of the celestial sphere may be achieved. Careful planning is required in order to optimise the observing program; both the rate at which the spacecraft is reorientated and the angular distance through which its attitude is changed between observations will influence fuel consumption. Whilst contingency planning is made for unscheduled repointing, many scientific missions inevitably cease only when the fuel has been exhausted (e.g. Ariel V). Other constraints which will be of concern in deriving mission profiles for scientific observatories include: minimum sun/telescope angle, earth limb/telescope angle, passage through the South Atlantic Anomaly in the Earth's magnetosphere and thermal balance of the spacecraft and its sensors. Analysis and planning activities associated with two specific types of mission, namely a geostationary communications satellite and a near polar remote sensing low earth orbit satellite, during their operational phases, are detailed in the following sections.

Decommissioning the spacecraft is the final event associated with a space mission. In GEO the demand for longitude slots is increasing and thus an uncontrolled satellite in this orbit is wasteful and also provides a collision probability [1]. It has therefore become standard practice to remove an obsolete spacecraft from GEO into a higher orbit by the use of residual propellant in the secondary propulsion system at the end of the life of the vehicle. This manoeuvre is frequently and appropriately called a 'graveyard burn'. It is also becoming the practice in LEO missions to provide a controlled re-entry through the

Earth's atmosphere. The reason for this is that uncontrolled re-entry can lead to the vehicle's breaking up, providing a collision hazard and adding to the problem of space debris.

A key aspect of mission analysis is to identify critical features of the mission which have an impact upon system and subsystem design. Any space vehicle is required to meet specific mission objectives and these thus provide *design requirements*.

These requirements will lead to design drivers which influence the whole nature of the vehicle. This may be seen by considering briefly the overall nature of the four types of orbit specified above.

GEO and LEO vehicles are generally in near circular orbits, resulting in a uniform performance from the communications links used by the vehicle. Similarly the thermal environment (excluding the difference between eclipse and sunlight phases) is near constant. If the Earth is used as a reference direction (as is often the case) attitude sensing for the vehicle is simplified. For HEO and non-geocentred orbits none of these features are present, thus impacting on subsystem design.

The majority of missions are either GEO or LEO, and thus the majority of this chapter is directed to these types. Rather specialized features are associated with non-geocentric missions, and these are not covered here. HEO orbits have some particular advantages; they can offer large periods of time when the vehicle motion relative to a given ground location is not large. This may be advantageously used for provision of communications links (as in the 12-hour Molniya orbit) or for providing downlink of data during continuous observations from scientific satellites (such as EXOSAT).

It is of course important to appreciate that to execute any given mission, the payload must be pointed in a specific direction at a specific time. The attitude history of the vehicle is thus inextricably linked with the mission analysis, which will then link to subsystem design. The control of attitude behaviour is left until Chapter 10.

5.2 GEOSTATIONARY EARTH ORBITS (GEO)

The utility of the geostationary orbit for providing global communications was first noted by Arthur C. Clarke [2]. Its primary attribute is that the subsatellite point is fixed at a selected longitude, with 0° latitude. It does not have dynamic tracking problems. GEO spacecraft may therefore provide fixed-point to fixed-point communications to any site within the beam of their antennas. Figure 5.2(a) shows the horizon as viewed from GEO, and the region over which the satellite appears with an elevation in excess of 10°; Figure 5.2(b) demonstrates that only three satellites are required to provide almost a global communications network. These figures may be compared with both Table 5.1 where concurrent visibility between New York and Paris is indicated for polar orbits of various altitudes, and Table 5.2 which indicates the number of satellites required for different links in a variety of circular orbits. Three important features emerge:

(a) At low altitudes, polar orbits offer considerable advantages over equatorial ones for links between centres at similar, though modest, latitudes—as illustrated by centres at approximately 40°N.

(b) the position under (a) is reversed for near-equatorial centres—as illustrated by the trans-equatorial sites (25°N and 35°S).

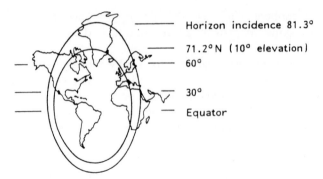

(a)   Coverage of one satellite

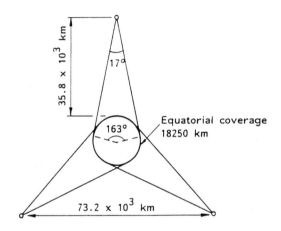

(b)   3 Geostationary satellites – Earth coverage

Figure 5.2   Geostationary views of the Earth

Table 5.1   Concurrent visibility: New York/Paris from polar orbits

Period (hr)	Altitude (km)	Maximum visibility (minutes)
2	1 700	7
4	6 400	56
6	10 400	112
12	20 000	270
24	36 000	594

Table 5.2 Number of spacecraft needed for continuous communication between specified sites ($e = 0$)

Sites (latitude)	Altitude (km)	Polar orbit	Equatorial orbit
Tokyo/Los Angeles	14 000	8	24
(35°41'N)/(34°00'N)	20 000	6	15
	36 000	7	1
New York/Paris	14 000	6	13
(40°45'N)/(48°51'N)	20 000	5	10
	36 000	4	1
Miami/Buenos Aires	14 000	9	4
(25°46'N)/(34°38'S)	20 000	10	4
	36 000	4	1

Finally and crucially

(c) the GEO is clearly most desirable from the coverage standpoint. The 24-hour geosynchronous orbit clearly offers unique advantages, providing almost complete global coverage (except for the immediate polar regions) from merely three satellites, and with no need for the ground antenna to switch between satellites.

For a service of high quality and reliability it is necessary to consider the control and replacement policy for a GEO spacecraft. The failure of a space segment would cause substantial financial penalties to the system operator, and since the ability to replace it is dependent on launcher availability, with a substantial time being needed, a philosophy of having an in-orbit spare is frequently adopted. This spare is at a slightly different longitude, and offers the advantage of extra capacity should it be required. As an example, for the Pacific Ocean coverage in March 1980 the Intelsat primary spacecraft was located at 173.87°E, with a contingency vehicle at longitude 179.13°E.

Maintaining a spacecraft's orbit is an essential requirement for maintaining a communications link. The capability and method of achieving GEO and maintaining a specific location will now be discussed and fuel requirements for station-keeping indicated.

5.2.1 Geostationary orbit acquisition

The final stage of the launch vehicle will place a satellite in a nominal orbit. The Space Shuttle for example takes typically two or three satellites into a near circular orbit of a few hundred kilometres altitude with an inclination of $\sim 28°$. An orbit transfer propulsion unit is required to expand this into an elliptical orbit whose apogee is near a geostationary altitude ($\sim 35\,786.4$ km), whilst perigee remains at the Shuttle altitude. The unit is frequently a solid rocket payload assist module or PAM. Precision in the timing of the firing of this motor is essential in order to obtain the correct transfer orbit; apogee must occur at a nodal crossing so that subsequently a single firing of an apogee kick motor may both circularize and change the orbit plane to become equatorial.

The Ariane launch scenario is similar in philosophy although the details differ. The third stage places the satellite directly into a transfer orbit with a perigee of ~200 km and an apogee near geostationary altitude. Separation from the launch vehicle occurs after ~15 minutes of powered flight. The transfer orbit phase is inclined to the equator by only 8°, significantly less than for Shuttle.

Several transfer orbit revolutions occur before injection of the satellite into a near-circular, near-GEO position. This period is essential for tracking the satellite and determining its orbit before the apogee motor is fired. This motor increases the velocity of the satellite from ~1.6 km/s to ~3 km/s at apogee. Most satellites will be spin-stabilized

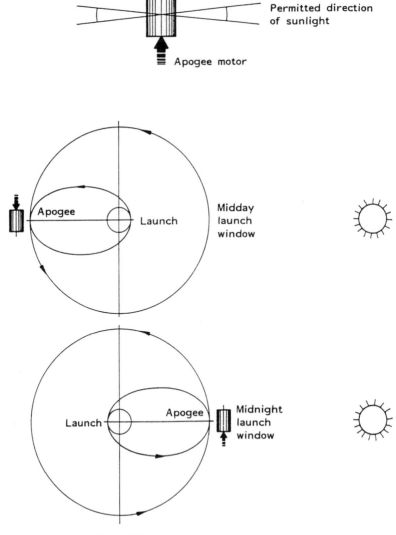

Figure 5.3   Vehicle orientation for injection

during the entire transfer and orbit acquisition phase, although vehicles using liquid apogee engines (low thrust) such as Olympus are three-axis stabilized. Orientation of the spin axis, and its control during motor firings, is particularly crucial for reasons of both orbit attainment and thermal control. This latter constraint typically requires a local midday or local midnight apogee firing to take place since the spin axis will be oriented along the thrust vector (see Figure 5.3).

For vehicles which utilize a liquid apogee motor (LAM), a single firing at apogee is insufficient to transfer the vehicle into the desired near GEO orbit. The low thrust of such motors ($\sim 10^2$ N rather than $\sim 10^4$ N for a solid rocket) would result in the need to rotate the thrust axis during firing in order to deliver the required impulse. Instead typically three firings of the LAM are utilized during successive apogee transits.

The final apogee firing places the satellite into a slightly elliptical orbit, termed a drift orbit, whose apogee is at GEO altitude. Its period is less than the GEO period, the precise value being chosen to minimize the ΔV requirements, whilst maintaining a reasonable GEO attainment strategy. As a consequence, the satellite progressively drifts eastward. Through a sequence of manoeuvres in both latitude and longitude it is eventually brought to rest at the required GEO location. These manoeuvres are similar to those described below for stationkeeping. Final station acquisition may be up to a month after satellite launch.

5.2.2 GEO station-keeping requirements

The primary disturbances on a satellite orbit were described in Chapter 4. The dominant effects for GEO are luni-solar perturbations, Earth triaxiality and solar radiation pressure.

Solar radiation perturbation is complex to model and depends critically upon both the ratio of projected-area to mass of the vehicle and the surface characteristics. Its predominant influence is upon the orbit eccentricity vector e. Generally an effective area-to-mass ratio which includes the reflectivity coefficient to solar illumination ε is used. When this is less than ~ 0.005 m^2/kg the effects of radiation pressure are significantly less than other perturbations. However, for larger values (>0.01 m^2/kg) radiation pressure can cause significant perturbation. Indeed it has been used to help control some spacecraft (e.g. Marecs A) by balancing other longitude influences.

The generalized techniques for examining the influence of perturbations, which were noted in Chapter 4, are not necessarily the optimum methods for analysis of GEO. This is particularly so since the orbit is ideally circular, and hence ω is undefined; the inclination is also zero, resulting in Ω being undefined. A method which is applicable is to linearize the equations of motion for small perturbations of magnitude δa in the semi-major axis a. For small values of δa, i and e, a linearized solution to Kepler's equation yields

$$r = A + \delta a - Ae \cos\left((t - t_0)\sqrt{\frac{\mu}{A^3}}\right) \tag{5.1}$$

$$\lambda = \Omega + \omega - \sqrt{\frac{\mu}{A^3}}t_0 - \frac{3}{2}\frac{\delta a}{A}\sqrt{\frac{\mu}{A^3}}(t - t_0) + 2e \sin\left((t - t_0)\sqrt{\frac{\mu}{A^3}}\right) \tag{5.2}$$

$$\theta = i \sin\left(\omega + (t - t_0)\sqrt{\frac{\mu}{A^3}}\right) \tag{5.3}$$

where A is the semi-major axis of a truly geostationary orbit (42 164.5 km), λ is the satellite longitude and θ its latitude. The evolution of r, λ and θ with time are shown in Figure 5.4. The simple sine and cosine dependence of r and θ indicate libration about single values, whilst the time dependence of the fourth term in equation (5.2) suggests that small deviations lead to a longitude drift rate. It is worth noting that for circular orbits whose semi-major axes are given by A and whose inclinations are non-zero, the ground track will be a figure of eight of the form shown in Figure 5.5.

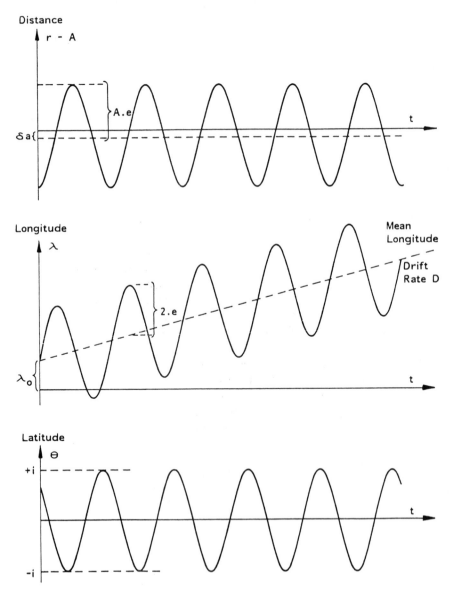

Figure 5.4   Linearized unperturbed spacecraft motion, given as distance (r), longitude (λ) and latitude (θ)

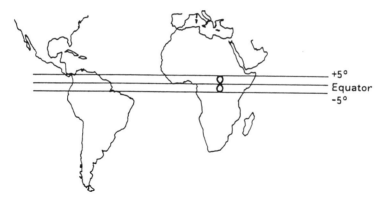

Figure 5.5 Ground track of geosynchronous vehicle having an inclination of 5°

Earth triaxiality perturbation is due to the J_{22} dominant Earth tesseral harmonic which produces a force on the spacecraft whose direction is shown in Figure 5.6. From Chapter 4 it is evident that a positive ΔV will expand an orbit, changing both the semi-major axis and the eccentricity. Since an expanded orbit relative to GEO has a longer period, a positive ΔV (i.e. an eastward impulse) will lead to westward drift in satellite longitude. Hence, referring to Figure 5.6, the longitude drift rates are opposed to the acting force direction, and it is apparent that 105.3°W and 75.1°E are stable equilibria against

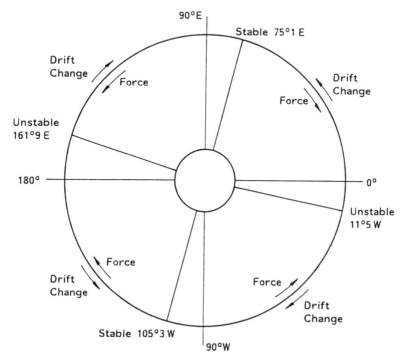

Figure 5.6 Force direction arising from J_{22} on a GEO vehicle

Table 5.3 Acceleration and station-keeping requirements for geostationary vehicles

Longitude (degrees east)	Acceleration (m/s²)	Fuel: ΔV (m/s/year)
− 160	-5.3×10^{-8}	1.67
− 140	-4.75×10^{-8}	1.50
− 120	-2.30×10^{-8}	0.73
− 100	$+8.10 \times 10^{-9}$	0.26
− 80	$+3.39 \times 10^{-8}$	1.07
− 60	$+4.61 \times 10^{-8}$	1.45
− 40	$+3.99 \times 10^{-8}$	1.26
− 20	$+1.48 \times 10^{-8}$	0.47
0	-2.10×10^{-8}	0.66
20	-5.09×10^{-8}	1.61
40	-5.73×10^{-8}	1.81
60	-3.30×10^{-8}	1.04
80	$+1.17 \times 10^{-8}$	0.37
100	$+5.21 \times 10^{-8}$	1.64
120	$+6.49 \times 10^{-8}$	2.05
140	$+4.44 \times 10^{-8}$	1.40
160	$+4.08 \times 10^{-9}$	0.13
180	-3.40×10^{-8}	1.07

equatorial drift. The magnitude of the acceleration as a function of longitude is shown in Table 5.3.

Luni-solar perturbations mainly cause out-of-plane forces acting on the spacecraft, leading to changes of inclination of the orbit. However, since a component of this force necessarily lies in the equatorial plane, then a, e and hence λ are also influenced. This equatorial component is nearly cyclic with the Earth orbit period, with magnitude less than that caused by the J_{22} term. It therefore does not lead to a significant secular evolution of the orbit, but it ultimately limits the closeness to a geostationary orbit. Figure 5.7 shows this near cyclic variability for both spring and neap tides during one sidereal day.

Returning to the out-of-plane components due to Moon and Sun, the periodic change of 18.6 years of the lunar orbit plane, plus precession of the Earth's spin axis, results in a 54-year period for the evolution of an uncontrolled geostationary orbit. The net force on the orbit plane evolution must clearly be opposed by an opposite ΔV, and is shown in Figure 5.8. The magnitude of the ΔV arising from luni-solar perturbation is typically ~ 50 m/s/yr and thus the propulsion requirements for the control of orbit inclination (i.e. north/south station keeping) is at least a factor of 20 larger than that required to overcome triaxiality (i.e. east/west station keeping) effects (see Table 5.3.). This results in the need for particularly careful control and planning of manoeuvres in order to avoid disadvantageous coupling between north/south and east/west control. This interaction is discussed in the following two sections.

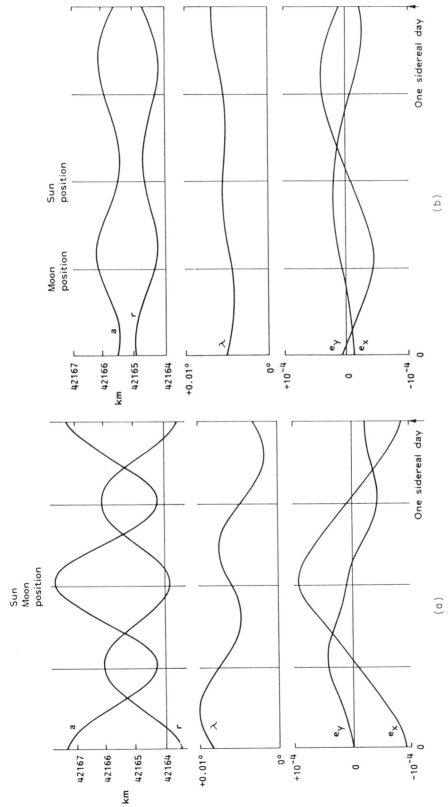

Figure 5.7 (a) Short-term variability during one sidereal day at spring tide. (b) Short-term variability during one sidereal day at neap tide

In summary the nature of station-keeping is to locate the spacecraft under the influence of the perturbations described within a specified range of longitude and latitude. This situation is shown in Figure 5.9. Whilst the apparent position of the spacecraft projected on the celestial sphere provides information to the ground control centre indicating instantaneously whether the spacecraft is within the correct longitude/latitude region, the range and range-rate are also required so that the actual orbit may be determined. This nominal orbit may then be used to schedule station-keeping manoeuvres to preclude departure from the required longitude/latitude location box. The size of this box necessarily impacts on the station-keeping schedule: the tighter the requirements the greater the impact. Requirements for $\sim 0.1°$ are typical, but this will become less when allowance is made for sensor errors.

5.2.3 Longitude station-keeping

Using the linearized approach noted above the subsequent change of longitude of a spacecraft following a small impulsive velocity increment ΔV is

$$\lambda = \lambda_0 + \frac{\Delta V}{r}\left\{-3t + 4\sqrt{\frac{A^3}{\mu}}\,\sin\left(t\sqrt{\frac{\mu}{A^3}}\right)\right\} \tag{5.4}$$

where t is the time since the impulsive burn. This is shown in Figure 5.10.

Longitudinal control can be effected by using an impulsive burn when the spacecraft reaches one extreme of the required error box. The subsequent longitude drift due to J_{22} (see Figure 5.6) brings the spacecraft back to the same side of the box as shown in Figure 5.11, and the process is repeated as a limit cycle. This neglects the short period fluctuations caused by luni-solar perturbations.

If ΔV is sized so that the spacecraft just reaches the opposite side of the box following a burn, and a constant acceleration f is assumed (as listed in Table 5.3), then the ΔV and

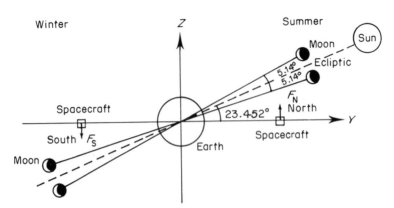

Figure 5.8 Out of plane forces F_W, F_S on a geostationary orbit. The Y-axis lies in the equatorial plane and the Z-axis is the Earth's spin axis. The diagram shows the relative positions of the Sun and Moon in summer and winter, relative to the GEO spacecraft

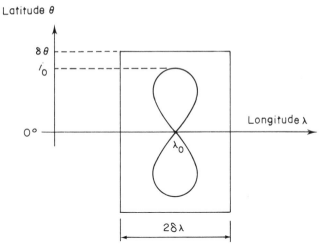

Figure 5.9 Dead-band of inclined geosynchronous vehicle. The figure of eight demonstrates the apparent ground track of a synchronized GEO satellite inclined at an angle i_0 to the equator

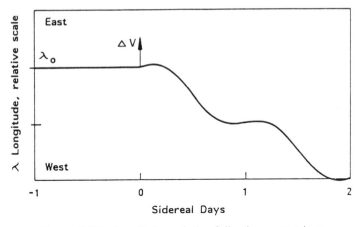

Figure 5.10 Longitude evolution following an east burn

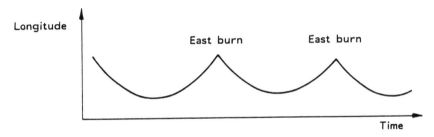

Figure 5.11 Parabolic shape of mean longitude free drift with station-keeping by east burns

the corresponding period T of the limit cycle are

$$\Delta V = \sqrt{(rf \lambda_{max})} \quad \text{and} \quad T = 4\sqrt{(r\lambda_{max}/f)} \tag{5.5}$$

where r is the orbit radius ($\sim 42 \times 10^6$ m) and the $\pm\lambda_{max}$ represents the box size.

Thus, for example, a satellite at a nominal longitude of $-160°$ ($f = -5.3 \times 10^{-8}$ m/s^2) requiring an error box of $\lambda_{max} = 0.1°$ would require $\Delta V = 0.25$ m/s and the period of the limit cycle would be $T = 54$ days.

5.2.4 Latitude station-keeping

Latitude drift may usefully be approached by introducing the two-dimensional vectors **i** and **e**, given by [3]:

$$\mathbf{e} = \begin{bmatrix} e_x \\ e_y \end{bmatrix} = \begin{bmatrix} e\cos(\Omega + \omega) \\ e\sin(\Omega + \omega) \end{bmatrix}$$

$$\mathbf{i} = \begin{bmatrix} i_x \\ i_y \end{bmatrix} = \begin{bmatrix} i\sin\Omega \\ -i\cos\Omega \end{bmatrix}$$

The natural evolution of inclination during a five-year period is shown in Figure 5.12. This wavy drift shows the half-yearly effect caused by the motion of the Earth about the Sun. The lunar periodicity of 14 days is smoothed out in this representation.

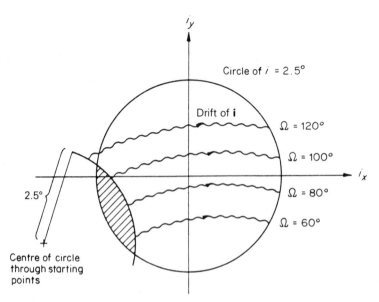

Figure 5.12 Uncontrolled evolution of inclination vector **i** in component form i_x, i_y, where $i_x = i\sin\Omega$ and $i_y = i\cos\Omega$

Table 5.4 GEO burn directions

	North burn	South burn
Spring	Morning	Evening
Summer	Midnight	Noon
Autumn	Evening	Morning
Winter	Noon	Midnight

The philosophy of inclination control is similar to that of longitude, a correcting nodal burn taking place just before the inclination drifts to the maximum permitted by the error box, $\pm i_{max}$ say. If the spacecraft's speed is V, then the directional change needed is $2i_{max}$, and this will require a ΔV of $2Vi_{max}$ (see equation 3.17).

For example, if $i_{max} = 0.1°$ and $V = 3\,075$ m/s, then $\Delta V = 10.7$ m/s.

Clearly the ΔV impulses for controlling inclination are appreciably greater than those needed for longitude, and their errors due to cold starts will therefore represent a smaller percentage error.

Latitude manoeuvres imply a change of the orbit plane. Such manoeuvres must be performed on the line of nodes, using north/south burns at times of day indicated by Table 5.4. Ideally the thrust vector should be perpendicular to the plane bisecting the initial and required orbit planes (see equation (3.18)), and should produce no change in the spacecraft's speed. The thrust direction should be accurate since a directional error ε will lead to a speed change of approximately $2Vi_{max}\varepsilon$, and this would influence the longitude drift rate; errors of only a few degrees could swamp the natural longitude drift rates noted above.

5.3 POLAR LEO/REMOTE SENSING SATELLITES

5.3.1 Viewing conditions/ground tracks

For viewing all parts of the Earth's surface at close quarters it is necessary to adopt a low-altitude polar obit. With the orbit fixed in space and the Earth rotating underneath it the result is that the ground tracks of successive orbits cross the equator at points which move westwards, as shown in Figure 5.13.

An Earth-synchronous orbit results when the subsatellite point follows a ground track identical to some previous orbit after a certain period of time. The repetition occurs on a regular basis, and it can be achieved in a variety of ways. (A geosynchronous orbit is of course another example of an Earth-synchronous orbit.) Between successive orbits the subsatellite point on the equator will change in longitude by $\Delta\phi$ radians, this angle being determined by two effects. The first of these is due to the rotation of the Earth beneath the orbit, and the second is caused by nodal regression. It will be assumed that a positive $\Delta\phi$ means a move towards the east.

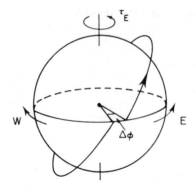

Figure 5.13   Orbit plane motion

The Earth rotates through one revolution in its *sidereal* period of τ_E, where $\tau_E = 86\,164.090\,55 + 0.015T$ seconds. T is measured in centuries from 1900. If the satellite's period is τ then the contribution to $\Delta\phi$ which is caused by the Earth's rotation will be given by

$$\Delta\phi_1 = -2\pi \frac{\tau}{\tau_E} \text{ rad/orbit} \tag{5.6}$$

The regression of the line of nodes (equation (4.35)) contributes

$$\Delta\phi_2 = -\frac{3\pi J_2 R_E^2 \cos i}{a^2(1-e^2)^2} \text{ rad/orbit} \tag{5.7}$$

The total increase in longitude at the equator is

$$\Delta\phi = \Delta\phi_1 + \Delta\phi_2 \text{ rad/orbit} \tag{5.8}$$

Clearly if we wish to have an Earth-synchronous orbit, then we will require that some integral number of orbits later the accumulated value of $\Delta\phi$ will equal 2π. In general we may, therefore, write

$$n|\Delta\phi| = m\,2\pi \tag{5.9}$$

where n is the total number of orbits performed and m is the number of revolutions (equivalent to days) before an identical ground track will occur.

Sun-synchronization occurs when the orbit plane rotates in space at the same rate as the Earth moves round the Sun—at one revolution per year, or roughly one degree per day. Figure 5.14 illustrates this over a period of about three months, during which the orbit clearly needs to rotate through 90° in order to be synchronized.

The required rotation rate is

$$\Delta\phi_2 = -2\pi \frac{\tau_E}{\tau_{ES}} \frac{\tau}{\tau_E} \text{ rads per orbit} \tag{5.10}$$

where $\tau_{ES} = 3.155\,815 \times 10^7$ s and is the orbital period of Earth round the Sun, and τ is the period of the satellite's orbit.

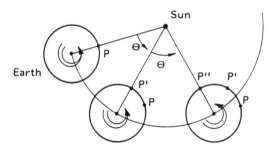

Figure 5.14 Sidereal and solar motion

Sun-synchronous orbits have an advantage for some Earth-viewing missions in that the Earth is always viewed at one of two times of day, as shown in Figure 5.15.

In general configuration (a) is unattractive for Earth viewing since the Sun is always low on the horizon at the subsatellite point, resulting in long shadows and low illumination. However, the orbit is attractive for the power subsystem as it provides lengthy eclipse-free periods, making energy storage minimal, and the array may be fixed relative to the body of the vehicle. Further, since the array is aligned along the direction of flight, a minimum projected area to the velocity vector of the satellite is obtained, thus reducing the influence of drag and increasing satellite lifetime.

The other feature to note about the variation in local solar time coverage is that since the orbit plane is fixed relative to the solar vector it is possible to cant the array relative to the spacecraft body, providing optimal illumination conditions for the solar array.

Regression of the line of nodes may be used to achieve synchronization without the use of fuel, and equation (5.7) indicates that an inclination in excess of $90°$ will be needed.

For Sun and Earth synchronization equations (5.8)–(5.10) apply, and these lead to the condition

$$nt\left(1 - \frac{\tau_E}{\tau_{ES}}\right) = m\tau_E \qquad (5.11)$$

The right-hand side of this equation indicates the number of days between successive identical ground tracks, and τ_E/τ_{ES} is the reciprocal of the number of days in a year.

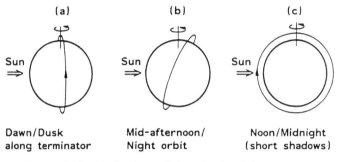

Figure 5.15 Implications of changing local time coverage

The angular displacement between successive orbits in a westward direction is given by

$$\Delta\phi = 2\pi\tau\left(\frac{1}{\tau_E} - \frac{1}{\tau_{ES}}\right) = 7.27 \times 10^{-5}\tau \text{ rad} \tag{5.12}$$

From equation (4.12) we have

$$\tau = 9.952 \times 10^{-3}a^{3/2} \text{ sec} \tag{5.13}$$

(where a is in kilometres). Since for remote sensing satellites in low Earth orbit the altitude is 550–950 km, then $\tau \sim$ 95–100 min, and so $\Delta\phi \sim 4.3 \times 10^{-1}$ rad. At the equator this translates into a distance of $\sim$2800 km between ground tracks, although this decreases at higher latitudes. It is clear from this that wide instrument swath widths will be required if complete Earth coverage is to be obtained. By extending the period between repetitions of a given set of ground tracks a more densely packed set may be achieved. For example, if the requirement is for a daily repeat ($m = 1$) of a set of ($n =$) 14, 15 or 16 tracks, then from equations (5.11) and (5.13) it may be deduced that the corresponding orbit altitudes

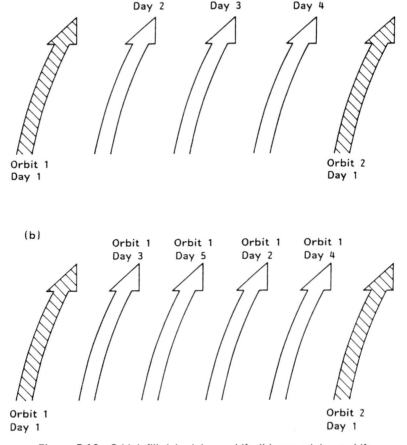

Figure 5.16   Orbit infill: (a) minimum drift, (b) non-minimum drift

are 894, 567 and 275 km. This type of orbit is known as a *zero drift* orbit, and has no infilling between ground-tracks in subsequent days. A greater density of tracks may be achieved by infilling the tracks of Day 1 on subsequent days. The cycle repeats itself over a number of days in excess of one ($m > 1$), with a number (n) of tracks which is not a multiple of m.

If the condition $(n \pm 1)/m = k$ is satisfied, where k is an integer, then a *minimum drift* orbit is obtained. In this situation two successive tracks of a given day are in-filled sequentially on subsequent days as in Figure 5.16(a). Non-minimum drift orbits, $(n \pm 1)/m \neq k$, infill in a nonsequential manner, as in Figure 5.16(b).

Since τ is a function of altitude (equation 5.13) it is possible to plot the repeat period of Earth synchronization versus altitude. This, for minimum drift orbits, is shown in Figure 5.17.

As an example, for Landsat 1/2 an 18-day repeat period of a minimum drift orbit was achieved with the following parameters:

- $\tau \sim 103.3$ min

- inclination $99°$

- descending node 9.38 local time

- Apogee height 920 km, $e = 0.002$

At the equator the orbital separation between successive orbits is ~ 2875 km. However, by considering the complete 18-day sequence the distance between adjacent ground tracks is reduced to 160 km. For Landsat D a different orbit philosophy was adopted, in part due to the need for a Shuttle retrievable system. Again, an 18-day repeat period was

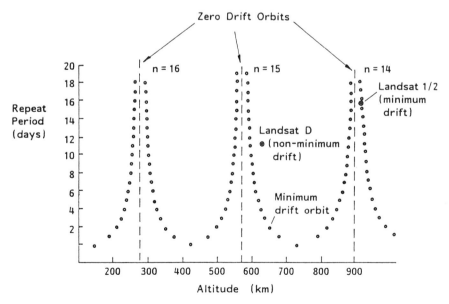

Figure 5.17 Locus of points for minimum drift orbits as a function of repeat period in days (m) and altitude

selected; however, a non-minimum drift orbit was adopted so that a suitable orbital altitude could be obtained. The orbit was nominally circular, having an altitude of 709 km,
period 99 mins, with an equatorial crossing time of 9.30 local time.

5.3.2 Ground station visibility

A satellite is visible at all points on the Earth's surface within a circle which is centred on
the subsatellite point, and whose diameter increases with satellite altitude (see Table 5.5).
However, signals from satellites at the horizon limit are considerably attenuated by the
atmosphere, and so for practical purposes the surface coverage is restricted to the region
in which the satellite elevation above the horizon is greater than $\sim 5°$.

Figure 5.18 shows the geometry associated with a satellite S in a circular orbit of height
h, passing immediately over a ground station at G. If it is visible only down to elevations
equal to θ, typically $5-10°$, then the semi-angle ϕ over which it is visible is given by

$$\phi = -\theta + \cos^{-1} \left\{ \frac{R_E}{(R_E + R)} \cos \theta \right\} \tag{5.14}$$

At the extremes of its visibility the satellite range is the slant range s, where

$$s = (R_E + h) \sin \phi / \cos \theta \tag{5.15}$$

Table 5.5 Relationship between altitude and Earth coverage angle

Altitude (km)	100	500	1 000	5 000	36 000
ϕ_0 ($\theta = 5°$)	6.2	17.5	25.5	51.0	76.3
ϕ_0 ($\theta = 0°$)	10.1	22.0	30.2	56.0	81.3
Slant range (km)	1 134	2 574	3 709	9 422	41 698

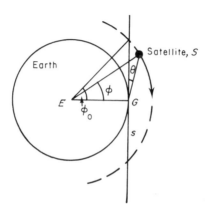

Figure 5.18 Satellite visibility θ, and
slant range S

The duration of a pass with a semi-angle ϕ is

$$\tau = 2\phi/\omega_{ES} \tag{5.16}$$

where ω_{ES} is the orbital angular velocity relative to Earth. ω_{ES} may be obtained from the Earth's angular rate ω_E (7.3×10^{-5} rad/s) and the satellite's orbital angular velocity ω, using

$$\omega_{ES}^2 = \omega_E^2 + \omega^2 - 2\omega_E\omega \cos i \tag{5.17}$$

For LEO it is sufficient to use ω in place of ω_{ES} in equation (5.16), where $\omega = 632(R_E + h)^{-3/2}$ and R_E, h are in kilometres. For example, a satellite in a 500 km orbit will have an elevation $>5°$ for 9.2 min and $>10°$ for 7.4 min during an overhead pass. Most passes will be shorter than this and a useful pass should be of more than four minutes duration.

5.3.3 Eclipse and launch windows

A spacecraft in a low-altitude orbit will experience eclipse, the frequency and duration of which is determined by orbit inclination—for example, in low-altitude equatorial orbit the satellite is eclipsed for about 40% of every orbit, whilst for dawn/dusk Sun-synchronous orbits, even at low altitude, several months of wholly sunlit operation may be obtained. These latter orbits, though nominally synchronous, are eclipsed as a result of the inclination of the ecliptic to the equatorial plane and the attendant seasonal variations in solar vector. Such concerns identify a suitable launch season for sunlit operation, as outlined below.

The launch season for wholly sunlit dawn/dusk sun-synchronous low altitude Earth

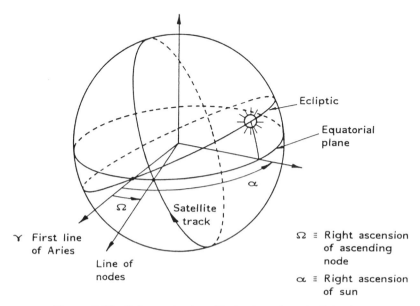

Figure 5.19 Solar motion/satellite track on the celestial sphere

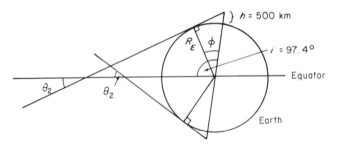

Figure 5.20   Eclipse condition for a space vehicle

orbits may be considered in the following way. With reference to Figure 5.19, it is clear that a daily launch opportunity will occur twice a day, when the line of nodes is perpendicular to the solar vector (Earth–Sun line). Thus for a right ascension of the Sun angle α, we require:

$$\alpha - \Omega = \pm \frac{\pi}{2} \qquad (5.18)$$

The season for the launch is influenced by both the inclination of the orbit (for a Sun-synchronous orbit this is related to the orbit altitude) and the fact that the apparent solar motion is from latitude 23.4°N (Summer solstice) to 23.4°S (Winter solstice). For a 500 km circular Sun-synchronous orbit, solving equations (5.7) and (5.10) yields an inclination i of 97.4°. From Figure 5.20 it is clear that $\cos \phi = R_E/(R_E + h)$, giving a value for ϕ of ~22°.

Hence, due to the inclination angle $i = 97.4°$, then $\theta_2 = 14.6°$ and $\theta_1 = 29.4°$. Thus the satellite will be eclipsed if the Sun's latitude falls below 14.6°S or increases above 29.4°N. Clearly the latter situation never occurs, but the satellite will be partially eclipsed during mid-winter. Approximating the solar motion to a sinusoidal motion projected on the equator, the duration of the eclipse season is ~2.6 months. Figure 5.21 is a plot of the totally sunlit life as a function of month of launch.

The above analysis is clearly simplified, in that it assumes a circular motion of the Earth about the Sun and also that a 1°/day motion in the ecliptic translates to a 1°/day motion on the equator. The relationship between ecliptic motion and equatorial motion is

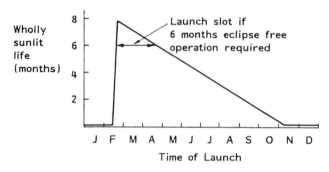

Figure 5.21   Launch season for eclipse-free operation of a remote sensing vehicle

$$\sin \theta_{eq} = \frac{\sin \theta_{ec} \cos \varepsilon}{\sqrt{((1 - \sin^2 \theta_{ec} \sin^2 \varepsilon))}} \qquad (5.19)$$

where θ_{eq} is equatorial angle measure, θ_{ec} is the corresponding ecliptic angle and ε is the obliquity of the ecliptic, $\sim 23.4°$.

A similar analysis may be performed for other types of orbit.

5.3.4 Satellite lifetime

For low Earth orbit vehicles, aerodynamic drag will eventually result in re-entry as described in Chapter 4. To the accuracy of a few percent the lifetime of an uncontrolled space vehicle will be

$$\tau \sim \frac{e_0^2}{2B} \left(1 - \tfrac{11}{6} e_0 + \tfrac{29}{16} e_0^2 + \tfrac{7}{8} \frac{H}{a_0} \right) \qquad (5.20)$$

where e_0 and a_0 are the initial values of eccentricity and semi-major axis once control has ceased, H is the scale height of the atmosphere near perigee and B is given by

$$B \sim \sqrt{\left(\frac{a_0^3}{\mu} \right) \frac{AC_D}{M} \rho_{p0} a_0 e_0 I_1 \left(\frac{a_0 e_0}{H} \right) \exp \left(-e_0 \left(1 + \frac{a_0}{H} \right) \right)} \qquad (5.21)$$

This expression neglects the effects of atmospheric rotation. I_1 is the Bessel function of the first kind and order 1 and ρ_{p0} is the atmospheric density at the initial perigeee. For large space vehicles and for military vehicles re-entry may cause problems due to both geographical position of the re-entry corridor and the size of individual spacecraft elements on ground impact. Most notably the effect on the public of Skylab re-entry over Australia, and Cosmos 943B over Canada, provide historical evidence of some of the problems that may be encountered in re-entry.

5.4 ELLIPTIC ORBITS

Elliptical orbits are of interest in that they offer the prospect of increased duration of ground station pass if the apogee point of the orbit is situated above the desired ground station. Whilst in the general case inclined elliptical orbits will not be stable due to precession in the line of apsides, it is possible to avoid this by choosing a suitable orbit inclination. In this way a near-stable orbit is obtained. The equation giving the rate of apsidal precession is, neglecting harmonic terms in the Earth potential function greater than 2:

$$\Delta \omega \sim \frac{3\pi J_2 R_e^2 (5 \cos^2 i - 1)}{2a^2 (1 - e^2)^2} \text{ rad/rev} \qquad (5.22)$$

Solving for $\Delta \omega = 0$ yields $i = 63.4°$. This expression also ignores any additional perturbative effects, such as luni-solar perturbation, which can become a significant disturbance for apogee heights which are beyond geostationary orbit. These orbits do, however, provide a useful operational environment for communications at high latitudes, since the sub-satellite point at apogee may be set to a latitude of $63.4°$. It transpires, however, that

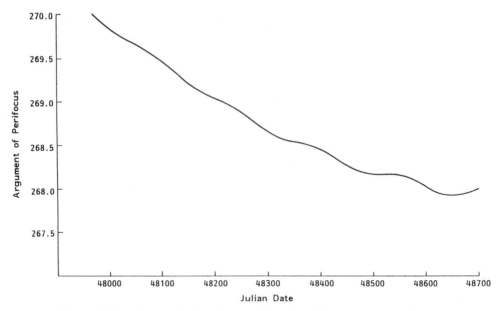

Figure 5.22   Uncontrolled evolution of Molniya orbit: argument of perifocus

even including higher harmonics in the Earth's gravitational field and luni-solar perturbations, only a small change in ω results for critical inclination orbits. Figure 5.22 shows the evolution of ω when the effects of lunar and solar gravitational fields include Earth's gravitational harmonics (tesseral and zonal) to order 5, and solar radiation pressure. In this case a 12-hour synchronous Molniya orbit has been modelled having $a = 26\,447$ km, $e = 0.72$ with $\omega = 270°$. After two years ω has changed by less than 2°.

The main application for this type of orbit is to provide a communication capability at high latitude. Appropriate orbital parameters (similar to those of the orbit displayed in Figure 5.22) can provide continuous, high-elevation coverage for typically 8 hours in a 12-hour orbit period. It is then possible to provide a 24-hour communication system with a minimum of three satellites. The notable disadvantage of this orbit, in comparision with GEO orbits, is that orbital placement is costly in propulsion terms. This is particularly noticeable for Ariane with its attendant launch site close to the equator; indeed the Ariane vehicle mission profile is optimized for GEO. One further disadvantage for a Molniya communication system arises from the 'spares' philosophy. For a GEO vehicle a nearby longitude slot may be used for an in-orbit spare. However, in Molniya orbit the three satellites required to establish a 24-hour link are each in a different plane. This results in it being impracticable to have an in-orbit spare satellite.

REFERENCES

[1]   Hechler, M. and Van der Ha, J.C. (1981) *J. Spacecraft and Rockets* **18**, 361.
[2]   Clarke, A.C. (1949) Extraterrestrial relays, *Wireless World*, October, 305.
[3]   Soop, E.M. (1983) *Introduction to Geostationary Orbits*, ESA SP 1053.

6 PROPULSION SYSTEMS

J. Barrie Moss

School of Mechanical Engineering, Cranfield Institute of Technology

6.1 SYSTEMS CLASSIFICATION

The broad classes of propulsion systems are distinguished in Figure 6.1. Of the several alternatives identified, exploitation in practical devices has focused largely on thermal and electric rockets. Primary propulsion for launch vehicles is further restricted— currently to solid- or liquid-propelled chemical rockets. We review these in detail in subsequent sections and refer only briefly to the more speculative options.

The solar radiation pressure at 1 AU from the Sun is roughly 5×10^{-6} N m^{-2} and therefore the surface area which must be deployed in order to produce significant thrust for primary solar sailing is extremely large. Space vehicles of more modest dimensions may however be subject to perturbing torques resulting, for example, from asymmetries in deployed surfaces.

Nuclear propulsion has been the subject of very detailed studies over many years, although these have not been pursued to flight demonstrations. The NERVA programme of the 1960s (Nuclear Engine for Rocket Vehicle Applications) resulted in a ground-tested solid core (graphite) U235 fission reactor-powered engine delivering approximately 300 kN thrust with a specific impulse of 825 s. With the emphasis on near-Earth operations, direct thrust nuclear rockets do not now appear to be cost effective and more recent studies focus on lower thrust orbit raising and manoeuvring propulsion as aspects of wider programmes for nuclear space power raising.

The continued development of high-power lasers and increasing interest in the establishment of complex orbiting platforms for power raising and communications in space have encouraged conceptual studies of laser propulsion. Recent thinking in relation to these propulsion strategies is reviewed by Caveney [1].

Crucial parameters which distinguish between the developed systems, and introduce important characteristics of systems generally, are readily identified. We show later that the beam (or kinetic) power of a rocket exhaust, P, is given by

$$P = \tfrac{1}{2}\dot{m}V_e^2 \tag{6.1}$$

where $\dot{m}$ is the exhaust mass flow rate and V_e the exhaust velocity.

Spacecraft Systems Engineering. Edited by P.W. Fortescue and J.P.W. Stark
© 1991 John Wiley & Sons Ltd

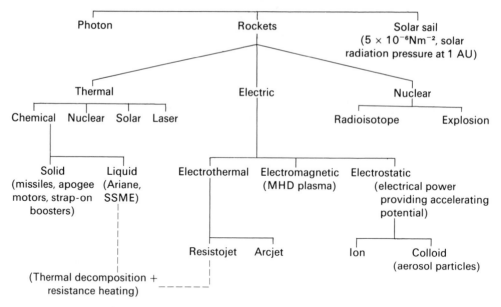

Figure 6.1   Propulsion systems classification

Since the thrust delivered by the rocket engine may be written

$$F = \dot{m}V_e \qquad (6.2)$$

then

$$P = \tfrac{1}{2}FV_e \qquad (6.3)$$

If we introduce the vehicle acceleration (or vehicle thrust-to-weight ratio), αg_0, where g_0 is the Earth's surface acceleration due to gravity, then we may write

$$F \approx M\alpha g_0$$

(M = vehicle mass) and, from (6.2) and (6.3),

$$P/M = \tfrac{1}{2}\alpha V_e g_0 \qquad (6.4)$$

Propulsion systems are readily distinguished by comparisons between their specific power P/M, exhaust velocity, V_e, and thrust acceleration, α as shown in Figure 6.2. The separately powered electric rocket is characterized by high exhaust velocity and low specific power or thrust acceleration. Nuclear or chemical rockets on the other hand offer high powers but with relatively poor propellant utilization through limited exhaust velocity. Booster operation from planetary surfaces is evidently restricted to these latter systems which are said to be *energy limited*—by the chemical energy stored in the propellants—but admit high thrust for comparatively modest engine weight. Although the specific impulse of electrically powered systems is high, they are *power limited*. Whilst the energy available from the Sun is unlimited, or from a radioactive source introduces a negligible fuel mass, the effective energy density is low and the necessary weight of the accompanying systems for electrical conversion is unacceptably large in high-power applications. Electrically propelled space vehicles are thus restricted to very small accelerations. As we shall

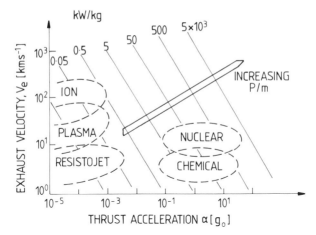

Figure 6.2 Comparative rocket performance

describe in Section 6.3 the more efficient expellant utilization possible with low thrust systems thus appears most naturally suited to orbit-raising manoeuvres, interplanetary missions and spacecraft attitude and orbit control.

Whilst the sound and fury of rocket propulsion is identified with launch vehicles there is a wide range of propulsive roles associated with space missions. We may illustrate these distinctive roles as follows:

- launch vehicles—main engines and 'strap on' boosters—developing high thrust for extended periods (approximately 2×10^6 N for 8 minutes in the case of each Space Shuttle Main Engine (SSME) for example)

- apogee motors for spacecraft orbit circularization and inclination removal (typically, 75 kN thrust for approximately 60 s, producing a velocity increment of approximately 2 km/s); perigee motors for orbit raising from low-altitude parking orbit (similar performance to that of apogee motors required by Payload Assist Modules (PAM) or Inertial Upper Stages (IUS) offering geostationary altitude acquisition for Shuttle-launched payloads).

- spacecraft attitude and station-keeping orbit control (thrust levels, ranging from milliNewtons to the order of 10 Newtons, pulsed operation over the duration of the complete mission).

Since propulsive requirements are frequently specified in terms of ΔV (cf. Chapter 5: Mission Analysis) it is useful to identify and compare some typical values:

- ΔV into low Earth orbit
  (including drag and gravity loss) ≥ 9.5 km/s

- additional ΔV to equatorial geostationary orbit
  from a 30° inclined parking orbit ~ 4.2 km/s

- additional ΔV for Mars fly-by ~ 3.4 km/s

- additional ΔV for solar System escape
  (without gravitational assist) ~ 8.5 km/s

- geostationary orbit station-keeping (Intelsat V)
  north–south (inclination $< 0.1°$ throughout 7 year life) $\Delta V \sim 347.5$ m/s
  east–west ($\pm 0.1°$ of desired longitude) $\Delta V \sim 29.0$ m/s

6.2 CHEMICAL ROCKETS

The rapid growth in rocket propulsion technology following World War II is largely based on chemical rockets. Simple reaction systems, in which the propulsive force exerted on a vehicle results from changes in system momentum through a high velocity exhaust, have proved uniquely successful in high-thrust atmospheric and space applications. The combustion of chemical propellants, solid or liquid, at high pressure liberates large quantities of energy in a small volume. The subsequent expansion of these high tempera-ture products of combustion through a nozzle, converts thermal energy to directed kinetic energy for rocket propulsion. Since the rocket engine carries both fuel and oxidizer the specific fuel consumption is substantially higher than that of an air-breathing reaction system such as the turbojet. It is, however, mechanically less complex than an aircraft power plant, since moving parts are confined to auxiliary systems such as propellant feed.

We show later in this section that chemical rockets are conveniently characterized by the Tsiolkovsky equation in field-free space

$$\Delta V = V_e \ln R \tag{6.5}$$

where ΔV denotes the rocket velocity increment and R is the mass ratio, initial mass to mass at burn-out. ΔV is typically prescribed by the mission whilst V_e is essentially fixed by the choice of propellant. Only by increasing the mass ratio can the shortfall in propellant energetics be accommodated and mission goals be attained.

6.2.1 Basic principles

We first review briefly those aspects of rocket motor performance, gas dynamics and thermochemistry which most directly influence design and operation. For further details of the analysis the reader is referred to the excellent texts by Barrère et al. [2] and Sutton and Ross [3].

Performance parameters

Consider the rocket illustrated schematically in Figure 6.3. Applying equation (3.15) from Chapter 3 leads to

$$M \frac{dV}{dt} = \dot{m} V_e + A_e(p_e - p_a) + F_{ext} \tag{6.6}$$

The rocket thrust F comprises two contributions, from the exhaust momentum flux and the exhaust plane pressure difference:

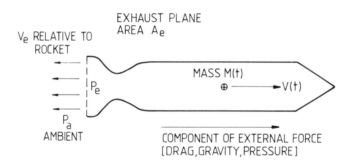

Figure 6.3 Schematic rocket

$$F = \dot{m}V_e + A_e(p_e - p_a) \tag{6.7}$$

where $\dot{m}$ is the propellant mass flow (kg/s)

V_e is its exhaust speed (m/s)

$A_e(p_e - p_a)$ is the resultant force on the rocket due to the pressure difference between nozzle exit and ambient

F_{ext} denotes the extra force in the direction of motion due to external forces (for example, aerodynamic drag or gravity).

We show in the following section that nozzle performance is maximized by complete exhaust expansion to ambient pressure, whence $p_e - p_a = 0$. In space applications the ambient pressure is either continuously varying with altitude or is zero, outside the Earth atmosphere. The effects of under-expansion, $p_e > p_a$, are thus partially offset by the thrust increases accompanying this pressure difference.

From equation (6.7) the thrust at sea level is

$$F_{SL} = \dot{m}V_e + A_e(p_e - p_{SL})$$

and at altitude h it may be expressed as

$$F_h = F_{SL} + A_e(p_{SL} - p_h) \tag{6.8}$$

In vacuo, when $p_h = 0$, it becomes

$$F_0 = F_{SL} + A_e p_{SL}$$

Typically $A_e p_{SL}/F_{SL}$ approaches 20%.

It is convenient to define an effective exhaust velocity

$$V_e^* \equiv V_e + A_e(p_e - p_a)/\dot{m} \equiv I_{sp}g_0 \tag{6.9}$$

where g_0 is the acceleration due to gravity at the Earth's surface.

I_{sp} is the specific impulse, the total impulse per unit propellant weight consumed, and is given by

$$I_{sp} = \frac{I}{M_p g_0} = \frac{\int_0^t F(t)\, dt}{g_0 \int_0^t \dot{m}(t)\, dt}, \tag{6.10}$$

$I_{sp} = F/\dot{m}g_0$ for constant thrust and exhaust mass flow rate.

Equation (6.6) may be expressed in terms of V_e^*, and when this is constant it may be integrated over the duration of the rocket motor firing giving

$$\Delta V = V_b - V_0 = V_e^* \ln\{M_0/M_b\} + \int_0^{t_b} \frac{F_{ext}}{M}\, dt \tag{6.11}$$

and hence the Tsiolkovsky equation (6.5), where the mass ratio

$$R \equiv M_0/M_b,$$

is the ratio of initial to burn-out mass.

The maximization of V_e^* for a specified velocity increment ΔV is essential for efficient design. We now describe the nozzle flow characteristics necessary to realize high exhaust velocity given a particular propellant selection.

Nozzle flows

We analyse the flow through a convergent–divergent nozzle, downstream from the combustion chamber and as illustrated in Figure 6.4, with the aid of the following simplifying assumptions:

1. The combustion products are homogeneous and of constant composition.

2. The products at temperature T and molecular weight W obey the perfect gas law relating pressure p and density ρ:

$$p = \rho T(R_0/W) \tag{6.12}$$

   where R_0 is the universal gas constant.

3. The specific heat of the mixture is invariant with temperature and pressure.

4. The flow is one-dimensional, steady and isentropic.

The conservation equations for mass and energy may then be written

$$\dot{m} = \rho V A \tag{6.13}$$

$$\tfrac{1}{2}V^2 + C_p T = \text{constant} = \tfrac{1}{2}V_c^2 + C_p T_c \tag{6.14}$$

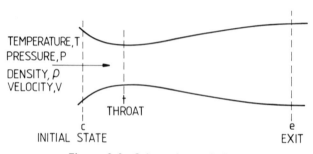

Figure 6.4   Schematic nozzle flow

whilst for an adiabatic flow process

$$p\rho^{-\gamma} = \text{constant}$$

and

$$\frac{T}{T_c} = \left(\frac{\rho}{\rho_c}\right)^{\gamma-1} = \left(\frac{p}{p_c}\right)^{(\gamma-1)/\gamma} \tag{6.15}$$

If the initial velocity V_c is negligibly small, then

$$\frac{\dot{m}}{A} = \left\{\frac{2\gamma}{\gamma-1}\, p_c\rho_c\left(\frac{p}{p_c}\right)^{2/\gamma}\left[1 - \left(\frac{p}{p_c}\right)^{(\gamma-1)/\gamma}\right]\right\}^{1/2} \tag{6.16}$$

Clearly (6.16) exhibits a maximum value for $(\dot{m}/A)$ corresponding to a critical throat condition, subscript t, at which

$$\frac{p_t}{p_c} = \left(\frac{2}{\gamma+1}\right)^{\gamma/(\gamma-1)}$$

and

$$\frac{T_t}{T_c} = \frac{2}{\gamma+1}$$

and

$$\frac{\dot{m}}{A} = \{\gamma\rho_t p_t\}^{1/2} = \rho_t V_t$$

where the critical throat velocity V_t is given by

$$V_t = \left\{\gamma\frac{p_t}{\rho_t}\right\}^{1/2} = \left\{\gamma\frac{R_0}{W}\, T_t\right\}^{1/2} = a_t, \quad \text{speed of sound at the throat.}$$

In convergent–divergent (de Laval) nozzles the velocity continues to increase downstream of the throat ($p_{\text{exit}}/p_c < p_t/p_c$) but the nozzle is choked—that is, the mass flow is simply determined by throat conditions, independent of the exit flow condition.

Mass flow can be expressed as a function of combustion chamber conditions (p_c, T_c) and throat area A_t,

$$\dot{m} = \sqrt{\gamma}\left\{\frac{2}{\gamma+1}\right\}^{(\gamma+1)/[2(\gamma-1)]}\frac{p_c A_t}{\sqrt{(R_0 T_c/W)}} \tag{6.17}$$

We may identify a characteristic velocity

$$c^* = \sqrt{(R_0 T/W)}\left/\left\{\sqrt{\gamma}\left[\frac{2}{\gamma-1}\right]^{(\gamma+1)/[2(\gamma-1)]}\right\}\right. \tag{6.18}$$

whence

$$\dot{m} = p_c A_t/c^* \tag{6.19}$$

From the energy conservation equation (6.14) we may determine the exhaust velocity V_e

$$\tfrac{1}{2}V_e^2 + C_p T_e = C_p T_c \quad (V_c = 0)$$

whence, after some manipulation

$$V_e = \sqrt{\left\{\frac{2\gamma R_0 T_c}{(\gamma - 1)W}\left[1 - \left(\frac{p_e}{p_c}\right)^{(\gamma-1)/\gamma}\right]\right\}} \tag{6.20}$$

We note that the exhaust velocity, V_e, increases with the following:

1. increasing pressure ratio p_c/p_e—though such benefits are limited by accompanying increases in motor weight;

2. increasing combustion temperature T_c—to be set against the adverse effects of higher temperatures on nozzle heat transfer and increased dissociation losses (see page 118);

3. low molecular weight; and

4. to a limited extent, by reducing the ratio of specific heats, γ—this is of limited practicality given the other influences.

It is convenient to identify a characteristic thrust coefficient C_F^0 such that

$$V_e = c^* C_F^0$$

where

$$C_F^0 = \sqrt{\left\{\left[\gamma\left(\frac{2}{\gamma + 1}\right)^{(\gamma+1)/(\gamma-1)}\right]\frac{2\gamma}{\gamma - 1}\left[1 - \left(\frac{p_e}{p_c}\right)^{(\gamma-1)/\gamma}\right]\right\}} \tag{6.21}$$

The exit-to-throat area ratio, A_e/A_t, can be determined from the continuity equation (6.13) such that

$$\frac{A_e}{A_t} = \frac{\rho_t V_t}{\rho_e V_e} = \gamma\left(\frac{2}{\gamma + 1}\right)^{(\gamma+1)/(\gamma-1)}\left(\frac{p_c}{p_e}\right)^{1/\gamma}\bigg/C_F^0 \tag{6.22}$$

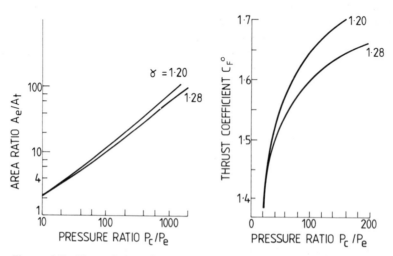

Figure 6.5 The variation of area ratio and thrust coefficient with pressure ratio

The variation of thrust coefficient and area ratio with pressure ratio are illustrated in Figure 6.5.

Nozzle performance and design

We now combine the rocket motor performance characteristics of Section 6.2 with the gas dynamics of the preceding section.

For given combustion chamber pressure p_c and mass flow rate $\dot{m}$ (and hence throat area) the motor thrust can be optimized. From equation (6.7)

$$F = \dot{m} V_e + A_e(p_e - p_a)$$

and incremental parameter changes are related by

$$\delta F = \dot{m}\delta V_e + \delta A_e(p_e - p_a) + A_e\delta p_e$$

But conservation of momentum insists that

$$\dot{m}\delta V_e + A_e\delta p_e = 0$$

whence

$$\frac{\mathrm{d}F}{\mathrm{d}A_e} = p_e - p_a = 0 \quad \text{for maximum thrust}$$

The thrust is thereby maximized when $p_e = p_a$ and the nozzle flow is said to be ideally expanded.

The exhaust flow patterns accompanying departures from this ideal behaviour are sketched in Figure 6.6. If the nozzle flow is over-expanded, $p_e < p_a$, pressure recovery to ambient conditions is effected through a series of shock waves. Penetration of these shock waves into the nozzle leads to separation of the wall boundary layers, enhanced viscous losses and reduced expansion ratio. Such behaviour is characteristic of operation at high ambient pressure, typical of sea-level or test-bed firing. Under-expanded flows are characterized by incomplete nozzle expansion, $p_e > p_a$, and reduced exhaust velocity. The low ambient pressures which give rise to such flows are typically realized in space vacuum operation.

Figure 6.7 illustrates the thrust improvement which would result from continuous adaptation, leading to $p_e = p_a$, with increasing altitude, and hence reducing p_a, in comparison with the thrust of an ideally expanded nozzle at sea level. The expansion ratio (A_e/A_t) of the nozzle becomes very large, however, and significant weight penalties may result. Careful optimization is necessary to reconcile these competing factors.

The thrust coefficient in (6.21) can be modified to include the effects of non-ideal expansion in the form

$$C_F = \frac{F}{p_c A_t} = C_F^0 + \frac{A_e}{A_t}\left(\frac{p_e}{p_c} - \frac{p_a}{p_c}\right) \tag{6.23}$$

The thrust coefficient for an ideal expansion, given p_c, p_a and A_t, is, from (6.21),

$$\{C_F^0\}_{\max} = \sqrt{\left\{\frac{2\gamma^2}{\gamma - 1}\left(\frac{2}{\gamma + 1}\right)^{(\gamma+1)/(\gamma-1)}\left[1 - \left(\frac{p_a}{p_c}\right)^{(\gamma-1)/\gamma}\right]\right\}}$$

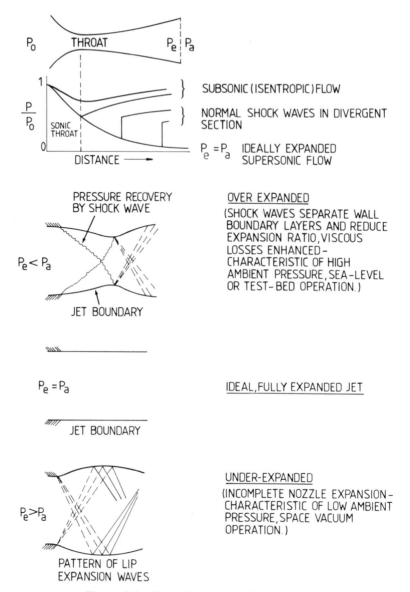

Figure 6.6   Nozzle flows: non-ideal expansion

and with (6.22) we write

$$\frac{C_F}{\{C_F^0\}_{max}} = \frac{C_F^0}{\{C_F^0\}_{max}} + \frac{\gamma(2/(\gamma+1))^{(\gamma+1)/(\gamma-1)}(p_c/p_e)^{1/\gamma}}{C_F^0\{C_F^0\}_{max}}\left(\frac{p_e}{p_c} - \frac{p_a}{p_c}\right) \qquad (6.24)$$

As Figure 6.8 indicates, the departure from an ideal expansion is less severe with under-expansion than with over-expansion. Flow separation from the nozzle, which contracts the jet in the over-expanded situation, leads to an increase in thrust over that which would result in the absence of such separation.

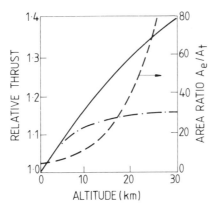

Figure 6.7 Comparative thrust: continuous and sea-level adaptation. ——— continuous adaptation —·— sea-level adaptation

As indicated earlier, nozzle adaptation has important implications for nozzle size and weight. From the manufacturing point of view, convergent–divergent nozzles which are conical represent the simplest designs. Two specific problems arise:

1. The exhaust is not directed in the axial direction and the flow divergence implies some loss of thrust.

2. In high-altitude applications the nozzle tends to be very long and correspondingly heavy.

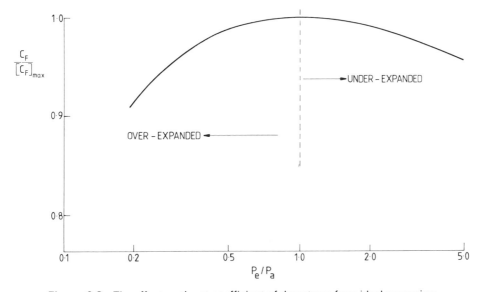

Figure 6.8 The effect on thrust coefficient of departures from ideal expansion

The axial thrust may be shown to be approximately

$$F = \tfrac{1}{2}\dot{m}V_e\,(1 + \cos a)$$

for cone included angle 2α. For $\alpha = 15°$, the multiplicative factor $\tfrac{1}{2}(1 + \cos \alpha)$ is 0.986 and the effect of divergence is small.

Designs to reduce nozzle length, and hence weight, resulting from small divergence must expand the flow from the throat more rapidly and subsequently turn the exhaust in the axial direction. Bell-shaped nozzles are designed to achieve this—see, for example, the established procedure due to Rao [4].

Thermochemistry

We identified from (6.20) that the basic elements in maximizing exhaust velocity (or specific impulse) were high combustion chamber temperature and pressure and low molecular weight. Some representative propellant properties will be illustrated later.

In developing equations for the ideal rocket motor we assumed that, in addition to isentropic flow (no viscous or heat losses), the combustion products were of known constant composition throughout the expansion. At the high temperatures of interest, however, a significant proportion of combustion products are dissociated and the extent to which the energy of dissociation can be recovered in the expansion has a substantial impact on nozzle performance.

In the expansion process, the variation of composition with position in the nozzle depends both upon local thermodynamic state—pressure and temperature—and chemical kinetic rates, in particular upon recombination rates. A complete description is complex and beyond the scope of this chapter but two limiting cases are readily distinguished:

1. frozen flow, in which the composition remains constant throughout the expansion irrespective of the variation in pressure and temperature;

2. equilibrium flow, in which the equilibrium composition corresponding to local conditions of pressure and temperature prevails along the nozzle.

The former implies that the chemical rates are vanishingly small whilst the latter implies that they are infinitely fast.

As the temperature decreases along the nozzle, species dissociated in the combustion chamber recombine, releasing energy and changing the composition. Whilst specific impulse (or thrust coefficient) is predicted to be larger for equilibrium rather than frozen flow at a particular pressure ratio (p_c/p_e), the accompanying area ratio A_e/A_t is also larger.

6.2.2 Propellants

In principle any chemical system producing heat release in a gas flow through exothermic reaction could be used as a propellant. In practice consideration of a range of factors such as the specific energy content, rate of heat release, ease of storage and handling, significantly restricts the choice. Heat release in a liquid-propellant rocket may be effected through the separate injection, mixing and combustion of two liquids—fuel and oxidizer—

comprising a bi-propellant or the exothermic decomposition of a monopropellant such as hydrazine or hydrogen peroxide. A bi-propellant is said to be hypergolic if the fuel and oxidizer react spontaneously on contact with each other. More generally, the requirements of separate propellant storage, pumping and feed to the injector, mixing and ignition mean that the high specific impulse is achieved at the expense of considerable complexity in design and manufacture. Significant flexibility in operation is however achieved, including stop–start options and throttling of thrust levels. By contrast the solid-propellant rocket is of comparatively simple design. The propellant is stored within the combustion chamber in the form of shaped grains bounded by the walls of the chamber. Once ignited combustion will generally proceed until all the propellant is consumed; the thrust–time relationship is then fixed by the grain configuration. Limited thrust regulation is possible.

We shall review briefly the key features of some commonly encountered propellants.

Liquid propellants

Table 6.1 illustrates properties of some representative propellants. We recall from equation (6.10) that in order to maximize exhaust velocity we seek propellants which give high combustion temperatures and low molecular weights. The typical variation of adiabatic flame temperature with equivalence ratio exhibits only a modest decline in T_c for richer-than-stoichiometric mixtures. It is thus possible to maintain high combustion temperatures with hydrogen as fuel, for example, and capitalize on the low fuel molecular weight by operating fuel rich.

The attraction of fluorine as an oxidizer (high combustion temperature, $W_F = 19$) is tempered by its highly corrosive properties which constrain the choice of liner materials for the combustion chamber and nozzle. Both oxygen and fluorine necessitate cryogenic cooling for storage as liquids since they have boiling points of 90 K and 85 K respectively. Long-term storage is therefore difficult and their application is restricted to launch vehicles. Nitrogen tetroxide is finding increasing applications in space vehicle propulsion despite its high molecular weight ($W_{N_2O_4} = 92$) since its boiling point is 294 K.

The traditional, high-thrust, bi-propellant combinations of LOX/LH_2, employed on Saturn V, Space Shuttle main engines and the upper stage of Ariane; $RFNA/UDMH$ employed on the Delta launcher; $N_2H_4/UDMH/N_2O_4$ on Titan and Ariane are increasingly supplemented in apogee motors and orbital manoeuvring systems generally by monomethyl hydrazine ($MMH–CH_3NHNH_2$)/N_2O_4 combinations. The latter offers a specific impulse of approximately 310 seconds, a value which is substantially greater than that available from monopropellant hydrazine decomposition, the readily stored alternative (boiling point 387 K). In a slightly different context, that of the provision of thermal control for propellant storage, we should note that both hydrazine and nitrogen tetroxide, have melting points in the neighbourhood of typical spacecraft ambient temperatures, 275 K and 262 K respectively.

Solid propellants

Solid propellants are typically of two types: either double-base, comprising homogeneous colloidal mixtures of nitrocellulose and nitroglycerine, or composite, comprising mixtures

Table 6.1 Liquid propellants

Fuel	Oxidizer	Molecular weight W products	Combustion temperature T_c (K)	Ideal specific impulse I_{sp} (s)	Mean density ρ (kg/m^3)
H_2 (hydrogen)	O_2 (oxygen)	10	2980	390	280
	*F_2 (fluorine)	12.8	4117	410	460
Kerosene	O_2	23.4	3687	301	1020
	F_2	23.9	3917	320	1230
	RFNA (red fuming nitric acid)	25.7	3156	268	1355
	N_2O_4 (nitrogen tetroxide)	26.2	3460	276	1260
	H_2O_2 (hydrogen peroxide)	22.2	3008	278	1362
N_2H_4 (hydrazine)	O_2 (oxygen)	19.4	3410	313	1070
	*HNO_3 (nitric acid)	20	2967	278	1310
UDMH	O_2 (oxygen)	21.5	3623	310	970
$(CH_3)_2NNH_2$ (unsymmetrical dimethyl hydrazine)	*HNO_3 (nitric acid)	23.7	3222	276	1220
	*hypergolic				
Monopropellants					
N_2H_4		10.3	966	199	1011
H_2O_2		22.7	1267	165	1422

All quoted values are for $p_c = 7$ MPa with an ideal expansion to $p_e = 0.1$ MPa. Higher chamber pressures admit increases in I_{sp}—for example, at 20 MPa, LH_2–LOX yields a specific impulse of ~460 s.

of an organic fuel and crystalline inorganic salt. Ammonium perchlorate, NH_4ClO_4, is the principal oxidizer used in composite propellants with a polymer fuel-binder, typically polyurethane or polybutadiene.

In comparison with the liquid propellants described in Table 6.1, the specific impulses for solid propellants are substantially lower—in the range 200–260 seconds. These values are not strongly influenced by the particular fuel binder, although high hydrogen to carbon ratios are favoured. The performance of composite propellants is improved by the addition of metals such as aluminium or beryllium in the form of finely ground particles ($\sim 10~\mu$m).

Solid propellant charges are typically cast or extruded. The components are mixed in the form of a dough, perhaps in the presence of a plasticizer, and cast or pumped into a mould or directly into the combustion chamber. The grain geometry is fixed by a mandrel which is removed after curing and solidification.

Further information on propellant energetics and physical properties is available from Barrère *et al.* [2] and Sutton and Ross [3]; some more advanced liquid propellant systems are reviewed by Weber in Caveney [1]; the complexities of solid propellant combustion are extensively discussed in Kuo and Summerfield [5].

6.2.3   Chemical rocket design

In this section we seek to identify some of the more important features of motor design in relation to high-thrust launch vehicles or booster applications. A detailed discussion of such a major topic is inappropriate to this book which seeks to familiarize the non-specialist with the diversity of disciplines and technologies at the heart of spacecraft engineering. From the standpoint of spacecraft design the involvement with the launch vehicle lies essentially in the role of procurement and the clear identification of the constraints which are imposed on the spacecraft by the launcher. These form the basis of Chapter 7, whilst the design of secondary propulsion systems, of more direct concern to the spacecraft engineer, is described in Sections 6.3 and 6.4.

Clearly if we distinguish the principal rocket components to be

- the thrust chamber;

- the propellant feed system;

- the propellant storage tanks,

then significant differences are evident between solid and liquid systems. For the solid propellant rocket, offering high thrusts of short duration, key features of the design are the choice of propellant grain to give the appropriate thrust law and materials selection for the nozzle and casing to combine low weight with reliable operation. The greater complexity of the liquid-propellant rocket requires that consideration be given to thrust-chamber design in respect of such components as the fuel injector and cooling system, and the propellant feed system in respect of the gas generator and turbopumps, for example.

Liquid propellant rockets

A typical liquid propellant rocket is illustrated in Figure 6.9. Liquid fuel and oxygen are pumped into the combustion chamber by turbopumps driven by a separate gas generator. The starter cartridge is typically a solid propellant gas generator.

Two important classes of engine in which fuel and oxidant are supplied by turbopumps are distinguished as those involving open and closed cycles. In the open cycle the turbine exhaust is discharged into the nozzle downstream of the combustion chamber at modest pressure in the expanding section. By contrast this exhaust is injected at high pressure into the combustion chamber, contributing significantly to the energy of the system, in the closed cycle. The improved performance of the latter cycle must be set against the disadvantage of more complex turbopump design necessary to operate at much higher discharge pressures.

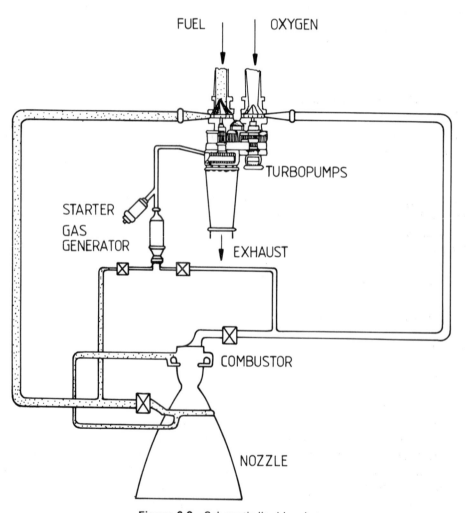

Figure 6.9   Schematic liquid rocket

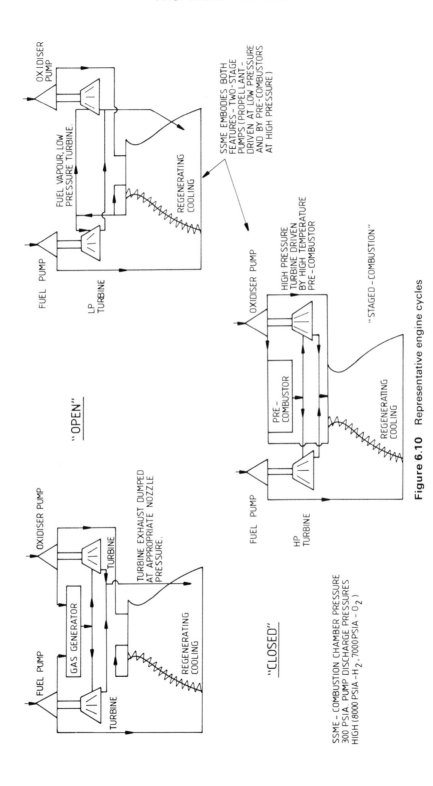

Figure 6.10 Representative engine cycles

Table 6.2   Illustrative comparison of closed- and open-cycle engines

Space shuttle main engine		Advanced gas generator cycle	
Thrust (kN) vacuum	2090	(e.g. as projected for Ariane HM60)	
Sea level	1700	Thrust (kN) vacuum	900
Specific impulse(s) vacuum	455	Sea level	715
Sea level	363	Specific impulse(s) vacuum	445
Mixture ratio		Sea level	349
(cf. 8:1 stochiometric	$2H_2 + O_2 \rightarrow 2H_2O$)	Mixture ratio	5.1
Chamber pressure (bar)	207	Chamber pressure (bar)	100
Nozzle area ratio	77	Nozzle area ratio	110
Flow rates (kg/s) Engine	468	Flow rates (kg/s) Engine	206
Gas generator	248	Gas generator	7.06
Pump discharge pressure (bar)		Pump discharge pressure (bar)	
LOX	319	LOX	126
LH2	426	LH2	150
Length (m)	4.24	Length (m)	4.0
Nozzle exit diameter (m)	2.39	Nozzle exit diameter	2.52
Burn time(s)	480	Burn time(s)	291
Mass (kg)	3022	Mass (kg)	1300

Two open-cycle configurations are illustrated in Figure 6.10; a separate gas generator cycle of the kind employed in the HM-7 engine of the Ariane launcher and coolant tap-off cycle in which vaporized hydrogen fuel from the nozzle coolant jacket drives the turbine. The turbine power is comparatively low but the complexity of the design is much reduced. Also illustrated is the closed, staged-combustion cycle employed in the Space Shuttle main engines. All of the fuel and part of the oxidizer is there supplied to a precombustor at high pressure. The high-energy, fuel-rich exhaust first drives the turbopumps and is then injected into the main combustion chamber with the remaining oxidizer.

An illustrative comparison between open- and closed-cycle engines is presented in Table 6.2, where SSME performance characteristics are summarized together with those of a projected European cryogenic engine for Ariane (HM60), incorporating an improved gas generator cycle (Pouliquen [6]). The gas generator consumes approximately 3.5% of the available propellant and the closed-cycle engine, also incorporating a much greater chamber pressure (two-fold increase), yields a vacuum specific impulse some 2% higher. The necessarily higher turbopump discharge pressure for both fuel and oxidizer in the staged-combustion cycle imply significantly more extensive design and development of turbines, pumps, and ancillary equipment. A simpler closed cycle is the expander cycle in which vaporized fuel drives the turbopumps but, in contrast to the coolant bleed cycle illustrated, all of the fuel is so employed and subsequently passes into the combustion chamber. Whilst particularly suitable for LH2, the fuel flow rate is limited by vaporization rate and hence effectively by heat transfer. Higher chamber pressures and increased thrust are thereby restricted in turn (cf. Brown [7]).

The configurations shown all employ regenerative nozzle cooling. Fuel or oxidizer is used as a coolant, flowing through a jacket surrounding the thrust chamber. The heat absorbed in this way, necessary in prolonged firings, enhances the initial energy content of the propellants prior to injection into the combustion chamber.

Solid propellant rockets

By contrast, solid propellant rockets are comparatively inflexible in their design. The gross classification of solid rockets is made on the basis of the propellant grain geometry since, once ignited, erosive burning proceeds to propellant exhaustion. The variation of burning surface area with time then determines the thrust-time history.

A propellant grain is said to be neutral if the thrust remains broadly constant throughout the firing with a burning surface area which is independent of time. Such behaviour would be characteristic of 'cigarette' burning as illustrated in Figure 6.11. In practice a convenient cylindrical geometry would insist that the burning surface area be small and hence of limited thrust.

A large surface area for burning combined with a roughly neutral thrust performance is provided by star-shaped cylindrical grains as shown. Different interior profiles give rise to two-step burning. These reflect the combined performance of progressive grains in which thrust increases with time, as in the case of internal cylindrical burning, and regressive grains in which thrust decreases with time, as with external rod-shaped burning.

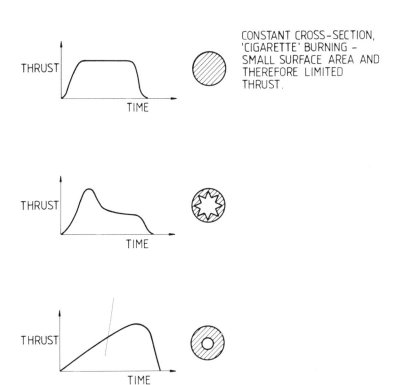

Figure 6.11 Solid propellant grain geometries

IGNITER   NOZZLE

TAPERING GRAIN CROSS-SECTION

- PROPELLANT MASS 5.0×10^5kg, INERT MASS 8.2×10^4kg
- VACUUM THRUST 11.8×10^6 N
- SPECIFIC IMPULSE ~ 260 SECONDS.

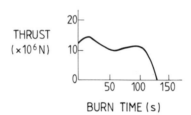

Figure 6.12   Space Shuttle solid rocket booster

Variations in grain geometry along the length of the propellant charge further permit the tailoring of the thrust–time relationship to the requirements of the mission. The configuration and performance of the solid rocket boosters used on the Space Shuttle are illustrated in Figure 6.12.

Solid propellant ignition is typically effected by a pyrotechnic or pyrogen igniter. A small quantity of heat-sensitive powdered explosive is ignited electrically and the heat released in turn ignites the propellant charge within the igniter. Typical igniter compounds are aluminium, boron, or magnesium with potassium perchlorate or nitrate oxidizers. In lateral burning cylindrical grains, the igniter is placed at the end of the chamber furthest from the nozzle so that the hot products of combustion from the igniter sweep across the grain. Main charge ignition occurs through convective and radiative heat transfer from these products, the balance between these processes being determined by the detailed igniter design.

6.2.4   Alternative high-speed air-breathing propulsion

Review of the first-stage Ariane performance described in Chapter 7 reveals that approximately 120 tonnes of propellant (55% of the initital mass) is consumed in accelerating the vehicle to 1500 m/s (20% of orbital speed) at 30 km altitude. The bulk of this propellant, which is consumed within the Earth's atmosphere (approximately 2/3 by mass), is N_2O_4

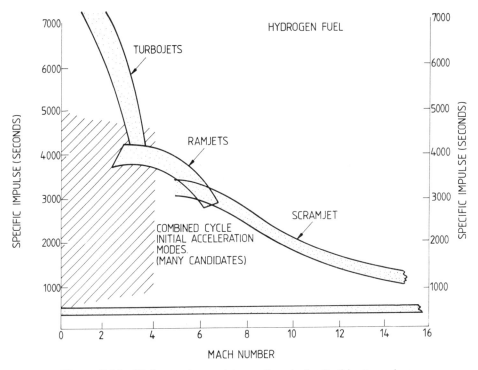

Figure 6.13 High-speed propulsion options in the Earth's atmosphere

oxidizer. Such bare statistics provide a prima-facie case for the investigation of air-breathing propulsion.

Figure 6.13 distinguishes the key propulsive options for high-speed air-breathing with hydrogen as fuel. Current turbomachinery, familiar in the aircraft jet engine, becomes increasingly unattractive thermodynamically at Mach numbers (ratio of flight speed to local sound speed) in excess of 3 and relatively high specific impulses can then be achieved much more simply by capitalizing on ram compression. In the conventional ramjet, however, the ingested air is diffused down to subsonic Mach numbers prior to mixing with the fuel and combustion. Isentropic (shock-free) subsonic diffusion leads to a substantial temperature rise, from T to T_0, which increases quadratically with Mach number, M:

$$T_0/T = 1 + \tfrac{1}{2}(\gamma - 1)M^2$$

For $M > 4.5$ heat transfer and dissociation losses in the combustor begin to erode subsonic-combustion ramjet performance significantly. However, technologies required by these engines are well-established and their range of application may be further extended by the judicious use of the high fuel heat capacity in a range of thermal protection and heat exchanger strategies, even embracing air liquefaction. The intimate relationship between vehicle trajectory, kinetic heating and combustion stability are illustrated in Figure 6.14. The target condition identified earlier at 30 km altitude corre-

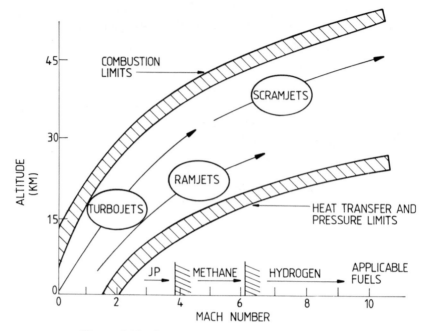

Figure 6.14   Constraints on air-breathing trajectories

sponds to a Mach number of 5.5, however, and, at such speeds and beyond, the supersonic combustion ramjet (SCRAMJET) is the more attractive prospect. Diffusion to subsonic Mach numbers for combustion is avoided but the necessary technologies for efficient high-speed mixing, ignition and stable burning remain poorly understood.

By comparison with rockets, air-breathing engines offer relatively modest thrust-to-weight ratios and composite engines which seek to use common components over a wide flight regime are crucial to single-stage-to-orbit operation. Cycle performance analyses and engine simulations continue to attract research interest (Bendot *et al.* [8]) and such imaginative vehicle studies as Star-Raker (Read *et al.* [9]) and HOTOL (Parkinson [10]).

6.2.5   Propellant management

The operation of liquid propellant rockets for space vehicle applications may expose the propulsion system to a dynamical regime not usually encountered in terrestrial applications, namely that of free-fall or low residual acceleration. We shall briefly review here some aspects of the problems encountered which do not have a ready analogue in $1g_0$ environments.

The near free-fall situations typical of space manoeuvres or residual drag in Earth orbit correspond to

$$|g_{\text{local}} - a_{\text{vehicle}}| = O(10^{-6}g_0)$$

The very small vehicle weight associated with such levels of acceleration invites consideration of forces such as surface tension, which might usually be considered to be insignificant except on very small scales.

Liquids may be characterized as wetting (surface spreading) or non-wetting. Liquid to solid surface contact angles may approach zero for perfectly wetting liquids, typical of cryogenics, but may exceed $90°$ for non-wetting liquids such as mercury. The surface tension of LOX (at 90 K) is approximately 13×10^{-3} N/m, compared with 460×10^{-3} N/m for mercury and 72×10^{-3} N/m for water.

It is convenient to review the significance of the several forces prevailing in liquid propellants by introducing the following dimensionless groups:

$$\text{Bond number, } Bo \equiv \frac{\text{gravitational}}{\text{surface tension}} \text{ forces}$$

$$= \rho L^2 g / \sigma$$

and

$$\text{Weber number, } We \equiv \frac{\text{inertia}}{\text{surface tension}} \text{ forces}$$

$$= \rho V^2 L / \sigma$$

We distinguish the respective inertia, capillary and gravity dominated regimes as shown in Figure 6.15. For $Bo > 1$ gravitational forces predominate and surface tension may be neglected and vice versa. We note that for surface tension to be significant in a $1g_0$ Earth surface environment for a liquid such as water then $Bo > 1$ implies a characteristic length scale of approximtely 2 mm and effects like capillary rise are only significant at such scales. For liquid oxygen in a residual microgravity environment, however, the $Bo > 1$ boundary corresponds to a length scale of 1 m.

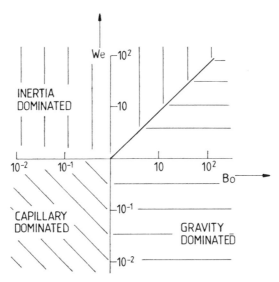

Figure 6.15 Distinctive regimes for forces acting on liquids

The equilibrium configuration of a liquid propellant in a partially filled tank under microgravity conditions is determined essentially by the minimization of free energy which is proportional to surface area. Active measures must be adopted to ensure that liquid propellant is available at the tank outlet for rocket motor restart, for example. The principal options in relation to propellant storage and delivery systems—supercritical storage, inertial and rocket bottoming systems, positive expulsion and capillary or surface tension systems—are reviewed by Ring [11].

In addition to the problem of propellant configuration within the tank, the response to dynamic excitation in flight in the form of propellant sloshing may also be important. The viscosity of cryogenic propellants is in general low, typically by more than an order of magnitude in comparison with water (or hydrazine). The damping of free surface oscillations in the fluid which would otherwise give rise to substantial fluctuating forces and moments on the tank walls may also require active provisions in the form of turbulence-generating baffles.

6.3 SECONDARY PROPULSION

The typical functions of spacecraft propulsion, as distinct from launcher operations from the Earth surface, may be summarized as follows in order of reducing thrust level:

- final orbit acquisition from the nominal orbit established by the launch vehicle;
- station-keeping and orbit control;
- attitude control.

General principles in relation to both solid and liquid propellant rockets have been reviewed in earlier sections. Here we shall focus on factors influencing the choice of system appropriate to the secondary propulsion roles set out above. The principal options are cold gas systems, monopropellant hydrazine, bi-propellant nitrogen tetroxide/ monomethyl hydrazine, solid propellants and electric propulsion. We shall consider each of these in turn.

6.3.1 Cold gas systems

Such systems comprise an inert gas, typically nitrogen, argon, freon, or propane, which is stored at high pressure and fed through a regulator to a number of small thrusters. A schematic of the system employed on the European EXOSAT spacecraft, a cosmic X-ray observatory, is shown in Figure 6.16. Propellants are selected for the simplicity of their storage and compatibility with other facets of spacecraft operation such as the effect of exhaust plume impingement on sensitive surfaces, solar cells, sensors and detectors.

Thrust levels are small, typically in the neighbourhood of 20 mN, but the emphasis of the role for cold gas thrusters is for the provision of small impulse bits in order to achieve high pointing accuracies. Minimum impulse bits of approximately 10^{-4} N s are often necessary for better than 0.1° accuracy in attitude control on some of the larger modern scientific satellites.

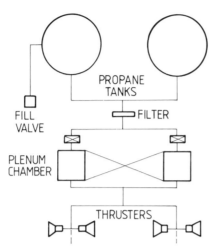

Figure 6.16 Schematic cold gas propulsion system

The specific impulses from cold gas propulsion systems are comparatively small (≈ 50 s) but the expellant mass is relatively unimportant in the fine-pointing role. Missions which require larger ΔVs for secondary propulsion do however necessitate higher-performance propellants in order to reduce the system mass.

6.3.2 Monopropellant hydrazine

As we described in Section 6.2, the decomposition of anhydrous hydrazine (N_2H_4), either thermally or catalytically, to the products nitrogen, ammonia, and hydrogen is exothermic. Expansion of the hot product gas through a nozzle will yield specific impulses in the range 200–250 s. The propellant is readily stored as a liquid, freezing point 275 K and boiling point 387 K, in tanks under the pressure of an inert gas such as nitrogen or helium.

A typical electrothermal hydrazine thruster configuration is sketched in Figure 6.17. The low-temperature monopropellant decomposition is enhanced by a resistively heated

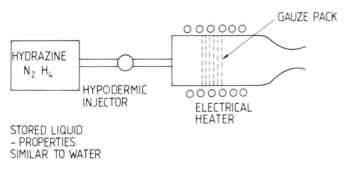

Figure 6.17 Schematic electrothermal hydrazine thruster

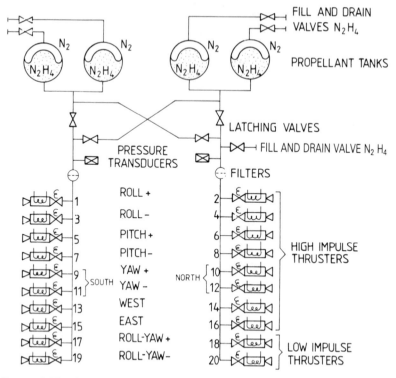

Figure 6.18 OTS propulsion system (reproduced by permission of European Space Agency)

metal catalyst, commonly platinum/iridium dispersed on a large surface area, porous substrate of aluminium oxide (e.g. Shell 405). Thruster performance is enhanced by higher-temperature operation but accompanying heat transfer losses and materials integrity problems increase.

Thrust levels of $\geq 10\,N$ are required for orbit control duties and Figure 6.18 illustrates the propulsion system employed on the Orbital Test Satellite. Such systems are extensively employed in the present generation of geostationary communications spacecraft for which the principal requirement is that of north–south station-keeping.

Progress has been reported with power-augmented hydrazine thusters which embody a more significant resistance heating component and which are of the tandem design of exothermic decomposer followed by a heat exchanger. Nozzle temperature limitations on materials, given the corrosive nature of the product gases, restrict the extent of this heating but specific impulses in the vicinity of 300 s appear achievable.

6.3.3 Bi-propellant MMH/nitrogen tetroxide

The increased interest in placing larger spacecraft into geostationary orbit for telecommunications missions and as orbiting platforms has meant that the greater technical complexity of bi-propellant engines is worthwhile. The incorporation of higher-performance

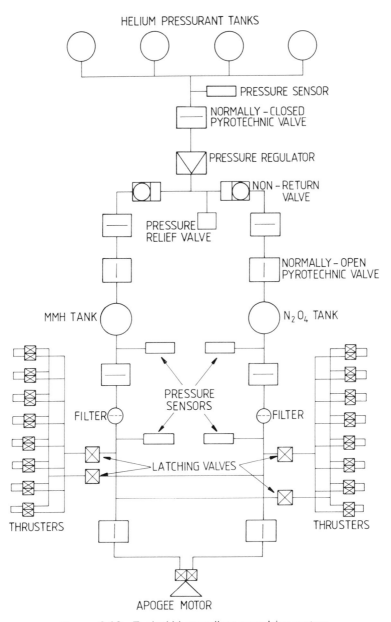

Figure 6.19 Typical bi-propellant propulsion system

propulsion systems in such spacecraft can yield significant propellant mass reductions. The N_2O_4/MMH combination will deliver specific impulses in excess of 300 s.

The configuration illustrated in Figure 6.19, capable of unifying the functions of orbit raising and AOCS, indicates the significant increase in propellant feed components required for the safe handling of these highly reactive, hypergolic propellants. A more detailed view of system trade-offs is given by Sackheim *et al.* [12].

6.3.4   Solid propellant apogee motors

In the emplacement sequence for a communications spacecraft into geostationary orbit
using an expendable launch vehicle such as Ariane, the final stage of the launcher injects
the satellite into an elliptical transfer orbit with apogee at geostationary height. The final
boost into circular orbit at apogee is typically achieved by a high-thrust, short-duration
burn from a solid propellant apogee boost (or kick) motor. For a satellite with an
on-station mass of approximately 1000 kg, of Intelsat V size, the necessary apogee motor
fuel approaches 900 kg and the propulsive ΔV is roughly 2 km/s.

Launch systems for similar missions using the Space Shuttle necessitate an additional
perigee motor to boost the spacecraft into geostationary transfer obit. The payload assist
modules (PAM) based on the solid propellant Minuteman III motor currently perform
this role and offer a level of performance not too dissimilar from that required of the
apogee motor. Launch vehicle operations are described more fully in Chapter 5, and we
focus here on the motor design.

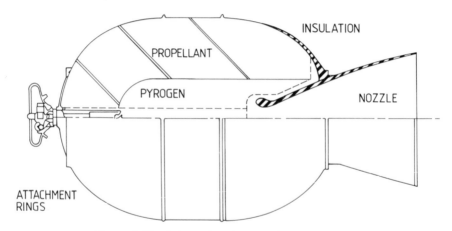

Figure 6.20   Schematic solid propellant apogee motor

Table 6.3   Performance data STAR 37E rocket motor

Total impulse	29.11×10^5 N s
Motor weight	1122 kg
Propellant weight	1039 kg
Burn-out weight	75.8 kg
I_{sp} (Propellant)	2800 N s/kg
Burn time	42.0 s
Mass fraction	0.926
Length	168.4 cm
Basic diameter	109.5 cm
Average thrust	68 800 N
Max. thrust	75 000 N

A typical solid rocket motor for secondary propulsion duties is illustrated in Figure 6.20. Table 6.3 refers to the characteristics of the Star 37E motor manufactured by Thiokol. This motor, which weighs 1122 kg, burns for 42 s and delivers an average thrust of 69 kN.

Clearly the emplacement roles of the solid propellant ABM, with its high thrust of short duration, and the bi-propllant MMH/N_2O_4, with the advantages of higher specific impulse and more controlled burning, are in part interchangeable. The impulse burn requires that the spacecraft should spin for reasons of stability and thrust alignment. It is therefore inherently less accurate than the extended burn, lower thrust operation of the bi-propellant motor which admits the option of quite precise spacecraft attitude control during the thrusting phase. The trade-off is once more that of propulsion system complexity for improved performance.

6.4 ELECTRIC PROPULSION

The opening section of this chapter revealed some crucial differences in the performance and operation of separately powered rockets. Unlike the chemical systems which we have described thus far, the energy required for expellant acceleration in an electrically propelled rocket derives from a quite separate source. Whether this source is solar radiation or nuclear fuel, it may be effectively unlimited and the constraints on performance relate to attainable thrust levels and efficient energy conversion.

Whilst potential advantages for electric rockets have been recognized for many years in relation to both lengthy interplanetary flights and near-Earth orbital operations, flight experience remains very limited. In subsequent sections we review key features of their distinctive performance, projected roles and design.

6.4.1 Electric propulsion fundamentals

In view of the importance of the power-plant to the rocket configuration it is convenient to analyse the performance in terms of the component masses: M_W, the power-plant mass; M_e, the expellant mass and M_P, the payload mass. The expellant storage and feed system may be assumed to be part of the power-plant, whilst the power-plant fuel mass will be considered to be negligibly small (nuclear fuel) or inappropriate (solar powered). The configuration envisaged is shown schematically in Figure 6.21.

The power-plant supplies the exhaust kinetic energy whence the jet power is related to the exhaust velocity by an expression of the form

$$W = \tfrac{1}{2}\dot{m}V_e^2 \tag{6.25}$$

where W denotes power-plant output, $\dot{m}$ the expellant mass flow rate and V_e the exhaust velocity. (If thruster process losses are introduced, then $W_{jet} = \eta W$. For present purposes we suppose $\eta = 1$.)

We relate the power output W to the power-plant mass M_W linearly and introduce the inverse specific power α such that

$$M_W = \alpha W \tag{6.26}$$

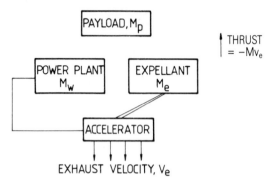

Figure 6.21 Schematic for separately-powered electric rocket

If we then suppose the exhaust mass flow rate to be constant throughout the burn time, t_b, such that

$$\dot{m} = M_e/t_b \qquad (6.27)$$

$$M_e = \frac{M_0 - M_p}{1 + \left(V_e^2 \middle/ \dfrac{2t_b}{\alpha}\right)} \qquad (6.28)$$

and

$$M_w = \frac{M_0 - M_p}{1 + \left(\dfrac{2t_b}{\alpha} \middle/ V_e^2\right)} \qquad (6.29)$$

where $M_0 = M_p + M_w + M_e$, the total rocket mass.

By identifying $\sqrt{(2t_b/\alpha)}$ as the characteristic velocity V_c and using the above relationships in the Tsiolkovsky rocket equation (6.5) it is possible to show that

$$\frac{\Delta V}{V_e} = \ln\left[\frac{1 + (V_e/V_c)^2}{\dfrac{M_p}{M_0} + (V_e/V_c)^2}\right] \qquad (6.30)$$

The rocket performance described by (6.30) is illustrated in Figure 6.22. For $V_e/V_c \ll 1$, increases in exhaust velocity at fixed characteristic velocity result in larger available propulsive ΔV. Such behaviour is essentially that observed with chemical propellants. Performance improvements accompany the better propellant utilization reflected in higher exhaust velocities. Unlike chemical propulsion, however, such benefits do not extend without limit and for $V_e/V_c > 1$, ΔV passes through a maximum and then decreases. Further increases in exhaust velocity require higher powers (cf. (6.25)) and therefore increased power-plant mass (cf. (6.26)). Beyond a certain stage these increases in power-plant mass outweigh the further reductions in expellant mass accompanying the higher exhaust velocities.

The presence of maxima in Figure 6.22 introduces a range of optimization studies. For a particular payload ratio, M_p/M_0, it is desirable to operate in the vicinity of the

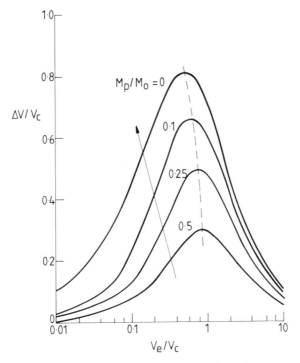

Figure 6.22 Separately-powered electric rocket performance

maximum, broadly corresponding to $V_e/V_c \simeq 1$, and high exhaust velocity then requires $2t_b/\alpha \gg 1$. Optimized electric propulsion usage will thus tend to imply lengthy burn times and small inverse specific power, α.

Consider the following illustration; for a payload ratio, $M_p/M_0 = 0.5$ and a mission ΔV of, say, 5 km/s, representative of an orbit-raising manoeuvre,

$$V_c \simeq V_e \simeq 16 \text{ km/s}$$

For $\alpha = 20$ kg/kW, representative of modern lightweight solar arrays, the burn time t_b would be 30 days. The mean acceleration over the thrust duration, $\Delta V/t_b$, is then $2 \times 10^{-4} g_0$. Clearly quite spectacular improvements in α would be necessary for such propulsion systems to operate from planetary surfaces. However, our example illustrates the potential for near-Earth missions in addition to its more traditional projected role in long-duration interplanetary flights. We explore these aspects in the following section.

6.4.2 Propulsive roles for electric rockets

The low thrust levels and extended burn times characteristic of electric rockets have tended to constrain their prospective application to interplanetary missions. If we recall that thrust acceleration levels of $2 \times 10^{-4} g_0$ are equivalent to approximately $0.35 \times$ solar gravitational acceleration at 1 AU, then continuous thrust at these levels has a pro-

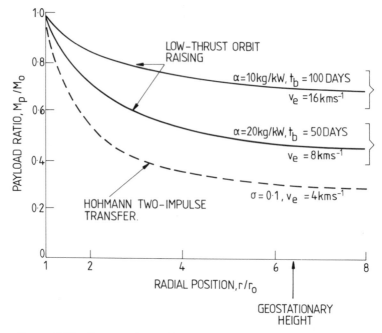

Figure 6.23 Comparative performance: chemical impulse versus low-thrust orbital transfer

nounced influence on interplanetary trajectories. Very significant savings in expellant mass and transfer time, in comparison with the minimum-energy Hohmann transfer, have been demonstrated for missions to Mars. See, for example Seifert [13].

More in keeping with the theme of this book, however, is the application to such near-Earth missions as orbit raising and station-keeping. Substantial benefits in terms of payload into final orbit may be achieved if the potential of high exhaust velocities to reduce propellant mass can be realized. Recalling Figure 6.22 the maxima are distinguished by $V_e/V_c \simeq 1$ and therefore high exhaust velocity must accompany extensive burn times, $V_c^2 = 2t_b/\alpha$. Figure 6.23 presents a simple comparison between a two-impulse Hohmann transfer and low-thrust orbit raising from low Earth orbit, radius r_0, to a higher orbit, radius r. The electric rocket is typically fired continuously and near-circumferential low thrust leads to a gradual spiralling expansion of the orbit radius. Projected burn times of months would probably be acceptable operationally for an unmanned space tug implying a modest extension of launch procurement schedules; the implications for thruster design are more important. Extended burn times do influence the propulsive ΔV required and low-thrust orbit-raising operations will in general require a larger ΔV. Figure 6.24 compares the impulsive and circumferential low-thrust requirements; transfer to geostationary orbit requires approximately 20% more ΔV for the low thrust mission. The requirements of accompanying manoeuvres for low thrust orbital operations such as changes in orbit inclination have been extensively analysed by Burt [14], for example.

It is important to appreciate that different electric propulsion techniques are appropriate to different missions, and that mission optimization is crucial to electric propulsion. For example, early investigation covered LEO/GEO transfers, whereas in more recent

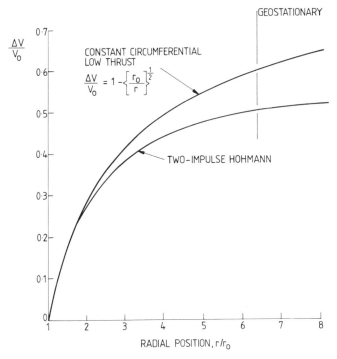

Figure 6.24 Comparative ΔV requirement for transfer between circular orbits r_0 to r

years attention has focused on more modest orbit raising such as from STS orbit to Sun synchronous altitudes. For this type of mission MPD arc jets show superior characteristics to other electric propulsion systems. For interplanetary trajectories, electrostatic-ion engines are superior as they are also for north–south station-keeping of GEO vehicles. In low Earth orbit operations resistojets may be used for orbit maintenance against aerodynamic drag effects—in this case bio-waste products (especially CO_2) are under active development for use on the international space station. A final objective for a particular type of electrostatic engine, FEEP, (see below), is for the very small impulse bit which may be achieved at high specific impulse—which makes this engine appropriate for fine pointing control.

6.4.3 Electric propulsion systems

The basic principles underlying electric thruster design are well established (cf. Stühlinger [15], Jahn [16], for example). Electrically powered expellant acceleration devices are of essentially three types: electrothermal, in which the enthalpy of the expellant is increased and converted into directed kinetic energy via a nozzle; electrostatic, in which charged particles, ions or colloids, are accelerated directly in an electric field; electrodynamic, in which crossed electric and magnetic fields induce a Lorentz force in a plasma. We illustrate briefly designs reflecting each of these distinctive approaches.

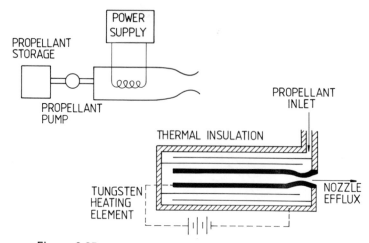

Figure 6.25 The simplest electrothermal thruster: the resistojet

Electrothermal thrusters

The resistojet is the simplest electrothermal thruster. Typically, the propellant is heated by passing it over a tungsten heating element. Broad features of the thruster are sketched in Figure 6.25. The more common propellants are hydrogen, nitrogen, ammonia and, in decomposition thrusters incorporating electrical heating and chemical heat release, hydrazine (power-augmented hydrazine thrusters, PAEHT). The exhaust velocity is a function of temperature and therefore materials' integrity and life considerations limit V_e to about 10 km/s. Efficiencies in excess of 70% and thrust levels ≤ 0.5 N have been demonstrated. Hydrogen is particularly attractive as a propellant since it is noncorrosive, with both high specific heat and thermal conductivity. It is, however, difficult to store, requiring cryogenic temperatures, and dissociation to atomic hydrogen at elevated temperature leads to frozen flow losses in the nozzle.

Ammonia is readily stored without refrigeration and is dissociated to lighter species on heating, but it is corrosive in both the heater and nozzle. Recent interest has focused on solid propellants, in particular Teflon, which is ablative. An electrical pulse creates a plasma which is augmented by further evaporation of the solid wall material and then exhausted through the nozzle. In this device self-induced electric forces are also important and it is therefore in reality a hybrid thruster.

Higher specific impulses are available from the arc jet (~ 2000 s) but at reduced electrical efficiency. The expellant itself is subject to ohmic heating by passing it through an arc discharge, thereby eliminating gas–solid heat transfer. Excessive electrode wear and the high plasma temperatures have inhibited flight operation of these devices.

Electrostatic thrusters

Electrostatic thrusters derive their thrust from the direct acceleration of positively charged propellant ions by an electric field. Devices are typically classified according to the mechanism by which the ions are extracted from the neutral propellant.

Surface contact ionization is perhaps the simplest mechanism. Ionization of the vapour of a material having a low ionization potential, for example caesium, is achieved by passing it through a heated porous tungsten plate. Radiative heat losses limit overall efficiencies.

Electron bombardment ion sources are the most extensively developed. Electrons emitted from an axially mounted thermionic cathode are attracted towards a concentric cylindrical anode. A weak, externally applied magnetic field causes the electrons to spiral within the chamber and propellant ionisation results from collisions between these electrons and propellant vapour. Although originally developed for use with mercury as propellant, recent research has focused on inert gases, argon and xenon for example, prompted by concern over the environmental impact in near Earth applications. A representative small thruster design for station-keeping is illustrated in Figure 6.26. Substantially larger inert gas thruster designs have been tested in the laboratory, delivering thrusts ≥ 1 N (cf. Kaufman and Robinson [17]). Alternative ionization mechanisms currently under investigation employ radio frequency or field emission sources. The attractions of propellants of large atomic weight which may be readily ionized is easily demonstrated.

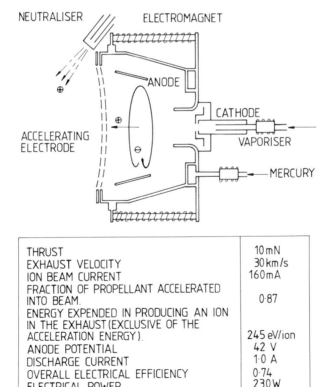

THRUST	10 mN
EXHAUST VELOCITY	30 km/s
ION BEAM CURRENT	160 mA
FRACTION OF PROPELLANT ACCELERATED INTO BEAM.	0·87
ENERGY EXPENDED IN PRODUCING AN ION IN THE EXHAUST (EXCLUSIVE OF THE ACCELERATION ENERGY).	245 eV/ion
ANODE POTENTIAL	42 V
DISCHARGE CURRENT	1·0 A
OVERALL ELECTRICAL EFFICIENCY	0·74
ELECTRICAL POWER	230 W

Figure 6.26 Typical 10 cm diameter station-keeping ion thruster

For negligible anode velocity we may balance particle kinetic energy and electrostatic work according to

$$\tfrac{1}{2}m_p V_e^2 = eV$$

whence the beam power

$$W = \tfrac{1}{2}FV_e$$

implies

$$F/W = \sqrt{\left(\frac{2m_p/e}{V}\right)} \tag{6.31}$$

Thus for given electrical power, and hence power-plant mass (6.26), the thrust is maximized for large mass-to-charge ratio (m_p/e).

A technology which has been actively developed in Europe is field emission electric propulsion—FEEP. In this sytem the ions are created directly from a liquid metal surface by means of a strong electric field. FEEP thrusters have been fabricated having a high efficiency and with the capability for high reproducibility of very short pulse times, leading to a reliable small impulse bit.

Electrodynamic thrusters

The magnetoplasmadynamic (MPD) arcjet evolved from the combination of electro-thermal arcjet and magnetogasdynamic technologies. A neutral plasma is accelerated by means of both Joule heating and electrodynamic forces. In the full MPD engine the self-induced magnetic field provides the dominant acceleration mechanism, with thrust being proportional to the current squared. At low pulse levels the propulsive efficiency is somewhat modest ($\sim 10\%$ for a 1 kW device [18]). As the power level increases into the megawatt range, significantly improved efficiencies result ($\sim 40\%$ at ~ 2 MW [19]). For such high power levels pulsed operation is essential and significant developments have taken place in Japan in this area, including flight experiments. As with electrothermal arc jets at present the major life limitation for these devices is due to cathode erosion. Typical operational characteristics are thrusts of a few newtons, with an $I_{sp} \geq 2000$ s.

REFERENCES

[1] Caveney, L.H. (1984) Orbit raising and manoeuvring propulsion: research status and needs. *Prog. in Astro. and Aero.* **89**. AIAA.
[2] Barrère, M., Jaumotte, A., De Venbeke, B.F., and Vanden Kerckhove, J. (1960) *Rocket Propulsion*, Elsevier, Amsterdam.
[3] Sutton, G.P. and Ross, D.M. (1976) *Rocket Propulsion Elements* (4th edn), Wiley.
[4] Rao, G.V.R. (1961) Recent developments in rocket nozzle configurations. *ARS J.* **31** (11), 1488.
[5] Kuo, K.K. and Summerfield, M. (1984) Fundamentals of solid-propellant combustion. *Prog. in Astro. & Aero.* **90**. AIAA.
[6] Pouliquen, M.F. (1984) HM60 cryogenic rocket engine for future European launchers. *J. Spacecraft and Rockets* **21** (4), 346.

[7]   Brown, J.R. (1983) *Expander Cycle Engines for Shuttle Cryogenic Upper Stages.* AIAA-83-1311.

[8]   Bendot, J.G., Piercy, T.G., and Brown, P.N. (1975) *Composite Engines for Application to a Single-Stage-to-Orbit Vehicle.* NASA CR-2613.

[9]   Reed, D.A., Ikawa, H., and Sadunas, J.A. (1979) *Star-Raker, an Airbreather/Rocket-powered Horizontal Take-off, Single Stage-to-Orbit Transportation System,* AIAA 79-0895.

[10] Parkinson, R.C. (1985) *HOTOL – A Third Generation Economic Launch Vehicle.* IAA-85-482.

[11] Ring. E. (1964) *Rocket Pressurisation and Propellant Systems,* Prentice-Hall, Englewood Cliffs, NJ.

[12] Sackheim, R.L., Fritz, D.E. and Macklis, H. (1979) *The Next Generation of Spacecraft Propulsion Systems.* AIAA 79-1301.

[13] Seifert, W.S. (1959) *Space Technology,* Wiley, New York.

[14] Burt, E.G.C. (1968) The dynamics of low-thrust manoeuvres. *J. Royal Aeronautical Society,* **72**, 925.

[15] Stühlinger, E. (1964) *Ion Propulsion for Spaceflight,* McGraw-Hill.

[16] Jahn, R.G. (1968) *Physics of Electric Propulsion,* McGraw-Hill, New York.

[17] Kaufman, H.R. and Robinson, R.S. (1983) Inert Gas Thruster Technology, *J. Spacecraft and Rockets,* **20** (1), 77.

[18] Uematsi, K. (1984) *Development of a 1 kW MPD Thruster.* AIAA-87-1023.

[19] Burton, R.L., Clark, K.E., and Jahn, R.G. (1983), *J. Spacecraft and Rockets,* **20**, 299.

7 LAUNCH VEHICLES

J. Barrie Moss

School of Mechanical Engineering, Cranfield Institute of Technology

7.1   INTRODUCTION

Launch vehicle performance and operations interact with spacecraft design and mission analysis in many important respects. The attainment of a minimum speed for insertion into orbit is clearly crucial to any mission but the procurement of the most appropriate launch vehicle also has profound implications for such diverse factors as spacecraft mass and configuration, launch window and mission profile, on-board propulsion requirements and overall mission cost. In this chapter we shall seek first to establish the basic principles which determine launcher design and hence constrain the spacecraft payload. Secondly, we shall outline some key features of the principal launcher alternatives in Europe and the United States, namely the unmanned, expendable Ariane and the manned, substantially reusable, Space Shuttle.

7.2   BASIC LAUNCH VEHICLE PERFORMANCE AND OPERATION

7.2.1   Vehicle dynamics

For purposes of illustration we specialize the equations of motion of the rocket to the vertical plane, parallel and normal to the flight direction, and to the motion of the centre of mass and the pitch rotation.

The configuration envisaged and nomenclature are shown in Figure 7.1.
Parallel to the flight direction

$$M\frac{dv}{dt} = F\cos(\alpha + \delta) - Mg\sin\gamma - D, \tag{7.1}$$

and normally

$$Mv\frac{d\gamma}{dt} = F\sin(\alpha + \delta) - Mg\cos\gamma + L \tag{7.2}$$

Spacecraft Systems Engineering. Edited by P.W. Fortescue and J.P.W. Stark
© 1991 John Wiley & Sons Ltd

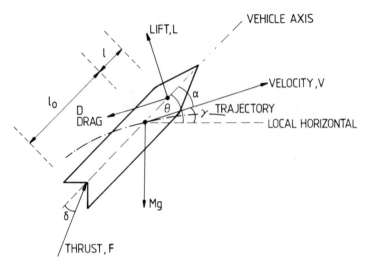

Figure 7.1   Configuration and nomenclature for rocket motion in the vertical plane

The displacement of the centres of pressure and nozzle from the centre of mass give rise to pitching moments. The accompanying angular equation of motion may be written

$$I_p \frac{d^2\theta}{dt^2} = (L \cos \alpha + D \sin \alpha)l - Fl_0 \sin \delta \qquad (7.3)$$

where I_p is the moment of inertia in pitch and

$$\theta = \alpha + \gamma$$

For small values of α and δ, equation (7.1) becomes

$$\frac{dv}{dt} = \frac{F}{M} - g \sin \gamma - \frac{D}{M} \qquad (7.4)$$

Recall from Chapter 6 that the thrust may be written

$$F = \left(\frac{-dM}{dt}\right) I_{sp} g_0$$

whence equation (7.4) becomes

$$\frac{dv}{dt} = -I_{sp} g_0 \frac{d}{dt}(\ln M) - g \sin \gamma - \frac{D}{M} \qquad (7.5)$$

On integration

$$v = v_0 + I_{sp} g_0 \ln\left[\frac{M_0}{M}\right] - \int_0^t g \sin \gamma \, dt - \int_0^t \frac{D}{M} dt \qquad (7.6)$$

and in addition to the ideal velocity increment at burn-out, say,

$$\Delta V_{ideal} = I_{sp} g_0 \ln \left[\frac{M_0}{M_b} \right]$$

we now distinguish propulsive losses associated with gravity and aerodynamic drag

$$\Delta V_g = \int_0^{t_b} g \sin \gamma \, dt$$

$$\Delta V_D = \int_0^{t_b} \frac{D}{M} \, dt,$$

whence

$$\Delta V = \Delta V_{ideal} - \Delta V_g - \Delta V_D \tag{7.7}$$

The relative magnitudes of these terms for an Earth surface launch are sketched in Figure 7.2. We note that the impulse requirement (minimum burn time) which might lead to reduced gravity loss is at variance with the requirement from the standpoint of drag loss, which might suggest a low-velocity ascent through the denser atmosphere since $D = \frac{1}{2} \rho V^2 S C_D$. In general this latter requirement is broadly satisfied, the first-stage propulsion of a multi-stage vehicle might typically have a thrust-to-weight ratio less than about 1.5, implying an initial vertical acceleration of about $0.5g$ and hence gradual ascent.

Vehicles are generally launched vertically and minimization of gravity loss, together with the eventual requirement of local horizontal payload injection, suggests that the flight trajectory be deflected from the vertical as rapidly as possible. The simplest manoeuvre to effect this is the gravity turn.

If we neglect the lift force in equation (7.2), then

$$v \frac{d\gamma}{dt} = -g \cos \gamma$$

and on integration

$$\sin \gamma(t) = \tanh \left\{ \tanh^{-1}(\sin \gamma_0) - \int_{t_0}^{t} \frac{g}{v} \, dt \right\} \tag{7.8}$$

More rapid pitching manoeuvres may be effected by thrust vectoring. A typical Ariane

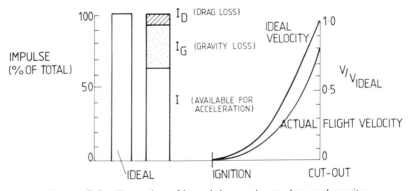

Figure 7.2 Illustration of launch losses due to drag and gravity

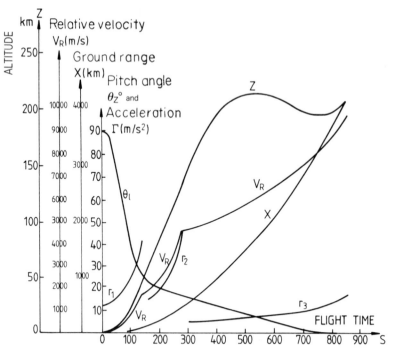

Figure 7.3 Typical Ariane 2 flight profile for transfer into GTO. Orbit: $a = 24\,731$ km, $e = 0.73$, $i = 9.65°$, $\omega = 180°$, $\Omega = -144.6°$, $Z_a = 35\,786$ km, $Z_p = 200$ km

II flight profile is shown in Figure 7.3. The local pitch angle decreases by 60° in approximately the first two minutes of the flight.

Analytic solutions are not available in general since gravity and drag are complex functions of altitude. We may illustrate typical features of the performance, however, by considering estimates for the simple case of vertical ascent *in vacuo* at constant thrust, specific impulse and gravitational acceleration.

From equation (7.6) the velocity at burn-out may be written

$$V_b = I_{sp}g_0 \ln R - gt_b$$

and, on further integration, the burn-out altitude is

$$h_b = I_{sp}g_0 \left\{ 1 - \frac{\ln R}{R - 1} \right\} t_b - \tfrac{1}{2}gt_b^2$$

whilst the burn-time t_b is given by

$$t_b = \frac{M_0 - M_b}{\dot{m}} = \left\{ \frac{M_0}{F} \right\} I_{sp}g_0 \left\{ \frac{R - 1}{R} \right\}$$

Figure 7.3 shows the full range of flight parameters for a typical Ariane II injection into geostationary transfer orbit.

It is appropriate at this point to review some representative mission requirements for spacecraft launches.

7.2.2 Mission requirements

The standard launch vehicle must provide an efficient means of boosting a spacecraft into the planned trajectory. The payloads may vary from low Earth orbit scientific satellites, through geostationary communications satellites, to Earth escape probes. Figure 7.4 distinguishes some representative launcher burn-out conditions for spacecraft emplacement. As the table of values shows, the minimum velocity at a typical injection height of 200 km is approximately 7.8 km/s, whilst that for geostationary transfer orbit must be 10.3 km/s.

In addition to accelerating the payload to these velocities the launcher must overcome the effects of aerodynamic drag and gravity. From equation (7.7)

$$\Delta V_{ideal} = \Delta V + \Delta V_g + \Delta V_D$$

and whilst detailed determination of the loss terms requires computation, estimates can be made for purposes of preliminary planning using such correlations as those described by White [1]. Representative losses for the first stage of Ariane using this approach are

$$\Delta V_g = 1.08 \text{ km/s}$$

$$\Delta V_D = 0.22 \text{ km/s}$$

The values quoted in Figure 7.4 embody a significant simplification relative to Earth rotation. The rotational velocity at the Equator is approximately 0.47 km/s and an eastwards launch from the Earth surface can evidently capitalize on this velocity. The

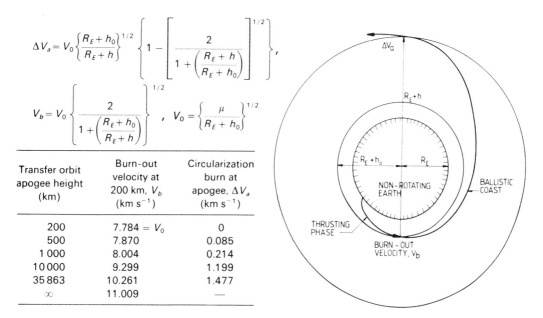

$$\Delta V_a = V_0 \left\{ \frac{R_E + h_0}{R_E + h} \right\}^{1/2} \left\{ 1 - \left[\frac{2}{1 + \left(\frac{R_E + h}{R_E + h_0} \right)} \right]^{1/2} \right\} ,$$

$$V_b = V_0 \left\{ \frac{2}{1 + \left(\frac{R_E + h_0}{R_E + h} \right)} \right\}^{1/2} , \quad V_0 = \left\{ \frac{\mu}{R_E + h_0} \right\}^{1/2}$$

Transfer orbit apogee height (km)	Burn-out velocity at 200 km, V_b (km s^{-1})	Circularization burn at apogee, ΔV_a (km s^{-1})
200	7.784 = V_0	0
500	7.870	0.085
1 000	8.004	0.214
10 000	9.299	1.199
35 863	10.261	1.477
∞	11.009	—

Figure 7.4 Launch vehicle burn-out velocities for spacecraft emplacement

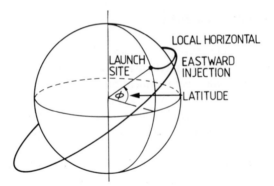

Figure 7.5 Schematic of the link between orbit
inclination and the latitude of the launch site

benefit reduces with increasing latitude of the launch site and vanishes completely for
launches into polar orbit.

The latitude of the launch site also has important implications for the subsequent
inclination of the spacecraft orbit. We have identified some significant benefits in respect
of launcher performance which accompany both rapid booster pitch angle changes to the
local horizontal and eastward injection. Figure 7.5 then illustrates a typical launch
sequence. The spacecraft orbit plane is essentially fixed by the velocity vector at burn-out
and the centre of the Earth. Without a very lengthy dog-leg manoeuvre prior to orbit
insertion, accompanied by increased losses both gravitational and aerodynamic, the
inclination of the orbit will then be approximately equal to the latitude of the launch site.

As we described in Chapter 5, the correct orbit inclination is a crucial orbital parameter.
In the case of the 24-hour synchronous communications satellite, for example, the

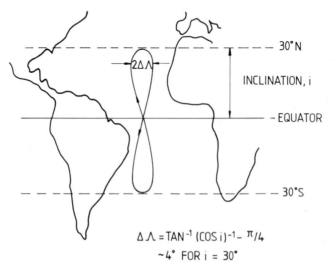

$$\Delta\Lambda = TAN^{-1}(COS\ i)^{-1} - \pi/4$$
$$\sim 4° \ FOR\ i = 30°$$

Figure 7.6 Drift of the subsatellite point for an inclined geosynchronous orbit

subsatellite point is only stationary relative to the Earth for equatorial orbits. Inclination leads to the subsatellite point describing a closed figure of eight as illustrated in Figure 7.6. The amplitude of the apparent drift of the payload, north and south, is $\pm i°$, where i denotes the orbit inclination. The control of orbit inclination, and specifically its reduction to zero in the case of geostationary communications satellites, is a further requirement of the launch phase.

Two guiding principles behind thrusting operations have been identified in Section 3.2.4 (see Figure 3.8). The maximum increase in speed, and hence kinetic energy, is clearly obtained by thrusting in the direction of motion, and the thrust is then used to best effect in orbit expansion. On the other hand the maximum deviation from the initial direction is obtained if the available increment ΔV is applied at right angles to the required direction. Furthermore the deviation is larger for a given ΔV the smaller the initial velocity. Relative to the problem of orbit raising illustrated in Figure 7.4, directional changes are evidently most economically effected at apogee when orbital speeds are least, for example, 7.8 km/s at 200 km altitude compared with 3.07 km/s at geostationary altitude of 36 000 km.

We shall return to these discussions in later sections when reviewing some specific applications but it is appropriate to note at this stage that the geographical position of the launch site may have important mission implications.

Multi-staged launch vehicles

It is evident from Figure 7.4 that the minimum burn-out velocity for orbital operations is approximately 7.8 km/s. Linking this requirement with equation (7.6) and neglecting for the moment drag and gravity losses, then from

$$V_b = I_{sp}g_0 \ln R \tag{7.9}$$

we may estimate the necessary rocket mass ratio R for a particular propellant combination, and hence I_{sp}. Suppose the initial rocket mass, M_0, comprises three components; payload M_p, structure M_s and fuel M_f, whence

$$M_0 = M_p + M_s + M_f \tag{7.10}$$

If we introduce the fractional payload ratio, $p = M_p/M_0$, and the propellant tankage structural efficiency, $\sigma = M_s/M_f$, then we may write

$$R = \frac{M_p + M_s + M_f}{M_p + M_s} = \frac{1 + \sigma}{p + \sigma} \tag{7.11}$$

Substituting (7.11) in (7.9) gives

$$V_b = I_{sp}g_0 \ln\{(1 + \sigma)/(p + \sigma)\} \tag{7.12}$$

For a typical structural efficiency, $\sigma \sim 0.10$, then the maximum value of the mass ratio, R, corresponding to zero payload is approximately $R = 11$. Rearranging equation (7.12), for given V_b

$$I_{sp} \geq (V_b/g_0) \ln\{p + \sigma)/(1 + \sigma)\}$$

and for the maximum value R together with the minimum burn-out velocity, this implies

$$I_{sp} \geq 331 \text{ s}$$

Thus, even under these ideal circumstances single stage to orbit could only be attained using high-energy propellant. If we incorporate a non-vanishing payload and make allowance for performance losses associated with gravity and aerodynamic drag, then low Earth orbit is scarcely attainable with the most energetic fuel/oxidizer combinations (cf. Chapter 6).

Since the mass of propellant tankage is large, typically comparable with the payload, significant performance benefits result from the progressive shedding of this mass by multi-staging. In these circumstances only a small fraction of the initial tankage mass is accelerated to the final speed.

If we identify the operation of each stage by an equation of the form given in (7.12), distinguishing the ith stage by subscript, then the velocity increment produced

$$\Delta V_i = V_{bi} - V_{b,i-1} = I_{sp,i} g_0 \ln\{(1 + \sigma_i)/(p_i + \sigma_i)\} \tag{7.13}$$

We observe that the payload for the i-stage comprises the whole of the $i + 1$ stage. Summing over n such stages, the final burn-out velocity is then given by

$$V_b = \sum_{i=1}^{n} I_{sp,i} g_0 \ln\{(1 + \sigma_i)/(p_i + \sigma_i)\} \tag{7.14}$$

An elaborate optimization procedure is required if we seek, for example, to maximize this velocity subject to constraints imposed by way of payload ratio and rocket performance. We should recall that the lower stages will also be subject to drag and gravity losses. The generalized problem has been extensively analysed, cf. White [1], for example. It is sufficient for purposes of illustration here to take the simplest case in which the stage specific impulses and structural efficiencies are equal, $(\sigma_i = s)$, whence

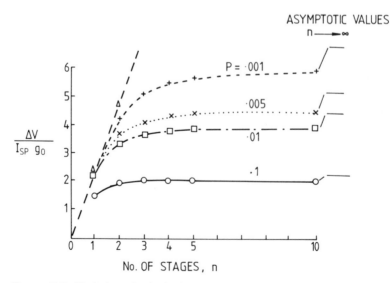

Figure 7.7   Variation of velocity increment with number of stages for fixed overall payload ratio P

Stage	Motor designation	Total impulse N − S vacuum	Average web thrust N vacuum	Burn time (s)	Total weight (kg mass)
1	ALGOL IIC	24 232 114	424 232	81.29	10 737
	ALGOL III	32 142 784	472 962	79.30	14 175
2	CASTOR II	10 304 114	291 514	38.08	4 433
	TX-354-3				
3	ANTARES II	3 218 217	97 550	36.69	1 272
	X259-B3				
4	ALTAIR III	762 972	23 745	35.00	301

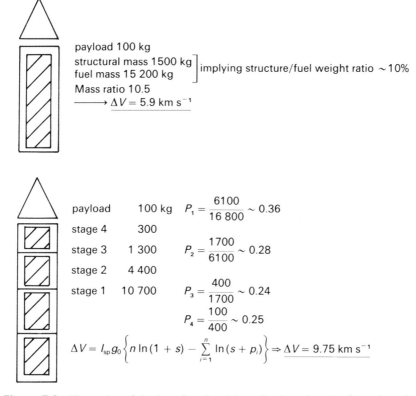

payload 100 kg
structural mass 1500 kg ⎤
fuel mass 15 200 kg   ⎦ implying structure/fuel weight ratio ~10%
Mass ratio 10.5
⟶ $\Delta V = 5.9$ km s^{-1}

payload   100 kg $P_1 = \dfrac{6100}{16\ 800} \sim 0.36$

stage 4   300

stage 3   1 300 $P_2 = \dfrac{1700}{6100} \sim 0.28$

stage 2   4 400

stage 1   10 700 $P_3 = \dfrac{400}{1700} \sim 0.24$

$P_4 = \dfrac{100}{400} \sim 0.25$

$\Delta V = I_{sp}g_0\left\{ n\ln(1+s) - \sum\limits_{i=1}^{n} \ln(s+p_i)\right\} \Rightarrow \Delta V = 9.75$ km s^{-1}

Figure 7.8   Illustration of the benefits of multi-staging based on the Scout launcher

$$V_b = \sum_{i=1}^{n} I_{sp}g_0 \ln\{(1+s)/(p_i+s)\}$$

Using the technique of Lagrange multipliers with the constraint imposed by specification of the overall payload ratio,

$$P = \prod_{i=1}^{n} p_i$$

it can be shown that V_b is maximized when the stage payload ratios are all equal, whence

$$p_i = P^{1/n} \qquad \text{(all } i\text{)} \tag{7.15}$$

and

$$V_{b\max} = I_{sp}g_0\{n\ln(1 + s) - n\ln(s + P^{1/n})\}$$

This expression has the asymptotic value

$$V_{b\max} \sim I_{sp}g_0 \ln P^{-1} \qquad \text{as } n \to \infty$$

The benefits introduced by multi-staging do not then increase without bound except for the impractical case of zero overall payload ratio, as shown in Figure 7.7. At the higher payload ratios of interest, $P \sim 0.10$, there is little advantage in further partitioning beyond three stages given the attendant increases in complexity and cost. The inclusion of the propulsive losses described earlier leads typically to a preferential loading of the first stage.

By way of illustration Figure 7.8 compares the staging of the four-stage, solid propellant Scout launcher with an equivalent single-stage vehicle having the same overall mass, payload ratio and common stage properties. The ΔV available is almost doubled as a result of the multi-staging, making near-Earth orbit accessible to small payloads.

7.3 SPACECRAFT LAUNCH PHASES AND MISSION PLANNING

In the preceding sections we have outlined some basic aspects of flight dynamics, the propulsive requirements for Earth orbit attainment and the response to the intrinsic mismatch between ΔV and chemical propellant performance reflected in multi-staging. Spacecraft missions invariably require the emplacement of the payload in an orbit having narrowly specified parameters—altitude, period, inclination and inertial orientation. The launch phase therefore embraces each of the propulsive manoeuvres necessary to achieve the initial emplacement and these are necessarily different for each mission. By way of illustration Figure 7.9 distinguishes the elements of geostationary orbit attainment.

7.3.1 Geostationary orbit emplacement

The three-stage expendable launcher—Delta or Ariane, for example—injects the satellite into geostationary transfer orbit (GTO) with perigee at 200 km and apogee at 36 000 km. Upper-stage burn-out occurs in the neighbourhood of the equatorial plane and the apogee at geostationary height, following a ballistic coast, coincides with the second crossing of the equatorial plane or node. The transfer orbit is inclined, the inclination being dependent upon the latitude of the launch site and the launch azimuth. As indicated earlier, a due-east launch (launch azimuth = 90°) will result in a transfer orbit inclination equal to the latitude of the launch site whilst any other launch azimuth must increase the orbit inclination.

Positioning of the geostationary apogee over the equator admits the possibility of emplacement in final equatorial geostationary orbit by a further single motor firing. The satellite itself is typically fitted with an apogee boost (or kick) motor (ABM/AKM) specifically to effect this combined manoeuvre of circularization and inclination removal.

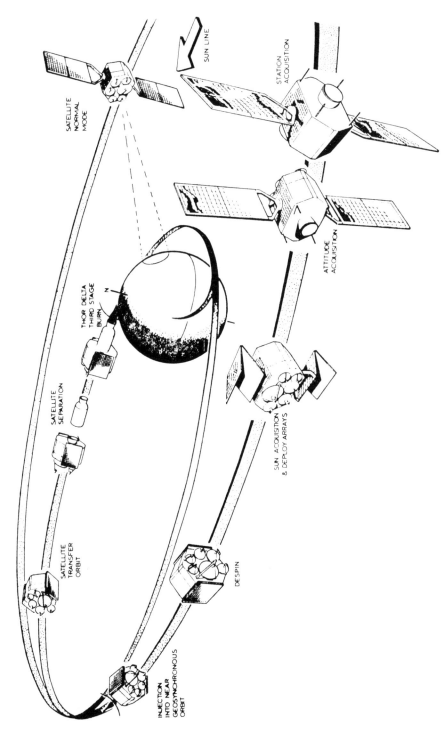

Figure 7.9 Typical sequence leading to a geostationary orbit attainment (reproduced by permission of European Space Agency)

Table 7.1 ABM impulse requirements for geostationary orbit emplacement

	Transfer orbit inclination (degrees)			
	0	10	30	50
ABM ΔV(km/s)	1.47	1.52	1.86	2.38

The transfer orbit apogee velocity is approximately 1.60 km/s whilst the equatorial geostationary orbital velocity is 3.07 km/s. Table 7.1 illustrates the impact of launch site latitude, and hence transfer orbit inclination, on ABM impulse requirement.

Near-equatorial launch sites such as the European Space Agency site at Kourou in French Guiana ($\sim 5°$N) or the San Marco floating platform off the Kenyan coast used by the small Scout launcher offer significant advantages over the eastern test range at the Kennedy Space Centre in Florida ($\sim 28°$N) or the Japanese and Russian sites at even higher latitudes.

Injection into the final orbit by the satellite ABM is not usually effected at the first transfer orbit apogee, approximately 5 hours after launch. The manoeuvre is constrained by the requirement to attain the operational station (longitude) within a defined period but also that of determining the satellite orbit and attitude, using ground station telemetry and tracking, in order to correctly orientate the motor prior to commanding its firing. Several apogees will then pass—elliptical transfer orbit period approximately 10 hours— before the ABM burn. The satellite must evidently be ground controlled during this phase and sufficiently powered electrically to maintain communications and some on-board systems, for example attitude and orbit control. If this phase is solely battery powered this will impose a further limitation on the acceptable length of time spent in transfer orbit.

In view of the possible accumulation of injection errors and the requirement to position the satellite precisely on longitude station, final positioning is achieved from a drift orbit. If the satellite orbit after ABM firing is arranged to be very slightly elliptical with perigee somewhat less then geostationary height, the orbit period will be slightly less than the Earth rotational period. As Figure 7.10 then illustrates, the subsatellite point at geosta-

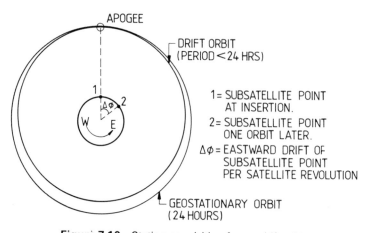

Figure 7.10 Station acquisition from a drift orbit

tionary height will drift gradually eastwards relative to the Earth. Short-duration firing of the satellite's station-keeping thrusters over several days then permits the precise positioning of the satellite.

7.3.2 Orbital mission performance—Scout launcher illustration

Similar, if less complicated, manoeuvres apply to most other near-Earth missions. The more energetic and accurate is the launch vehicle injection, the smaller is the secondary satellite propulsion requirement and the greater is the true payload fraction. Preliminary mission planning is essentially iterative as described in Chapter 5. The initial identification of broad mission objective—Earth observation platform, space-borne telescope, communications satellite, for example—will involve the specification of the satellite orbit. Launch vehicle performance summaries which identify the payload mass which a particular vehicle can inject into a prescribed orbit presents the first constraint. Detailed consideration of the distinctive launch capabilities of Ariane and Space Shuttle is presented in later sections and we shall here illustrate briefly some general principles in relation to the small

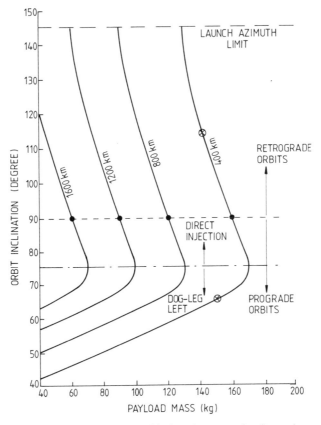

Figure 7.11 Illustrative orbital performance for Scout from Vandenberg, WTR

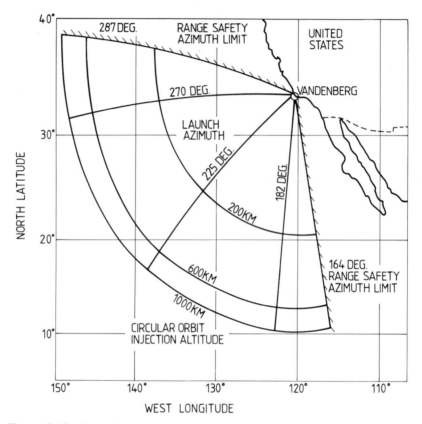

Figure 7.12 Ground track to injection for Scout from Vandenberg, WTR (re-produced by permission of LTU Missiles and Electronics Group, Missiles Division)

four-stage Scout vehicle used formerly for launching scientific payloads into low-Earth orbit.

Figure 7.11 illustrates typical users' manual information for the LTV Scout launcher into circular orbits of varying inclination from the Western Test Range in California [2]. The mission performance has been considerably simplified for ease of presentation, but some key features emerge.

As might be expected the payload capability into circular orbits of increasing altitude but fixed inclination, for example polar orbits (90° inclination achieved with a launch azimuth of 182°), decreases steadily: 158 kg into a 400 km orbit, 88 kg into a 1200 km orbit. Further reductions would accompany their expansion to elliptical orbits; 140 kg into an orbit having perigee at 400 km and apogee at 1200 km, for example. Changes in orbit inclination also have a substantial impact on mission performance. For retrograde orbits (inclination > 90°) modest decreases in payload accompany increases in inclina-tion. By contrast dramatic reductions must be accommodated for payload injected into prograde orbits (inclination < 90°). To explain these observations we must consider the geographical location of the launch site and in particular the azimuth limits set by considerations of range safety.

The location is illustrated in Figure 7.12. Situated on the west coast of the United States and in the neighbourhood of major centres of population, launches are limited to a sector bounded by azimuth limits of 164° and 287°. Recall that in the topocentric coordinate system, azimuth angle is measured from due north in an eastwards direction. Limited prograde but a wide range of retrograde orbits may be achieved by direct injection. The south-easterly heading of these prograde orbits leads to the launch benefiting from a small component of earth rotation, in contrast to the westerly headings of launches into retrograde orbits. Prograde orbits having inclinations of less than 76.5° can be achieved from the 164° azimuth by a dog-leg left manoeuvre. The necessary programmed yaw changes extend the flight profile and introduce propulsive losses which increase with the extent of the dog-leg. However, if we compare Figure 7.11 the directly injected retrograde orbit of 115° inclination with its counterpart, the prograde orbit of 65° inclination achieved by a dog-leg manoeuvre, then a payload mass improvement is recorded for the latter strategy.

7.3.3 Configuration interactions

Whilst payload mass may justifiably be considered to assume an overriding significance in determining the mission, in many circumstances the spacecraft configuration may also be seriously constrained by the size and shape of the available payload volume. Multi-stage expendable launch vehicles are essentially cylindrical, and aerodynamic considerations naturally restrict the payload fairing (or envelope) to a shape resembling a cone–cylinder combination. Spacecraft designed to be launched by such vehicles tend to be slender, at least in their stowed configuration—solar arrays, communications antennae, scientific instruments may have to be folded, furled, or telescoped to conform to the fairing and deployed on station. The inclusion of the mechanism necessary to effect this deployment then adds substantially to the complexity and vulnerability of the payload design.

Although the launch phase duration may be measured in hours in a mission with an operational timescale of years, the launch often produces the most demanding environment faced by the mechanical design. The longitudinal acceleration is high: in excess of $4g_0$ in the case of Ariane II at second-stage burn-out (cf. Figure 7.3). Sensitive elements of the payload and deployable equipment must withstand both the mean acceleration and structural vibration and resonances accompanying motor firing and stage separation.

As we indicated in Section 7.3.1, geostationary orbit emplacement also incorporates quite major satellite thrusting phases at apogee (ABM firing) and, in the case of Shuttle launches (cf. Section 7.5), a perigee burn in addition. In general these motors employ solid propellants and deliver relatively high thrusts in firings of short direction (cf. Chapter 6). In the absence of fine control of thrust, the spacecraft and motor are spun-up to an angular rate of ~ 100 r.p.m. This both provides a measure of gyroscopic stiffness for guidance and reduces the effects of any thrust misalignment. In consequence, however, the deployment of such lightweight, flexible structures as solar arrays is further delayed and substantial power-raising in transfer orbit is prevented. The enhanced mission flexibility which accompanies lower thrust, more readily controlled and higher specific impulse bi-propellant rocket motors is a major factor in their current development as secondary propulsion.

7.4 THE ARIANE LAUNCH VEHICLE

In the preceding section we have described some general features of launch vehicle performance and operation from the user or payload perspective. Here we describe the Ariane launch vehicle in more detail but from the same standpoint; the reader will be referred to other sources for launch vehicle design information.

7.4.1 Vehicle design summary

Whilst earlier expendable launch vehicles evolved from missiles and were adapted to a wide range of missions, the Ariane programme was specifically directed towards the task of geostationary communications satellite emplacement. Whilst the performance envelope has been expanded through Ariane 1 to 4 to embrace increased payload mass and larger dual-launched satellites, the target mission which determines the overall design remains that of injection into geostationary transfer orbit (GTO). Ariane comprises a three-stage liquid propellant rocket; the first and second stages burning a UDMH fuel-nitrogen tetroxide oxidizer combination whilst the third stage employs liquid hydrogen and liquid oxygen. The mass breakdown for the initial Ariane I configuration is given in Table 7.2 and sketched in Figure 7.13. The vehicle is 48 m in length with a lift-off mass of 210 tonnes and payload delivery into GTO of 1750 kg. By the tenth flight in August 1984 and the inauguration of Ariane 3, improvements in the performance of the third-stage HM7 engine, increased combustion chamber pressures on the first and second stages and the addition of two solid propellant strap-on boosters led to increases in lift-off mass and payload to 230 tonnes and 2500 kg respectively. This latter level together with a

Table 7.2   Mass breakdown of Ariane I

| | Lift-off mass (tonnes) | | | |
| | Structures | Propellants | Total | |
			Item	Cumulative
Fairing	0.84		0.84	0.84
VEB	0.32		0.32	1.16
Third stage	1.2		9.45	10.61
Liquid oxygen		6.72		
Liquid hydroger		1.53		
Second stage	3.8		37.85	48.46
UDMH		11.8		
N_2O_4		21.7		
Water		0.55		
First stage	13.3		159.65	208.11
UDMH		50.5		
N_2O_4		93.35		
Water		2.5		
Payload			1.75	209.86

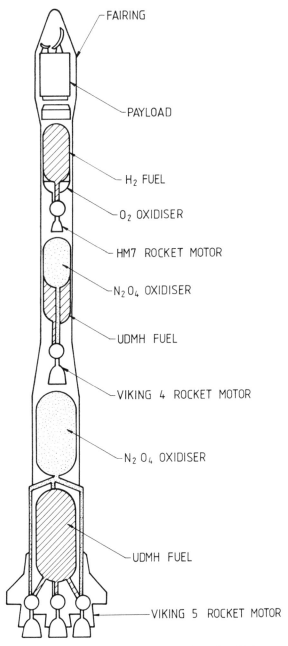

Figure 7.13 Schematic Ariane launch vehicle

Table 7.3 Ariane. Summary of engine characteristics

	Ariane 1			Ariane 3			
	Viking V	Viking IV	HM7	Viking V	Viking IV	HM7	First-stage strap-on booster
Vacuum thrust (kN)	660	700	62	720	760	63	700
Specific impulse (s)	278	294	442	278	294	446	240
Chamber pressure (bar)	53	53	30	58	58	35	50
Burn time (s)	138	131	563	142	140	731	30
Propellant type	$UDMH/N_2O_4$		LH_2/LO_2	$UDMH/N_2O_4$		LH_2/LO_2	Flexadyne powder

minor modification to the payload fairing permits two PAM-D class satellites to be accommodated using the SYLDA, dual launch system (see page 165). The designation PAM-D denotes satellites which are comparable with Delta-launched payloads and utilize this particular payload assist module as the perigee stage when Shuttle launched (see page 169).

The first stage (Drakkar) propulsion uses four Viking V engines, each producing a thrust of 550 kN, whilst the second stage uses a single Viking IV engine producing a thrust of 760 kN at altitude. The motor designs are substantially the same and employ a simple gas generator cycle (e.g. Chapter 6). A small fraction of the $UDMH/N_2O_4$ propellants are fed into a gas generator which drives the propellant turbopumps. The exhaust product gases are cooled using water from a toroidal tank. The combustion chamber pressure is 58 bars and duration of the burn is 140 s. Details of the engine development are reported by Souchier [3] and conveniently summarized by Velupillai [4].

The third stage HM7 engine uses liquid hydrogen and liquid oxygen and is a development by SEP of engines first envisaged for the second stage of the ill-fated Europa launcher. The thrust delivered is 62 kN and the motor fires for 730 s. A crucial distinction between the design of this engine and the Space Shuttle main engine (SSME) is again the use of a gas generator to drive the turbopumps. The turbine exhaust is then simply expelled whereas the staged combusion of the SSME admits its further use propulsively (cf. Chapter 6). The combustion chamber pressure is however only 35 bar and this is then reflected in a comparatively simple design [5].

The comparative engine characteristics of Ariane 1 and 3 are summarized in Table 7.3. The two solid propellant strap-on boosters, each containing 7350 kg of flexadyne powder moulded into a six-pointed star grain, enhance the performance very substantially, increasing take-off thrust from 2490 kN to 4040 kN. A further increased contribution from strap-on boosters came with the Ariane 4 programme inaugurated in 1988. Four boosters, all solid, all liquid or in combination, permit further expansion of the payload capability into GTO. For the most powerful AR44L combination (with liquid propellant boosters) this payload is approximately 4200 kg injected into geostationary transfer orbit and hence two Atlas sized satellites. This further Ariane first-stage development is described by Souchier and Pasquier [6]; the additional liquid propellant strap-on booster, designated Viking VI, carries 37 tonnes of $UDMH/N_2O_4$, burns for 135 s and delivers a thrust of 666 kN.

7.4.2 Mission performance

Ariane is launched from a near-equatorial site in Kourou, French Guiana at a latitude of 5.2°N. As described earlier such a location is particularly favourable for geostationary satellite emplacement but launches can be made from the CSG (Centre Spatial Guyanais) within azimuth limits −10.5° to 93.5°. A range of performance curves is shown in the appropriate Ariane users manual [7] for purposes of preliminary mision planning and embraces a variety of elliptical and Sun-synchronous orbits in addition to the primary programme objective of GTO emplacement.

Figure 7.14 illustrates the basic elliptical orbit performance of Ariane 3 for a due east launch ($i = 5.2°$). The single launch payload capability into GTO is identified to be 2450 kg. Alternatively this may be configured as two Delta 3920 or STS PAM-D sized

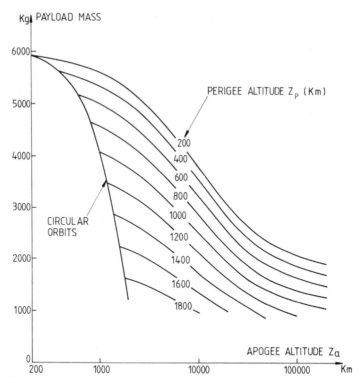

Figure 7.14   Ariane 3 performance in elliptical orbits. Launch azimuth 90°, inclination 5.2°

satellites of the following general form:

2 × 1150 kg Delta satellites	2300 kg
SYLDA dual launch system and adaptor	180 kg

A brief description of the SYLDA/SPELDA dual launch systems and payload envelopes is given shortly.

If we focus specifically on the dual launch into geostationary transfer orbit, both satellites are injected into the same orbit although they may be oriented and spun-up (≤ 10 r.p.m. as required) quite independently after third-stage engine cut-off. A minimum relative velocity of 0.5 m/s is established between the two satellites at separation and a safe distance is determined between them before apogee motor firing. The argument of perigee at injection is approximately 180° and the longitude of the first descending node occurs in the vicinity of 10°W, off the coast of West Africa.

The enhanced performance to be offered into GTO by Ariane 4 with the addition of alternative combinations of liquid and solid propellant strap-on boosters is illustrated in Table 7.4. The values quoted include the masses of spacecraft, dual-launch system (if employed) and appropriate adaptors.

The injection phase is inertially stabilized and the expected accuracy is high. Typical

Table 7.4 Ariane 4 performance into GTO

Launch vehicle configuration	Mass (kg)
A40	1900
A42P	2600
A44P	3000
A42L	3200
A44LP	3700
A44L	4200

Dual launch system		Standard adaptors	
Short SPELDA	350 kg	'937'	48 kg
Long SPELDA	400 kg	'1194'	43 kg
SYLDA (3990)	185 kg		
SYLDA (4400)	185 kg		

values of standard deviation for key injection parameters are: inclination 0.07°, argument of perigee 0.45°, perigee altitude 1 km, apogee altitude 100 km.

7.4.3 Dual launch system

The Ariane 3 launch vehicle can simultaneously place in GTO two independent satellites, each of a mass broadly comparable with that launched by Delta vehicle or Shuttle/ PAM-D combination. This dual launch system is given the acronym SYLDA (Système de Lancement Double Ariane).

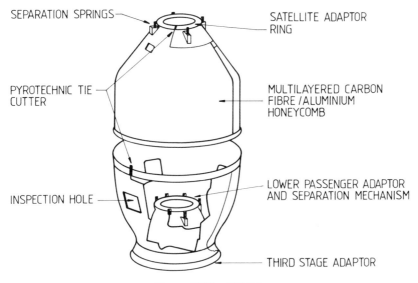

Figure 7.15 SYLDA

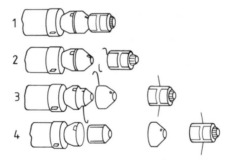

Figure 7.16 Ariane dual launch separation phases

An extended version of SYLDA incorporating a longer fairing and mounted on the more powerful Ariane 4 is designated SPELDA (Structure Porteuse pour Lancement Double Ariane).

The SYLDA is illustrated in Figure 7.15 and consists of a load-carrying, carbon-fibre reinforced shell which encapsulates the lower satellite and supports the upper one. The shell comprises two separable parts, retained in flight by a clamp band, and incorporating satellite adaptor rings. The two satellites are quite independent, without mechanical or electrical interfaces. The sequence of events leading to the release and injection into orbit of the two satellites is shown in Figure 7.16. Following third-stage engine cut-off, the attitude control system orientates the stage correctly and spins up. Pyrotechnic cutters are fired and release the spring-loaded upper satellite. The upper half of the shell is itself

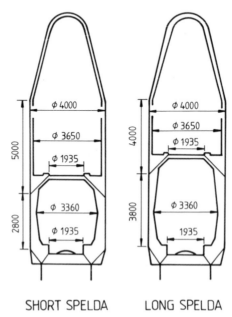

SHORT SPELDA   LONG SPELDA

Figure 7.17 Typical payload envelopes for enhanced dual launch capability (SPELDA)

then pyrotechnically released and the lower satellite is spring-released. The complete manoeuvre lasts approximately 30 s.

Illustrative payload envelopes and fairings are sketched in Figure 7.17. In addition to the envelope constraints introduced by the upper and lower passenger volumes, limitations are imposed on the satellite mass and inertia characteristics. These are specified in the relevant user's manual.

7.4.4   Future developments

Planned expansion of the Ariane series embraces Ariane 5 to be developed for the 1990s which will notionally offer a single launch payload capability of 7 tonnes into GTO and 18 tonnes into LEO (low earth orbit). The latter would represent approximately one half of the Shuttle payload and holds out the prospect of a launch capability for the small reusable manned vehicle, Hermes. The centre-piece of this launcher development is the HM60 Vulcain engine, comparable in size with the Rocketdyne J2 Saturn V second stage.

Studies have identified a need for an improved core stage motor in the 106 kN thrust range. This would be a cryogenic LOX/LH2 engine producing fifteen times the thrust of the present HM7 engine in the Ariane third stage with a combustion chamber pressure of 100 bar. The gas generator cycle would be retained, on the ground of reduced complexity and development cost for a modest performance compromise [8]. This stage is augmented initially by two solid propellant boosters (P230), each delivering 650 kN for a burn time of 125 s. Launches into geostationary transfer orbit will also incorporate a MMH/N_2O_4 second stage.

In operational terms first-stage reusability offers the prospect of significant savings in launch costs. Recovery and reuse of at least the first-stage strap-on boosters is envisaged for Ariane 4 and subsequent vehicles in the series.

7.5   THE SPACE SHUTTLE TRANSPORTATION SYSTEM

In contrast to the Ariane launch vehicle described in the preceding section, the Space Shuttle is both manned and substantially reusable. As its name implies its function is then more broadly defined than simply that of a spacecraft launch vehicle. For the purposes of this chapter, however, we shall neglect its important role in such manned near-Earth activities as Spacelab and focus on its orbital mission performance.

7.5.1   Vehicle design summary

The Shuttle primary propulsion comprises two basic elements; an external tank, 47 m in length and 8.7 m in diameter, which carries the LOX/LH2 propellants and feed system; the propellants are then fed to three main engines (SSME); two solid strap-on boosters augment the available thrust at launch, burning in parallel with the main engines for 120 s.

The solid rocket booster employs a composite propellant, a polybutadiene acrylic binder and ammonium perchlorate oxidizer, in a tapering eleven-point star grain configuration. The booster has an inert mass of 8.2×10^4 kg and propellant charge of

Table 7.5 Space Shuttle Main Engine and booster performance

	SSME Rocketdyne Space Shuttle Orbiter (each of three)	Solid rocket booster (each of two)
Thrust (10^6 N)	2.1 (vacuum)	11.8 (sea level)
Specific impulse (s)	455	260
Burn time (s)	480	120
Propellant	LOX/LH2	PBAN/AP

5.0×10^5 kg. The performance of the boosters and main engines is summarized in Table 7.5. The spent boosters separate from the external tank and are recovered by parachute approximately 300 km down range.

The SSME employs a staged-combustion closed cycle (cf. Chapter 6). The hydrogen fuel and a small fraction of the oxidizer are partially burnt in a pre-combustor. The efflux from this drives the high-pressure propellant turbopumps. The turbine exhaust, together with the remaining oxidizer, is then injected into the combustor chamber. The latter operates at 207 bars and therefore necessitates substantially higher turbopump discharge pressures—approximately 500 bars. The external tank is jettisoned before final orbit insertion.

Following external tank separation, the orbiter orbital manoeuvring subsystem (OMS) provides the thrust to perform orbit insertion, circularization and such on-station manoeuvres as rendezvous and de-orbit. The OMS engine employs MMH fuel and N_2O_4 oxidizer, delivering a specific impulse of 313 s and a vacuum thrust of 27 kN. The OMS tankage is sized to provide sufficient propellant for a total ΔV of 305 m/s when the vehicle carries a payload of 29.5 tonnes.

7.5.2 Mission performance

Shuttle launches presently take place from the Kennedy Space Centre in Florida, latitude 28.5°N. The payload capability into low Earth circular orbits in the range 200–400 km from this site, is large (≤ 29.5 tonnes) and therefore factors such as payload volume in-orbit operations usually impose more significant constraints on spacecraft design than does mass alone. Increases in parking orbit altitude can be achieved by a trade-off between payload mass and additional MMH/N_2O_4 OMS propellant (cf. Figure 7.18). In general, however, satellite orbit raising is most economically effected by tailoring specialist perigee stages to individual mission requirements.

Due eastwards Shuttle launches will characteristically lead to the deployment of satellites into a 200 km circular parking orbit inclined at 28.5° to the equator. A second Shuttle launch site at the Western Test Range in California would make polar orbits, which are more appropriate to Earth observation missions, more accessible. The payload capability is, however, reduced by a factor of approximately one third relative to a KSC launch.

The Shuttle cargo bay is roughly cylindrical, 18 m in length and 5 m in diameter.

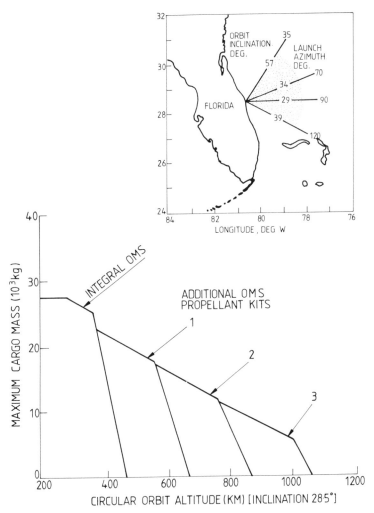

Figure 7.18 Typical STS orbital performance for KSC launch (NASA [18])

Spacecraft may be mounted vertically or horizontally, although the cost structure, based on the fraction of cargo bay length occupied by the payload, favours the vertical configuration. In the case of a mission necessitating substantial orbit raising, injection into GTO for example, the additional length of the propulsion module or perigee motor may then prove a significant cost factor.

The large number of geostationary communications satellites encourages the more detailed consideration here of the orbit-raising necessary for such missions. In launches to date the role of the Space Shuttle Upper Stage has been borne by the McDonnell-Douglas Payload Assist Module (PAM), broadly sized according to payload mass and, more specifically, the equivalent expendable US launcher, D for Delta-sized and A for Atlas-sized satellites—approximately 1100 kg and 2100 kg into GTO respectively

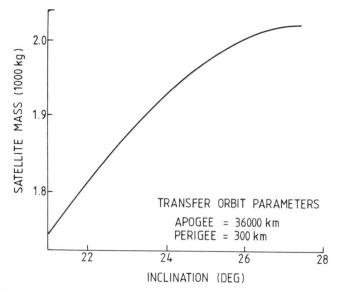

Figure 7.19 STS PAM-A GTO performance (Reproduced by permission of McDonnell Douglas Space Systems Company)

(McDonnell-Douglas Corp. [9, 10]). PAM-A performance is summarized in Figure 7.19. As illustrated earlier, substantial payload reductions accompany inclination changes in the neighbourhood of perigee.

The PAM-A incorporates a spin-stabilized, solid propellant motor derived from the Minuteman LIII. Several failures in the early Shuttle launches have been attributed to PAMs and considerable interest surrounds the development of higher-performance, more accurate orbital transfer vehicles.

7.5.3 Shuttle operations and satellite deployment

Payloads launched from the Kennedy Space Centre may be integrated into the cargo bay vertically or horizontally, depending upon the payload dimensions. As indicated earlier 'short and fat' designs, vertically mounted, are encouraged by the Shuttle pricing policy and the majority of launches to date have consisted of these designs. In these circumstances the mated satellite and solid spinning upper stage (SSUS) are carried in a reusable cradle assembly. This is mounted in the cargo bay in the KSC vertical processing facility.

Since Shuttle-launched satellites are deployed from low Earth parking orbit, in principle a greater measure of flexibility is introduced into the mission profile. Injection into GTO is possible at the ascending or descending nodes, for example, whilst Ariane direct injection occurs solely at the descending node. The satellite is, however, subject to a fluctuating solar input in parking orbit once the Shuttle cargo doors are opened. The thermal environment to which the satellite is then exposed differs substantially from that in its operational configuration and specific thermal protection is required.

The payload cradle carries thermal covers in the form of the movable clam shells illustrated in Figure 7.20.

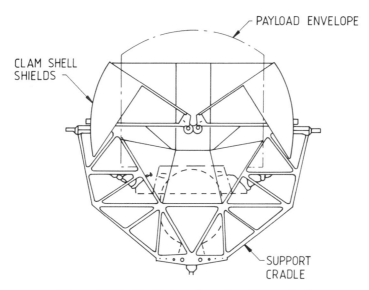

Figure 7.20 Shuttle launch cradle with sunshield

A typical satellite release sequence comprises the following steps: the satellite is spun up to 50 r.p.m. and checked out; the cradle retaining clamp is pyrotechnically freed and the satellite is ejected by springs at approximately 1 m/s; the PAM then fires some 45 minutes later. If two satellites are launched—for example, SBNS-3 and Anik C3 on flight STS-5 in November 1982—the planned deployments are roughly 24 hours apart.

In future, large satellites of the kind prompting the development of the Centaur upper stage ($\sim$4500 kg into GTO) will necessitate horizontal mounting in the cargo bay. For satellite deployment, the upper stage would be raised to an angle of approximately 45° before pyrotechnic separation and spring release.

An alternative deployment strategy centres on the Orbiter Remote Manipulator Systems (RMS). The robot arm is capable of both deployment and retrieval of large payloads of envelope and mass comparable with the cargo bay limits; the use to date has however been largely restricted to retrieval—the Solar Maximum Mission repair, for example. This demonstration of the Shuttle's role in satellite repair is important in that it encourages designs with a built-in measure of maintenance and reuse. The provision of sufficient propellant at end-of-life to permit the satellite to descend from operational orbit to a Shuttle accessible height then becomes important.

We have touched upon an important facet of Shuttle operation in promoting and developing near-Earth orbit activities. The requirement of improved performance and greater operational flexibility in terms of launch vechicles, in particular, is then reflected in the development of an Orbital Transfer Vehicle.

7.5.4 Future developments—Orbital Transfer Vehicles

We have identified in earlier sections the requirement for a Shuttle upper stage to boost payloads into those higher Earth orbits of commercial and scientific interest which are

not immediately accessible to the Shuttle. In the important case of injection into geostationary transfer orbit this role is presently assumed by the payload assist modules. The
performance of these spin-stabilized solid rocket motors is comparable to that of such
expendable launch vehicles as Delta and Atlas-Centaur. However, this represents only
an interim solution to a problem intrinsic to Shuttle-launched satellites destined for high-
energy orbits.

The concept of a space tug, operating exclusively in Earth orbit as a vehicle for orbit
raising, plane-change or space-probe boost to Earth-escape and incorporating significant
aspects of reusability, is as old as the Shuttle itself. The twists and turns in history of this
idea are comprehensively reviewed by Dooling [11]. The continued growth, for example,
in the mass of geostationary communications satellites and the Arianespace response have
certainly not lessened the requirement. A clearly defined alternative to designs based
upon existing solid rockets (IUS) or the Centaur, cryogenic LOX/LH2 upper stage, has
yet to emerge, however.

Expendable rocket stages naturally offer the prospect of reduced development costs
associated with their necessary adaptation to OTV missions. The 'short and fat' philosophy described in the previous section, however, applies equally to a high-performance,
horizontally mounted OTV, and therefore argues for the development of a specialist
vehicle.

The concept of an electrically propelled tug was quite widely canvassed in the 1970s
but the intervening years have not led to the development of suitable space-proven
hardware. Accordingly high-energy, chemical upper stages still predominate, although
the propellant economies which might accompany an aero-assisted de-boost to Shuttle
orbit height are under investigation. At perigee heights of 200 km or less, aerodynamic
drag is a significant force and trajectories which graze the atmosphere admit the prospect
of useful savings in propellant for controlled manoeuvres designed for Shuttle rendezvous
[12].

7.6 ALTERNATIVE EXPENDABLE LAUNCHERS

We have highlighted in brief case studies the major role played by Ariane and Shuttle in
satellite launches from Europe and the United States. It is important also to recognize
the continued launcher development in the Soviet Union and the more recent emergence
of such capabilities in Japan and China.

Comparatively limited information is available on the current family of Soviet launch
vehicles—Sapwood, Skean, Proton and Scarp—or the three launch sites, Tyuratam-
Leninsk, Kapustan Yar and Plesetsk. Information gleaned from launch logs and some
Soviet sources (e.g. Glushko [13], however, is usefully summarized in Wilson [14].
Plesetsk, situated at 62°N, serves near polar orbits in much the same way as the Vanden-
burg WTR whilst the other sites at roughly 48°N fulfil a role similar to the KSC. In the
light of earlier discussions these high latitudes have been one of the factors which has
encouraged the exploitation of 'Molniya' orbits for communications purposes (cf. Chapter
5). Booster evolution, larger payloads and higher-energy propellants—from kerosene/
LOX to UDMH/N_2O_4 and LH2/LOX, for example—appear to parallel the kind of
developments described in detail earlier, culminating in the Energia heavy lift launcher
unveiled in May 1987.

The Japanese N-series of launch vehicles, inaugurated by NASDA in 1975, have a performance which is similar to that of the McDonnell-Douglas Delta launcher and on which early designs were based. It is a three-stage vehicle with solid rocket augmentation, utilizing in turn kerosene/LOX, $UDMH/N_2H_4/N_2O_4$, and solid propellants for the respective stages. Launches are made from the Tanegashima space centre. Future extensions of the programme have focused on the development of a cryogenic LOX/LH2 engine for a fully domestic H-II launcher with a 2 tonne capability into GEO [15].

In April 1985 a Chinese three-stage launch vehicle, the Long March-3, successfully boosted an experimental communications satellite STW-1 into geostationary orbit, heralding their entry into the commercial exploitation of space. All these programmes indicate that expendable launch vehicles will continue to play an important role in spacecraft engineering for the foreseeable future. Furthermore an even greater diversity of launch vehicle options may become available to payload designers as confidence grows in these various national programmes.

7.7 TRENDS IN LAUNCH VEHICLE COSTS

The increased commercial exploitation of both low-Earth and geostationary orbits has focused attention on specific launch cost as a figure of merit. Koelle [16], in a review of launch vehicle evolution, points out that in little more than a decade specific launch costs have reduced fivefold. Table 7.6 provides a typical breakdown of the present costs involved in the launch of a geostationary communications satellite and a space platform by Shuttle.

The launcher, and related services, comprise a major component of these costs and the present levels—roughly $25 000/kg into GEO—remain high and must inhibit the further rapid exploitation of space. Although heavier-lift launch vehicles offer reduced specific launch costs into LEO—by an order of magnitude in the case of the Space Shuttle or Saturn V—their direct commercial significance is correspondingly smaller, however. Comparable reductions would be required even there for more massive programmes related to the industrialization of Earth orbit.

Given the widespread use of high-energy propellants, notably LH2/LOX, delivering a vacuum specific impulse of 455 s in the case of the SSME, and overall payload fractions

Table 7.6 Illustrative launch costs into Earth orbit

Projected EURECA μgravity		Communications satellite	
Platform	46 MAU	Satellite	50 MAU
(for five flights)			
Shuttle	18	Shuttle	16
Retrieval	8	PAM-DII	12
Launch Vehicle Services	3	Launch Processing	2
(optional)			
Insurance	26	Insurance	16
	101		96

$\geq 2\%$, there must appear limited scope for substantial further improvements in launch costs for expendable vehicles.

Concern for such levels continues to provide the stimulus for both feasibility studies and technology demonstrations of launch vehicles incorporating some of the air-breathing elements outlined in Chapter 6. Reusable single-stage-to-orbit vehicles, particularly those capitalizing on hybrid or composite engines such as the British Aerospace HOTOL concept [17], might then offer access to quite different generic trends and realize the goal of specific launch costs into LEO of, for example, $500/kg or less.

REFERENCES

[1] White, J.F. (1962) *Flight Performance Handbook for Powered Flight Opterations*, Wiley, London.

[2] Vought Corporation (1974) *Scout Planning Guide*.

[3] Souchier, A. (1983) Development of the thrust augmented Viking engine and first stage DRAKKAR propulsion systems for the Ariane 3 launcher. *Proc. 34th International Astronautical Congress*, Budapest, Hungary. IAF Paper 83-04.

[4] Velupillai, D. (1983) Ariane Uprates. *Flight International*, **123**, 1166.

[5] Souchier, A. and Kirner, E. (1983), Ariane 3 third stage propulsion systems and HM7B engine development. *Proc. 34th International Astronautical Congress*, Budapest, Hungary. IAF Paper 83-387.

[6] Souchier, A. and Pasquier, J. (1983) Ariane 4 liquid boosters and first stage propulsion system. *AIAA/SAE/ASME 19th Joint Propulsion Conference*. AIAA Paper 83-1192.

[7] *Arianespace Ariane Users Manual*—Ariane 2/3 (1980), Ariane 4 (1983), Evry, France.

[8] Pouliquen, M.F. (1984) HM60 Cryogenic Rocket Engine for Future European Launchers, *J. Spacecraft and Rockets* **21** (4), 346.

[9] McDonnell-Douglas Corporation (1980) STS PAM-A. *User Requirements Document*.

[10] McDonnell-Douglas Corporation (1985) PAM-D/DII *User Requirements Document*.

[11] Dooling, D. (1982) A Third Stage for Space Shuttle: What happened to Space Tug. *J. Brit. Interplanetary Soc.* **35**, 553.

[12] Walberg, G.D. (1982). A Review of Aeroassisted Orbital Transfer. *AIAA Conf. on Atmospheric Flight Mechanics*, San Diego, California. AJAA Paper 82-1378.

[13] Glushko, V.P. (1975) *Rocket Engines* GDL-OKB, Novosti Press,

[14] Wilson, A. (ed.) (1982) Soviet Astronautics, *J. Brit. Interplanetary Soc.* **35** (2).

[15] Godai, T. (1986) HII: A New Launch Vehicle in the 1990's *Acta Astronautica* **14**, 143

[16] Koelle, D.E. (1986). Launch Vehicle Evolution—from Multistage Expendables to Single Stage Reusables, *Acta Astronautica* **14**, 159.

[17] Parkinson, R.C. (1985) *HOTOL—A Third Generation Economic Launch Vehicle*. IAA-85-482.

[18] *NASA Shuttle Data for Payload Designers* (1980).

8 ATMOSPHERIC RE-ENTRY

R. A. East

Department of Aeronautics and Astronautics, University of Southampton
0703 - 595000

8.1 INTRODUCTION

Although many spacecraft are expendable on completion of their missions, a significant number of applications require recovery of either the complete vehicle, or some portion thereof, either from the considerations of reuse, or of the survival and/or recovery of scientific and industrial payloads and/or operating personnel. These requirements are relevant to both Earth return or to other planetary missions. Soft landing on to planets without atmospheres can only be accomplished by retro-braking, whereas aerodynamic deceleration and manoeuvre can be beneficially employed on those planets with significant atmosphere. However, concomitant with the advantageous effects of aerodynamic forces, the associated convective and radiative heating present problems which require solution if safe recovery is to be effected.

Problems associated with atmospheric re-entry from sub-orbital trajectories were first encountered by man-made objects with the German V-2 ballistic missiles in World War II. The Mercury, Gemini and Apollo manned vehicle programmes in the USA required considerable development of re-entry technology and the modest use of aerodynamic lift was used in these re-entry trajectories. The concept of a reusable lifting re-entry vehicle has been realized by the USA with the Space Shuttle Orbiter and the USSR with its Buran Orbiter which provides a space launch and return capability for low Earth orbit (LEO). The smaller European reusable space vehicle, Hermes, will also use this principle.

The aerodynamic forces encountered in lifting re-entry provide a means of spacecraft manoeuvre. This capability can be exploited to provide orbital plane changes and orbital transfer with a significant reduction in energy expenditure in comparison with trajectory changes effected using extra-atmospheric propulsive manoeuvres. An excellent survey of aeroassisted orbital transfer has been given by Walberg [1] in which three primary mission applications have been categorized:

- synergetic plane change;
- orbital transfer vehicle applications—studies having been primarily directed towards LEO missions;

Spacecraft Systems Engineering. Edited by P.W. Fortescue and J.P.W. Stark

● planetary mission applications involving aerobraking and aerocapture.

In the first of these, a combined aerodynamic manoeuvre/propulsive phase is proposed in the atmosphere after de-orbit. After a banked aerodynamic turn, the vehicle is re-boosted to the required orbital altitude, with a further propulsive phase to circularize the orbit in a different plane from the original.

Orbital transfer vehicle applications are primarily concerned with the application of aerodynamic braking as an alternative to propulsive braking in the context of return from high Earth orbit (HEO)—usually geosynchronous (GEO)—to low Earth orbit (LEO). The uses of aerodynamic lift (as for the synergetic plane change vehicle) in addition to drag can be applied to effect not only the required velocity decrement (~ 2.3 km/s) but also the orbital plane change (28.5°) needed to transfer from GEO to a typical Space Shuttle LEO. To avoid excessive mass penalties associated with the thermal protection system required to alleviate the aerodynamic heating problem, the atmospheric encounter occurs at high altitude in an essentially noncontinuum flow regime. The consequent low drag necessitates the use of multiple passes (up to 30) through the atmosphere to obtain the required velocity decrement. Greater operational flexibility in the form of shorter orbital transfer time (one or two passes) can be obtained from lower altitude atmospheric encounters, but at the expense of a heavier thermal protection system and the use of an ablative heat shield which requires refurbishment after every mission.

An example of such a vehicle concept was the Aeromanoeuvring Orbit-to-Orbit Shuttle (AMOOS) studied by the Lockheed Missiles and Space Co. [2]. A hypersonic lift-to-drag ratio in the range 0.5–1.0 provided a controlled manoeuvring capability which enabled a single-pass orbit transfer, with 7° of the required 28.5° plane change provided by aerodynamic forces.

The use of inflatible ballutes to obtain very low ballistic coefficients, in combination with a retro-propulsive atmospheric encounter, appears to provide a means of retaining operational flexibility (one or two passes) with a high-altitude, and hence low-density, atmospheric encounter.

Although the velocities of encounter are significantly higher than for Earth orbit applications, somewhat similar atmospheric braking manoeuvres have been proposed for aero-assisted Mars, Uranus and Saturn/Titan missions. Multi-pass aerodynamic braking can be effected by placing the vehicle in a highly eccentric orbit, which is followed by multiple high-altitude encounters through which the required circular orbit for planetary capture is obtained from aerodynamic braking. This technique requires small rocket burns at apoapsis in order to maintain the periapsis on each successive pass at an altitude such that the aerodynamic loads and heating are not excessive. For an alternative technique, known as aerocapture, the vehicle encounters the planetary atmosphere at hyperbolic velocity and descends to a lower level than for the aerobraking concept with consequently higher heating rates. After flying a roll-modulated, constant drag trajectory to achieve the required velocity decrement, a pull-up manoeuvre is used to exit the atmosphere, followed by a brief propulsive phase to circularize the orbit.

The aerodynamic features of the range of atmospheric braking and manoeuvring space vehicles are principally determined by the velocity–altitude regime of the atmospheric flight. Figure 8.1 shows a comparison of the flight regimes of several of the vehicles proposed for Earth atmosphere encounters. Also shown is a typical entry altitude where the 'top' of the Earth's atmosphere is encountered (~ 150 km). At the higher altitudes

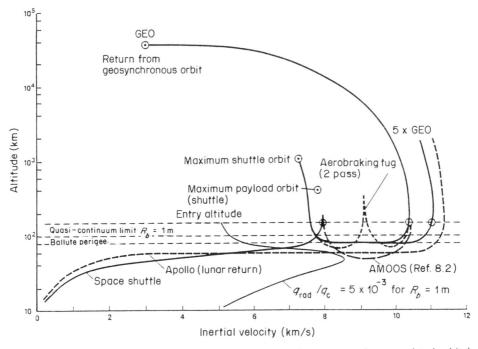

Figure 8.1 Altitude–velocity diagram for various types of re-entry and aero-assisted orbital transfer vehicles. (After Walberg [1], and Howe [12].)

within the atmosphere, rarefied flow effects dominate and the gas cannot be considered as a continuum. In the rarefied regime a slip condition occurs at the wall for both velocity and temperature in contrast to the no-slip condition for continuum flow. The upper limit of the continuum regime is given when the Knudsen number ($Kn = \lambda/R$ = mean free path/body radius) ceases to be small compared to unity; for a characteristic body radius of 10 m this occurs at an altitude of order 100 km. As shown in Figure 8.1, the bulk of the trajectories for the Space Shuttle orbiter, Apollo re-entry, ICBM and AMOOS lie within the continuum flow regime. In contrast, for aero-assisted orbital transfer vehicles (AOTVs) and plane change vehicles, a significant fraction of the atmospheric encounter occurs in a very energetic rarefied flow regime. A consequence of this is that chemical non-equilibrium effects, radiative heat transfer and air ionization are phenomena which have a major influence for AOTVs.

A further generic class of space vehicle is the orbit-on-demand type in which either a single-stage-to-orbit or air-launched vehicle takes off from the Earth's surface, orbits for a short period (perhaps accompanied by a plane-change manoeuvre), de-orbits and re-enters with significant cross-range to land horizontally and be capable of refurbishment within a short period. These requirements lead to vehicles with low wing loading with largely metallic thermal protection systems. High cross-range performance is achieved by high lift-to-drag ratios. Proposals for vehicles of this type include the transatmospheric space plane in the USA and the HOTOL vehicle in the UK. From the re-entry viewpoints, the low wing loading permits deceleration at higher altitudes than for the Space Shuttle orbiter, although the major portion of the trajectory is still within the continuum flow

Table 8.1   Re-entry specific kinetic energy and maximum energy flux

Vehicle type	Entry velocity (km/s)	Specific kinetic energy $(\frac{1}{2}V_e^2)$ (MJ/kg)	Approximate maximum energy flux $(\frac{1}{2}\rho_\infty V_\infty^3)$ (MW/m^2)
Ballistic $(W/C_D S = 4.5 \times 10^3 \text{ N/m}^2)$			
(5° entry angle)	7.9	31.3	20
Ballistic $(W/C_D S = 7 \times 10^4 \text{ N/m}^2)$			
(5° entry angle)	7.9	31.3	1000
Space shuttle	7.8	30.5	20
(370 km orbit)			
Apollo	11.3	64.3	65
Aerobraking tug (GEO return—			
30 pass)	10.3	53.0	6
AMOOS [2]	10.3	53.0	350
Transatmospheric space plane	7.8	30.5	0.5

regime. However, more of the trajectory occurs in a regime of chemical non-equilibrium and consequently effects of surface catalyticity will strongly influence the aerodynamic heating rates experienced.

With this wide diversity of vehicle types and flow regimes, the aerothermodynamic problems encountered vary considerably from type to type. Entry into planetary atmospheres other than the Earth's also requires a detailed knowledge of the relevant atmospheric physical properties. However, the essential principles involved in either re-entry prior to landing, or atmospheric encounter for orbital transfer, is that kinetic energy must be dissipated as heat in the surrounding gas. Table 8.1 provides a comparison of the specific kinetic energy $(\frac{1}{2}V_e^2)$ possessed by various space vehicle types at re-entry. Also shown are approximate values of the maximum specific energy flux $(\frac{1}{2}\rho_\infty V_\infty^3)$ encountered by the vehicle during its re-entry trajectory.

For a space vehicle returning to the Earth's surface from circular orbit (400 km) the kinetic energy per unit mass is 30.4 MJ/kg. The free stream energy flux for a vehicle of the Space Shuttle orbiter type, at a positon in the flight regime where peak aerodynamic heating occurs, is of order 2×10^7 W/m^2. Alternatively, we note that for a typical Space Shuttle orbiter re-entry mass of 8×10^4 kg and a re-entry duration of approximately 35 minutes, these quantities imply an average power dissipation rate during re-entry of order 1200 MW. Some of this power is dissipated as heat to the atmosphere and the remainder as heat to the vehicle.

The principal mechanisms which transfer heat to the atmosphere are the strong shock wave formed by the vehicle at hypersonic speeds and the viscous dissipation in the boundary layer and wake (see Figure 8.2). The shock wave increases the entropy of the atmosphere through which it propagates and influences a significantly greater cross-section of air than that in close proximity to the vehicle. The structure of the vehicle is heated by frictional dissipation and convection and, at higher velocities, by radiation from the high-enthalpy gas within the shock layer. Heat is also radiated away from the high-temperature vehicle structure to the surroundings. To maximize the ratio of the energy dissipated to the atmosphere to that to the vehicle structure requires the formation of the strongest

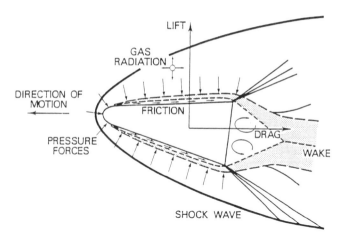

Figure 8.2   Blunt body hypersonic flow field features

shock wave with large lateral influence. This is accomplished by bluff body shapes for which the total heat load transmitted to the vehicle during re-entry will be a minimum. However, bluff shapes have poor aerodynamic lifting performance and their re-entry trajectories are determined principally by non-lifting ballistic considerations. A result is that the peak heating rates may be very high and ablative thermal protection systems are necessary. For slender vehicle shapes, by virtue of the weaker shock wave, a greater fraction of the total energy dissipated is transferred to the structure. This may not be disadvantageous since greater aerodynamic sophistication may be used to generate an efficient lifting shape which may glide and decelerate over relatively long periods at higher altitudes during re-entry, thereby reducing the peak heating rate below that for bluff non-lifting configurations. For slender lifting vehicles with low wing-loading the heating rates may be reduced to sufficiently low values that equilibrium between convective heat input and radiative heat loss can be maintained at temperatures low enough that metallic surface construction, without an ablating thermal protection system, can be tolerated.

Of central importance to all classes of re-entry vehicle or spacecraft involving atmosphere encounter is the thermal protection system. The particular type employed is a function of both the peak heating rate and the total heat load. The basic systems employed involve:

1. *heat absorption*, either by surface material through a rise in temperature, or by phase change, or chemical change, and

2. *heat rejection* of part of the heat input, by either mass efflux from the surface, or by thermal radiation from the heated surface.

The weight of the required thermal protection subsystem (TPS) is a vital consideration in assessing the performance of the complete spacecraft system. Assessed on an efficiency criterion of heat absorption per unit TPS weight, the direct use of thermally absorbing material is very inefficient in comparsion with phase or chemical changes in surface material, with vaporization or chemical reaction being rather better than melting. Ablative TPSs utilize vaporization with subsequent mass transfer to the vehicle boundary layer as the essential mechanism for thermal protection. The mass efflux into the boundary layer modifies the velocity and temperature gradients adjacent to the surface and reduces the incident heat flux. However, limitations on the total mass of ablative material which may be carried make this technique more suitable where the total heat load is of limited duration, as for rapid deceleration in the lower atmosphere.

8.2 EQUATIONS OF MOTION FOR PLANETARY ENTRY

To illustrate the essential physical principles which govern the flight mechanics of bodies encountering planetary atmospheres, the basic equations of motion of a point mass m with reference area S in motion about a spherical planet of radius R with angular rotation ω will be described. The formulation to be followed is that described by Vinh et al. [3] in which the vehicle position with respect to the centre of the planet is given in terms of the distance r, the angle of longitude and the angle ϕ of latitude. The vehicle moves at absolute velocity V in an atmosphere of density ρ along a path with an elevation angle γ and angle ψ of rotation northwards from the easterly local line of latitude to the projection of the vehicle's velocity vector on to the planet's surface (heading angle).

The vehicle may possess a bank angle σ and angle of attack α with respect to the trajectory. The components of the resultant of the aerodynamic and propulsive forces acting on the vehicle are given by F_T in the direction of the velocity vector $\mathbf{V}$ and F_N in the upwards plane of symmetry of the vehicle and perpendicular to the velocity vector $\mathbf{V}$.

Assuming that the atmosphere rotates with the planet at the same angular velocity ω about its axis, three equations of motion may be derived for the accelerations in the following directions [3]:

1. along the trajectory,

2. perpendicular to the tangent plane of the trajectory which passes through the centre of the planet, and

3. perpendicular to the trajectory in the tangent plane of the trajectory which passes through the centre of the planet.

The general equations of motion quoted by Vinh et al. may be simplified by assuming that the planetary, and hence atmospheric, angular velocity is small and that the Coriolis acceleration may be neglected. Although these assumptions are applicable to relatively short-range atmospheric encounters, the terms must be considered for long-range high-speed flight. With these assumptions, the equations may be written as

$$\frac{dV}{dt} = \frac{F_T}{m} - g \sin \gamma \tag{8.1}$$

$$V\frac{d\gamma}{dt} = \frac{F_N}{m}\cos \sigma - \left[g - \frac{V^2}{r}\right]\cos \gamma \tag{8.2}$$

$$V\frac{d\psi}{dt} = \frac{F_N \sin \sigma}{m \cos \gamma} - \frac{V^2}{r}\cos \gamma \cos \psi \tan \phi \tag{8.3}$$

In additon, for thrusting flight, the rate of change of mass of the vehicle is given by

$$\frac{dm}{dt} = -\frac{T}{c} \tag{8.4}$$

where c is a parameter characterizing the performance of the propulsion system and T is the thrust acting on the vehicle.

Equations (8.1)–(8.3) represent three coupled nonlinear equations which may, in general, be solved by step-by-step numerical integration once expressions for the aerodynamic coefficients, the atmospheric density and gravitational acceleration variation with altitude are given. The aerodynamic coefficients of lift and drag are given by

$$C_L = C_L(\alpha, M, Re)$$

$$C_D = C_D(\alpha, M, Re)$$

although at hypersonic Mach numbers they are only a weak function of M. The variation of the gravitational acceleration with altitude is given by

$$g = g_0 \left(\frac{r_0}{r}\right) \tag{8.5}$$

where g_0 is the gravitational acceleration at a reference radius r_0 from the centre of the planet.

Although for greatest precision numerical integrations require the use of data describing the accurate atmospheric density variation with altitude (see for example the US 1962 model Earth atmosphere [4]) useful results can be obtained from the assumption of density varying exponentially with altitude as given by

$$\rho = \rho_s e^{-\beta h} \tag{8.6}$$

where ρ_s is the atmosphere density at the planet's surface, h is the height above the surface given by

$$h = r - R \tag{8.7}$$

and β is the reciprocal of the scale height, which may be considered approximately constant.

Numerical integrations of the equations (8.1)–(8.4) can provide useful representations of both the launch trajectory (powered) and re-entry trajectory of lifting space launch vehicles (zero thrust) once the initial conditions have been specified.

8.3 BALLISTIC ENTRY AT LARGE ANGLES OF DESCENT

8.3.1 Aerodynamic deceleration for ballistic entry

For ballistic entry, the only aerodynamic force is assumed to be drag ($C_L = 0$) which is usually large compared with the vehicle's weight. If gravity force is ignored the result is that the trajectory remains straight. With these assumptions the equations of motion become

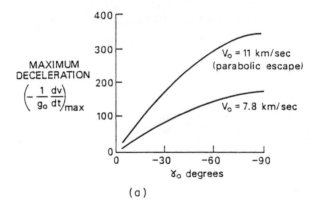

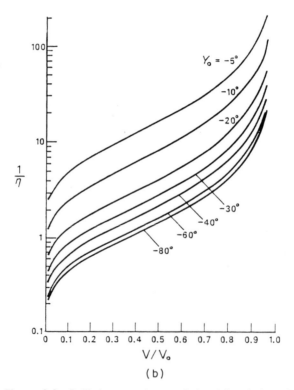

Figure 8.3 Ballistic entry characteristics: (a) variation of
the maximum deceleration with entry angle γ_0 and entry
velocity V_0 for ballistic entry; (b) ballistic entry at large flight
path angle: speed–altitude diagram for several values of the
initial angle (after Vinh *et al.* [3]) (reproduced by permission of The University of Michigan); (c) ballistic entry at
large flight-path angle: acceleration—altitude diagram for
several values of the initial angle (after Vinh *et al.* [3])
(reproduced by permission of The University of Michigan)

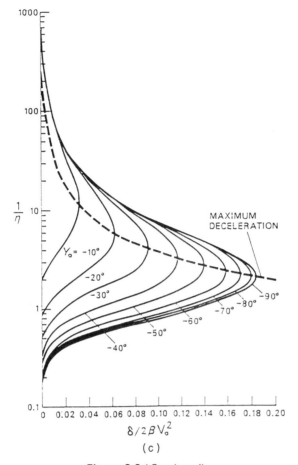

Figure 8.3 (*Continued*)

$$\frac{\mathrm{d}V}{\mathrm{d}t} = -\frac{\rho S C_D V^2}{2m} = -\eta \beta V^2 \qquad (8.8)$$

$$\frac{\mathrm{d}r}{\mathrm{d}t} = \frac{\mathrm{d}h}{\mathrm{d}t} = V \sin \gamma \qquad (8.9)$$

where a dimensionless height variable η is introduced which is defined by

$$\eta = \frac{\rho S C_D}{2m\beta} = \frac{\rho_s S C_D}{2m\beta} e^{-\beta h} \qquad (8.10)$$

The deceleration δ may be written as

$$\delta = -\eta \beta V_0^2 \exp(2\eta/\sin \gamma_0)$$

whence the maximum deceleration has the value

$$\delta_{\max} = \frac{\beta V_0^2}{2e} \sin \gamma_0 \qquad (8.11)$$

Table 8.2 Maximum deceleration (δ/g_0) for ballistic re-entry into planetary atmospheres at escape and low orbital velocities with $m/SC_D = 50$ kg/m² (after Gazely [5]).

Planet	Maximum deceleration (δ/g_0)					
	Entry at escape velocity			Entry at orbital velocity		
	$\gamma_0 = -5°$	$\gamma_0 = -20°$	$\gamma_0 = -90°$	$\gamma_0 = -5°$	$\gamma_0 = -20°$	$\gamma_0 = -90°$
Venus	28.6	112	326	14.3	56	163
Earth	28.3	111	324	14.2	55.5	162
Mars	1.6	6.3	18.3	0.8	3.2	9.2

where the velocity is

$$V = V_0 \exp(-\tfrac{1}{2}) = 0.607V_0$$

and the altitude is given by

$$\therefore (h)_{\delta_{max}} = \frac{1}{\beta} \ln\left(-\frac{\rho_s SC_D}{m\beta \sin \gamma_0}\right) \tag{8.12}$$

Equation (8.11) shows the maximum deceleration to be independent of drag coefficient and shape but eqn (8.12) shows that the altitude at which it occurs is dependent on the ballistic coefficient (m/SC_D); the higher the values of m/SC_D, the lower the altitude at which peak deceleration occurs. The maximum deceleration is a strong function of both entry angle and entry velocity as shown in Figure. 8.3(a). Generalized graphs for the variation of the velocity ratio and the deceleration are given in Figure 8.3(b) and (c). Table 8.2 gives values of the peak deceleration experienced in the atmospheres of Venus and Mars in comparison with those for Earth atmosphere re-entry.

8.3.2 Aerodynamic heating for ballistic entry

The essential features of aerodynamic heating relevant to planetary atmospheric entry were first considered by Allen and Eggers [6]. A typical expression for the convective heating rate $\dot{q}$, for example, is

$$\dot{q} = k'\rho^{1/2}V^3 \tag{8.13}$$

where, as shown by Lees [7], for stagnation point heating the quantity k' varies inversely with the square root of the nose radius of curvature and, more generally, is a function of the flow regime, the boundary layer state and the transport properties of the atmosphere.

Using the equations of motion the maximum heating rate is given by

$$\dot{q}_{max} = k'\left(\frac{m\beta \sin \gamma_0}{3SC_D}\right)^{1/2} V_0^3 \exp(-\tfrac{1}{2}) \tag{8.14}$$

when

$$V = V_0 \exp\left(-\frac{1}{6}\right) = 0.85V_0 \tag{8.15}$$

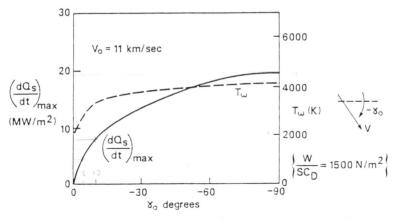

Figure 8.4 Maximum stagnation point heating rate and equilibrium wall temperature variation with entry angle

which occurs when the altitude is given by

$$(h)_{\dot{q}_{max}} = \frac{1}{\beta} \ln\left(\frac{-3\rho_s SC_D}{m\beta \sin \gamma_0}\right) \tag{8.16}$$

Unlike the peak deceleration, the maximum heating rate is shown by equation (8.16) to be dependent on drag coefficient, body shape (through k') and the ballistic coefficient. Equation (8.14) demonstrates that reductions in peak heating rate may be obtained by adopting high-drag, large-nose radius shapes with low ballistic coefficients. It is also noted that the altitude at which peak heating occurs is greater than that for maximum deceleration.

Heating rates typical of those for a blunt ballistic entry shape entering the Earth's atmosphere at 11 km/s are shown in Figure 8.4 together with estimates of the equilibrium wall temperature assuming radiative heat loss. Table 8.3 gives values of the equivalent radiation equilibrium temperature for various entry angles into Venusian and Martian atmospheres in comparison with those for Earth atmosphere re-entry.

Table 8.3 Maximum heating rate expressed as an equivalent surface radiation equilibrium temperature (K) at escape velocity for ballistic entry into planetary atmospheres with $m/SC_D = 50$ kg/m² (after Gazely [5]).

Planet	Peak radiation equilibrium temperature (K) Entry at escape velocity		
	$\gamma_0 = -5°$	$\gamma_0 = -20°$	$\gamma_0 = -90°$
Venus	2600	3050	3500
Earth	2800	3300	3800
Mars	1300	1550	1800

8.4 LIFTING ENTRY

In the early development of recoverable space vehicles, atmospheric entry was usually accompanied by purely ballistic or near ballistic trajectories in which very little use was made of aerodynamic lift. Currently lifting re-entry vehicles, of which the Space Shuttle orbiter is typical, permit greater trajectory flexibility in terms of manoeuvrability and lower heating rates together with controlled horizontal landing capability. Synergetic plane change and aero-assisted orbital transfer vehicles also require the significant use of aerodynamic lift during atmospheric encounters.

The exact treatment of lifting re-entry requires the numerical solution of the equations of motion of planar entry trajectories.

Although first-order solutions have only limited applicability, the essential physical features of lifting entry, in comparison with ballistic entry, may be illustrated by them. As an example, the results of first-order solutions for gliding entry at constant large lift-to-drag ratio and small trajectory inclination angles will be considered.

Figure 8.5 shows the variation in velocity along equilibrium glide trajectories for various values of $W/(SC_L)$ and demonstrates the higher altitudes at which entry takes place for vehicles with low wing loading (W/S) and high lift coefficient (C_L). The corresponding flight time along equilibrium glide paths is shown in Figure 8.6.

Although these first-order results are over-simple for application to accurate trajectory calculations they do include the basic features of lifting re-entry, i.e. the capability for lower maximum deceleration than experienced in non-lifting ballistic entry. As presented, they are limited to small flight-path angles. For medium inclination angles $(3° < \gamma < 15°)$, medium lift-to-drag ratios $(0.025 < L/D < 1)$ and entry at near orbital velocity $(V_0 \simeq (gr_0)^{1/2})$, the first-order solution of Lees et al. [8] is more appropriate. Values of the peak

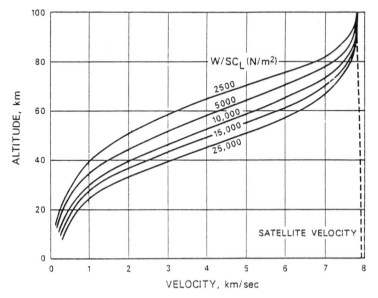

Figure 8.5 Altitude–velocity diagram for equilibrium glide paths from polar entry (Reprinted with permission © 1990 Society of Automotive Engineers, Inc.)

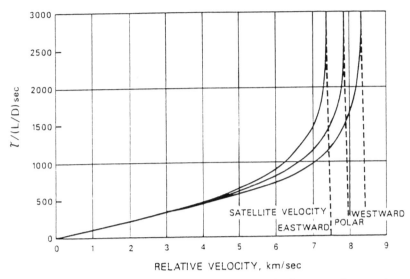

Figure 8.6 Flight time along equilibrium glide paths for entry from various orbital inclinations (Reprinted with permission © 1990 Society of Automotive Engineers, Inc.)

deceleration experienced by a lifting re-entry vehicle obtained using Lees' method are shown in Figure 8.7 and demonstrate the substantial reductions in maximum deceleration obtainable. The aerodynamic heating rates in gliding re-entry may be approximately estimated using first-order results.

The maximum heating rate $\dot{q}_{max}$ may be shown to be of magnitude

$$(\dot{q})_{max} = k' \frac{2}{3\sqrt{3}} g r_0 \left(\frac{2mg}{C_L S}\right)^{1/2} \qquad (8.17)$$

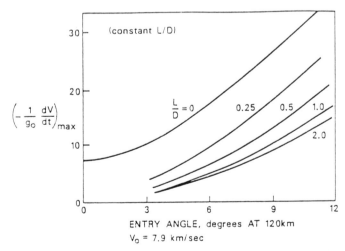

Figure 8.7 Variation of maximum deceleration with entry angle for various lift–drag (*L/D*) ratio (reproduced by permission from Loh, W.H.T. (1968) *Re-entry and Planetary Entry*, Springer-Verlag)

Table 8.4 Comparison of maximum heating rates for ballistic and shallow lifting entry into planetary atmospheres expressed as equivalent surface radiation equilibrium temperatures ($m/SC_D = 50$ kg/m²) (after Gazely [5]).

| Planet | Peak radiation equilibrium temperature (K) | | | | | |
| | Ballistic entry at near orbital speed | | | Shallow lifting entry at near orbital speed | | |
	$\gamma_0 = -5°$	$\gamma_0 = -20°$	$\gamma_0 = -90°$	$L/D = 1$	$L/D = 2$	$L/D = 5$
Venus	2050	2300	2700	1250	1150	1050
Earth	2150	2550	2890	1350	1250	1100
Mars	1050	1200	1200	800	750	650

For lifting re-entry, the maximum heating rate is reduced by using vehicles with low wing loading (W/S) and high maximum lifting capability (C_L). This implies a gliding entry at lower values of local density, and hence greater altitude.

The design of vehicle configurations to meet the requirement for the highest values of $C_{L_{max}}$, in order to reduce the peak heating rates and thereby minimize thermal protection system mass, has been addressed by Townend [9, 10]. Recessed under-surface shapes which support strong lifting surface shock waves and which minimize gas spillage from the lifting surface are potential candidates to meet the criterion for high $C_{L_{max}}$.

The effect of the reductions in peak heating resulting from lifting entry are demonstrated in Table 8.4, in which the equivalent surface equilibrium temperatures are compared for ballistic and lifting entry into planetry atmospheres.

Optimum lifting re-entry trajectories usually employ an initially high angle of attack (40–50°) flight regime during which peak heating occurs, followed by a hypersonic glide phase at lower angles of attack. Trajectory control is achieved by pitch and roll modulation.

8.5 THERMAL PROTECTION SYSTEMS (TPS)

Figure 8.8 (after Loh [11]) demonstrates the combination of peak aerodynamic heating and exposure time to which various generic classes of Earth atmospheric flight vehicles are subject. On the assumption that a balance is achieved between convective and radiative (at higher velocities) heat input and radiative heat loss from the surface, equilibrium surface temperatures are also given in Figure 8.8 and typical materials required to withstand these temperatures are quoted. For vehicles entering the Earth's atmosphere from LEO the heating arises predominantly from convective sources. For higher velocities, typical of those for AOTVs changing from GEO to LEO, or for vehicles returning from planetary missions, radiative heating from the gas within the shock layer becomes of considerable importance. Boundaries for radiative heating at which the fraction of the incident energy flux ($\frac{1}{2}\rho V^3$) which is transferred to the vehicle by radiation reaches 0.5%, are shown in Figure 8.1. The increase in radiative heat transfer with increase of velocity is shown in Figure 8.9 (after Howe [12]) for the assumptions of both chemical equilibrium and non-equilibrium in the shock layer. Considerable enhancement of the radiative heat

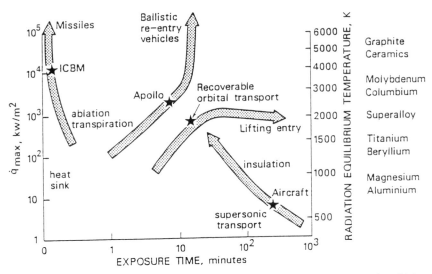

Figure 8.8 Variation of heating rates and exposure times for various generic vehicle types (reproduced by permission from Loh, W.H.T. (1968) *Re-entry and Planetary Entry*, Springer-Verlag)

transfer by non-equilibrium radiation can occur as demonstrated by Park [13]. Predictions of equilibrium and non-equilibrium radiative heating have been made by Menees [14] for both a Titan aerocapture vehicle and for two near-Earth AOTVs. For both applications it is concluded that non-equilibrium effects enhance radiative heating, which generally exceeds convective heating. For entry velocities in excess of about 12 km/s the combined effects of radiative and convective heating on the heat load experienced by the vehicle may reach about one fifth of the local vehicle kinetic energy.

The design of suitable thermal protection systems (TPS) to accommodate the heat loads

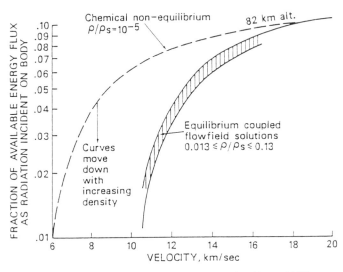

Figure 8.9 Radiative flux limitation (after Howe [12])

imposed by entry is of considerable importance in systems studies of potential vehicle candidates proposed for atmosphere encounters. The principal types of system used depend on either, or both, of the essential physical principles of

1. absorption of heat by surface material by temperature rise, phase change or chemical change, and

2. rejection of heat by mass efflux from the surface and/or surface radiation.

These takes the following forms:

(a) use of vehicle structure as a heat sink,

(b) radiative cooling from high-temperature structural materials,

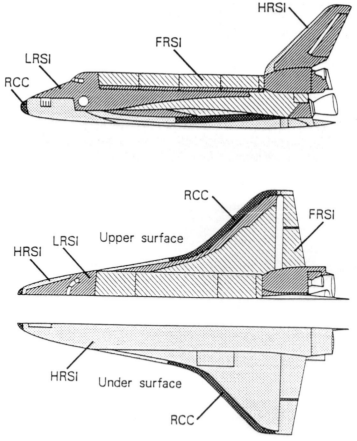

Figure 8.10   Space Shuttle Orbiter: thermal protection systems.
RCC: reinforced carbon–carbon.
HRSI: high-temperature reusable surface insulation.
LRSI: low-temperature reusable surface insulation.
FRSI: flexible reusable surface insulation

Table 8.5 Space Shuttle orbiter data

Thermal protection system (TPS) mass	7861 kg	(17 332 lb)
Thermal protection system area	1103 m²	(11 874 ft²)
Landing mass	82 275 kg	(188 000 lb)
Empty mass	68 040 kg	(150 000 lb)
Gross wing area	249.9 m²	(2690 ft²)
TPS/landing mass ratio	9.2%	
Re-entry wing loading	341 kg/m²	(69.9 lb/ft²)

(c) radiative cooling from the surface of insulating materials supported on relatively cool and conventional structures.

(d) ablative cooling materials supported on insulation, and

(e) transpiration cooling.

Each imposes a mass penalty on the vehicle and the important parameter which determines the effectiveness of the system is the heat absorption per unit mass of the TPS.

A reduction in the rate of heat conduction $\dot{q}_k$ into the vehicle structure can be effected by the use of surface insulation materials. Suitable surface coatings can optimize the radiative cooling properties and the thickness and type of the insulation can be matched to the local requirement at various positions on the vehicles. The TPS of the Space Shuttle orbiter is typical of this type, which is shown schematically in Figure 8.10. Supporting data for the mass and area of the TPS for the Shuttle Orbiter are given in Table 8.5.

With a TPS mass in excess of 8% of the landing weight, conservative design due to uncertainties in aerodynamic heating predictions can impose severe mass penalties.

Ablative heat shields possess a much higher energy-absorption efficiency than simple metallic heat sinks, although having the obvious disadvantage of non-reusability. Heat is primarily absorbed in vaporizing the surface material. Mass transfer of the vapour into the boundary layer perturbs the velocity and temperature profiles in a favourable manner which further reduces the heat transfer rate. A simple model of ablating surface protection is shown in Figure 8.11.

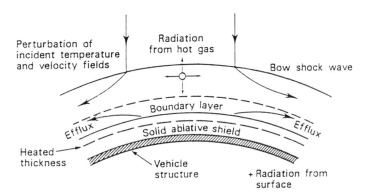

Figure 8.11 Schematic diagram of ablating surface protection system

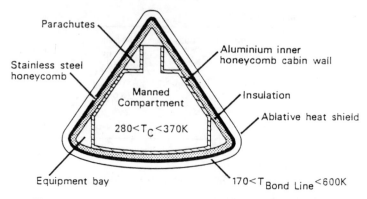

Figure 8.12 Apollo entry module thermal protection system

Figure 8.12 shows the Apollo re-entry module, together with the ablative TPS which was necessary to provide protection from the thermal loads encountered during the atmospheric entry from the lunar mission. The mass penalty associated with the TPS is well demonstrated and the overall system performance gains which can be obtained from increases in TPS efficiency are considerable.

8.6 THE ENTRY CORRIDOR

The final consideration in this chapter is concerned with the limiting boundaries which constrain the concept of safe entry into planetary atmospheres. Earlier in the chapter simple analyses have been presented for determining the deceleration and aerodynamic heating as functions of entry velocity and trajectory inclination. When a limit has been set for the maximum tolerable g for the spacecraft structure and its payload (Figure 8.13

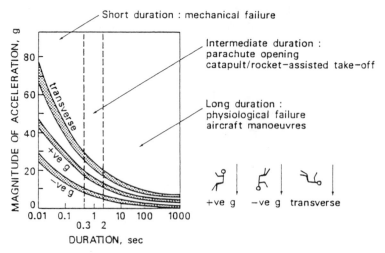

Figure 8.13 Permissible g-loading on man (reproduced by permission from Loh, W.H.T. (1968) *Re-entry and Planetary Entry*, Springer-Verlag)

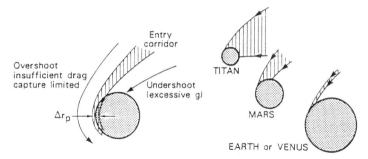

Figure 8.14 Definition of the entry corridor and comparison of entry corridors for Titan, Mars, Earth and Venus atmospheric entries (reproduced by permission from Loh, W.H.T. (1968) *Re-entry and Planetary Entry*, Springer-Verlag)

shows the permissible g-loading for man) or the maximum aerodynamic heating, this determines the maximum entry angle which can be tolerated for a given entry velocity. For angles greater than this, excessive g, or excessive heating, will result and such trajectories are described as 'undershoot' trajectories. Conversely, for sufficiently small entry angles the deceleration experienced in the atmospheric encounter may be too low for the vehicle to be captured within the atmosphere and it exits the atmosphere to complete a further elliptic orbit before a further atmospheric encounter. For lifting vehicles, atmosphere exit after the first entry may occur during skip trajectories but, in this case, re-entry into the atmosphere takes place before completion of a further orbit. All trajectories having entry angles below the limiting value are designated as 'overshoot' trajectories.

Using these concepts Chapman [15] has defined an entry corridor as shown schematically in Figure 8.14. The limiting overshoot and undershoot trajectories are shown as solid lines and hypothetical continuations of these trajectories in the absence of a planetary atmosphere are shown as dotted extensions. The periapsis of each of the hypothetical trajectories defines an entry corridor with Δr_p given by

$$\Delta r_p = r_{p_{\text{overshoot}}} - r_{p_{\text{undershoot}}}$$

where r_p is the hypothetical conic perigee.

Table 8.6 Entry corridor widths in kilometres for entry into planetary atmospheres at a velocity of $(2gr_0)^{1/2}$.

Planet	Corridor width (km)					
	5 g limit			10 g limit		
	$L/D = 0$	$L/D = 1$	$L/D = 1$ modulated	$L/D = 0$	$L/D = 1$	$L/D = 1$ modulated
Venus	0	43	58	13	84	113
Earth	0	43	55	11	82	105
Mars	338	482	595	644	885	1159
Jupiter	0	55	68	0	84	113

The widths of the entry corridors (Δr_p) for non-lifting ($L/D = 0$) and lifting ($L/D = 1$) entry into various planetary atmospheres, as calculated by Chapman [14], are given in Table 8.6 for entry at parabolic velocity ($V_0 = (2gr_0)^{1/2}$).

REFERENCES

[1] Walberg, G.D. (1985) A survey of aeroassisted orbit transfer *J. Spacecraft and Rockets*, **22** (1), 3–18.
[2] White, J. (1974) Feasibility and trade off study of an aeromanoeuvring orbit-to-orbit shuttle (AMOOS). Lockheed Missiles and Space Co., Huntsville, Ala., NASA CR-130431. ,
[3] Vinh, N.X., Busemann, A., and Culp, R.D. (1980) *Hypersonic and Planetary Entry Flight Mechanics*, University of Michigan Press.
[4] *US Standard Atmosphere, 1962* (1962) US Government Printing Office, Washington DC.
[5] Gazely, C. (1960) *Atmospheric Entry*. The RAND Corporation Report, P-2052.
[6] Allen, H.J., and Eggers, A.J. (1958) A Study of the Motion and Aerodynamic Heating of Missiles Entering the Earth's Atmosphere at High Supersonic Speeds. NACA TR 1381.
[7] Lees, L. (1956) Laminar heat transfer over blunt-nosed bodies at hypersonic flight speeds, *Jet Propulsion*, **26** (4) 259–69.
[8] Lees, L., Hastwig, F.W., and Cohen C.B. (1959) Use of aerodynamic lift during entry into the Earth's atmosphere, *American Rocket Society Journal*, September, 633–41.
[9] Townend, L.H. (1972) Some design aspects of space shuttle orbiters, *Progress in Aerospace Science*, **13**.
[10] Townend, L.H. (1979) Research and design for lifting re-entry, *Progress in Aerospace Science*, **19**, 1–80.
[11] Loh, W.H.T. (1968) *Re-entry and Planetary Entry*, Springer-Verlag.
[12] Howe, J.T. (1985) Introductory aerothermodynamics of advanced space transportation systems, *J. Spacecraft and Rockets*, **22** (1), 19–26.
[13] Park, C. (1985) Radiation enhancement by non-equilibrium in Earth's atmosphere, *J. Spacecraft and Rockets*, **22** (1), 27–36.
[14] Menees, G.P. (1985) Trajectory analysis of radiative heating for planetary missions with aerobraking of spacecraft, *Spacecraft and Rockets*, **22** (1), 37–45.
[15] Chapman, D.R. (1959) *An Approximate Analytical Method for Studying Entry into Planetary Atmospheres*. NASA Technical Report R-11.

9 SPACECRAFT STRUCTURES

R.F. Turner

Rutherford Appleton Laboratory

9.1 INTRODUCTION

Spacecraft design owes most of its generic development to the aircraft industries which fostered it. The principal factor driving structural design is that of minimizing mass. It must be extremely efficiently used and utterly reliable. The prime requirement is for the minimum structure which can achieve its goals—during the dynamic loading with which it is presented during the testing and launch phases and finally in the zero-gravity operational environment. What makes spacecraft structural engineering perhaps unique is that its goals are strongly dependent on other subsystems such as thermal design, attitude control, communications and power.

Structural design does not only encompass materials selection and configuration but must also include analysis and verification testing as part of the process, with an increasing reliance being placed upon analytical methods as experience grows.

9.2 DESIGN PHILOSOPHY

Spacecraft Design is an iterative process which starts with the initial concept and nominally ends at a predetermined freeze date at the start of the construction phase. (In practice the active role of design tends to end only shortly before shipment to the launch site.) With such multi-disciplinary projects it is convenient to specify a number of formal design reviews to assess the interaction of all the separate design factors and subsystems. This activity (see Figure 9.1) must proceed rapidly over very complex technical (and possibly geographical) interfaces and benefits considerably from the use of sophisticated computer-aided design methods.

The first step in structural design is to establish a specification based on the mission requirements with as much detail as can be expected at the concept stage. This may be quite mission specific, contrasting the thermal dissipation requirements of a communications satellite with the precision requirements of large antennas or telescopes. A gener-

Spacecraft Systems Engineering. Edited by P.W. Fortescue and J.P.W. Stark

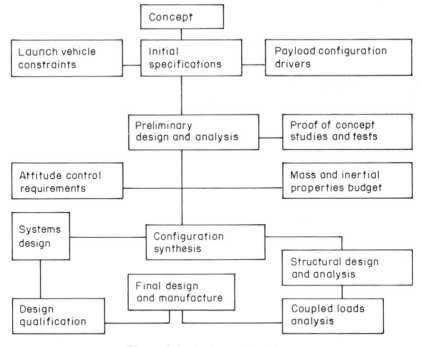

Figure 9.1   Design methodology

alized list of requirements for structures is given in Figure 9.2. At this stage the configuration will dominate as the distribution of masses becomes established. The concept of a load path from the interface with the launch vehicle (from which all accelerations are imparted) through the spacecraft structure to the mounting points of individual systems or units is followed. At each mounting point individual sets of interface requirements (alignment, thermal, field of view, screening, connections and accessibility) will be generated, thus setting some of the constraints to be applied to the design of the structure.

Thus far, the problem of providing a structure to meet the specification could be regarded as relatively straightforward. However, it is the need for extreme mass efficiency and reliability which presents a challenge. The selection of materials is often dominated by stiffness and structural dynamics rather than stress levels.

Testing as a means of validation is an inherent part of the design, by full test, or by limited testing supported with analysis and modelling. The final design stage requires a

```
* Accommodation for the payload and spacecraft systems
* Withstand launch loads
* Stiffness
* Provide environmental protection
* Alignment
* Thermal and electrical paths
* Accessibility
```

Figure 9.2   Requirement for a spacecraft structure

coupled loads analysis which combines the characteristics of the full assembly of the spacecraft with those of the launch rocket.

In the case of large structures such as the Space Station, reconfigurations will occur on orbit during its 25-year lifetime, well beyond the initial concept. Although launch accelerations are no longer a problem, very low-frequency dynamic characteristics during attitude manoeuvres and orbit boosting will present an additional design condition.

9.3 MATERIALS SELECTION

9.3.1 Properties of materials

The selection of an appropriate material for an application requires a knowledge of the way each property can best be used and where each limitation must be recognized.

Selection criteria can encompass the following:

- specific strength
- stiffness
- stress corrosion resistance
- fracture and fatigue resistance
- thermal parameters
- sublimation/erosion
- ease of manufacture/modification

A simple criterion based upon strength/density is not sufficient. For example, the directional characteristics of composites must be recognized, and also the low fracture resistance of some materials when they are considered as alternatives to metals. Glass is an extreme case.

Specific strength (f_y/ρ)

An elementary comparison of materials can be made by examining their proof strengths (load-carrying capability within a given fraction—usually 0.1 or 0.2%—of elongation at the elastic limit) with respect to density (see Figure 9.4, column G).

For metals, titanium alloys show up well but would have to be used in very thin sections to realize their advantage and could then be prone to buckling.

Stiffness (*the relationship between load and deformation*)

For the three common load cases shown in Figure 9.3 the stiffness has been expressed in terms of the material properties (Young's modulus and density). This then allows an examination and selection of materials to be made [1]. For example, in Figure 9.4 columns D, E, and F show the efficiencies of selected materials for the different load cases. Notice

Case	Deflection of a beam	Buckling of a strut	Buckling of a panel (w fixed) (t and L variable)
Characteristic equations	$\delta = \dfrac{PL^3}{3EI}$ $I = \dfrac{wt^3}{12}$ Weight $= L.w.t.\rho$	$P_e = \dfrac{\pi^2 EI}{L^2}$ $I = \dfrac{\pi d^4}{64}$ Weight $= \dfrac{\pi d^2}{4} L\rho$	$P = \text{const.}\dfrac{E}{1-\sigma^2}\left(\dfrac{t}{w}\right)^2 \cdot tw$ Weight $= w.t.L.\rho$
Structural efficiency $=\dfrac{\text{load}}{\text{weight}}$	$\dfrac{P}{WT} = \text{const.}\left(\dfrac{\delta t^2}{L^4}\right)\cdot\dfrac{E}{\rho}$	$\dfrac{P}{WT} = \text{const.}\left(\dfrac{P_e}{L^4}\right)^{1/2}\dfrac{E^{1/2}}{\rho}$	$\dfrac{P}{WT} = \text{const.}\left(\dfrac{P_e}{L^2}\right)^{2/3}\dfrac{E^{1/3}}{\rho}$
Structure loading coefficient	$\left(\dfrac{\delta t^2}{L^4}\right)$	$\left(\dfrac{P_e}{L^4}\right)^{1/2}$	$\left(\dfrac{P}{L^2}\right)^{2/3}$
Material efficiency criterion	$\dfrac{E}{\rho}$	$\dfrac{E^{1/2}}{\rho}$	$\dfrac{E^{1/3}}{\rho}$
Nomenclature	P = Load P_e = Critical buckling load (Euler theory) δ = Deflection L = Length σ = Poisson's ratios (0.3)	d = Diameter w = Width WT = Weight	ρ = Density t = thickness E = Young's modulus

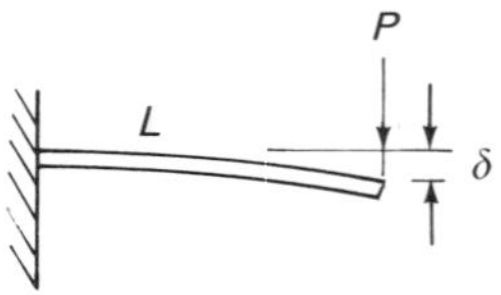

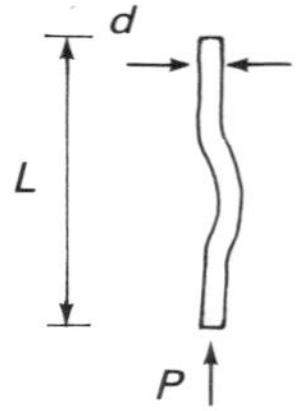

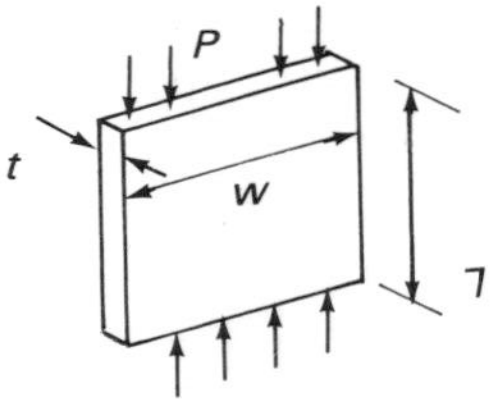

Figure 9.3 Materials selection criteria—stiffness

C.R.   corrosion resistance
S.C.C.   stress corrosion cracking

Caution: There is considerable variation of the properties of materials according to conditions (ageing, temper, form and structure orientation). Consult manufacturers data

Material	Density (kg/m³)	Young's modulus E (GPa)	Yield strength f (MPa)	Selection criteria $\dfrac{E}{\rho}$	$\dfrac{E^{1/2}}{\rho}$	$\dfrac{E^{1/3}}{\rho}$	$\dfrac{f_y}{\rho}$	Thermal expansion (μm/m K^{-1})	Fracture toughness (MPa m)	Fatigue strength (MPa)	Comment
Aluminium alloy											
6061 . T6	2800	68	276	24	2.9	1.5	98.6	23.6	186	97	Good C.R.
7075 . T6	2700	71	503	26	3.1	1.5	186.3	23.4	24	159	Prone to SSC in T6 Form
Magnesium alloy											
A2 31 B	1700	45	220	26	3.9	2.1	129.4	26			Prone to SCC
ZK 60 A.T5 Extrn	1700	45	234	26	3.9	2.1	137.6	26		124	
Titanium alloys											
T1 – 6A1 – 4V (annealed)	4400	110	825	25	2.4	1.1	187.5	9	75	500	
(solution treated and aged)			1035								
Beryllium alloys											
S 65 A	2000	304	207	151	8.7	3.4	103.5	11.5	42	690	Hot pressed sheet ⎱ Low fracture toughness
S R 200 E			345								
Ferrous alloys											
INVAR		150	275/415					1.66			Low expansion Ferromagnetic
Stainless steel											
AM 350 (SCT850)	7700	200	1034	26	1.84	0.8	134.3	11.9	40/60	550	Austentic
304L Ann	7800	193	170	25	1.8	0.7	21.8	17.2			
Composites											
KEVLAR 49 0° (Aramid fibre)	1380	76*	1379†	55	6.3	3.1	999.3	–4			Structure members Pressure vessels Rocket casings
90°	1380	5.5	29.6	4	1.7	1.3	21.4	57			
Graphite epoxy Sheets (undirectional) GY70/934	1620	282	586	174	10.4	4.0	361.7	–11.7 (Longitudinal) 29.7 (Transverse)			Sheet
Column ref. (See text)	A	B	C	D	E	F	G	H	I	J	

* Tensile modulus
† Tensile strength

Figure 9.4   Sample materials properties

how similar the efficiencies for titanium and aluminium alloys are, despite the greater specific strength of the former. Ashby and Jones give an explicit treatment of this subject together with illuminating case studies [2].

Materials show variations due to treatment and supply state, and therefore any application considered should be accompanied by a proper detailed examination of the properties from specialist sources.

Consideration needs to be given to the availability of materials, difficulties of manufacture, the method by which fasteners are attached and how late modifications would be conducted.

Stress corrosion resistance

In a terrestrial environment, most alloys are susceptible to stress corrosion cracking (SCC). Material under static or dynamic loading may exhibit minor cracking at grain boundaries, and in the presence of a corrosive medium or moisture these cracks can develop.

The strength of materials can be adversely affected prior to launch unless precautions are taken such as:

- Choose alloys less susceptible to SCC.

- Heat treat components to remove residual stresses due to manufacturing processes.

- Avoid material combinations which promote galvanic corrosion.

- Avoid exposure to atmospheric conditions which can induce corrosion.

Fracture and fatigue resistance

Column I of Figure 9.4 gives an indication of the fracture resistance of some materials, and in column J fatigue resistance is compared (see Section 9.3).

It is generally accepted that microcracking exists in all structures to some degree. Therefore it is incumbent upon the designer to demonstrate that under foreseeable circumstances failure due to such flaws in the material cannot occur. As structural efficiency steadily improves and safety factors are reduced, so fracture resistance becomes more important. Essentially, a crack tip is a point of stress concentration. If this region of stress rises above a certain level the crack will progress. The whole essence of fracture mechanics is coupled to strain energy, stress concentration and material state [3].

For a given stress level there is a critical crack length below which crack propagation will not occur, and above this the crack will propagate with increasing rapidity towards total failure of the section. It is therefore the job of the analyst to demonstrate:

1. by calculation, that the design and material can sustain non-catastrophic cracking up to a certain size;

2. by nondestructive testing (NDT) techniques that no such cracks exist prior to launch;

3. that in the case of manned flight one or more additional load paths exists to carry load in cases of failure, *or* that the failed item can be contained in such a manner that no primary function or structure is compromised.

Two issues arise from such judgement. If a critical crack length is too small to be readily inspectable, the factor of safety in that area must be increased. Further, a considerable knowledge of the local stress concentrations under all possible load conditions must be assumed—this leads to the application of finite element analysis (FEA) to the field of fracture machanics.

Particularly in the field of manned spacecraft, NASA will require that all designs are submitted with a fracture mechanics assessment. This has to demonstrate that the design is tolerant of such failures, either that another path can accommodate the additional load (in fact at least two more are often required) or that in the case of subsystems the failed part is contained, without danger, inside some acceptable housing.

Thermal parameters

A spacecraft will experience extremes of thermal loading from direct solar heating to viewing deep space and yet its mission may require the maintenance of focal lengths to very fine tolerances. Figure 9.4 column H compares some expansion coefficients. Note Invar and graphite epoxy.

Sublimation/erosion

Normally the sublimation of metals does not pose any major problems in the space environment, although for thin films the rate at which their thickness decreases may be significant. Erosion by atomic oxygen, particularly in respect of polymeric materials in low Earth orbit can be a significant factor in materials selection [4]. This is a subject of investigation and research in view of the possible long-term problems which could ensue in the 25-year life of a space station.

Ease of manufacture/modification

Some materials present individual problems during component manufacture; some forms of beryllium and the cutting fluids used for others present toxic material control problems which preclude late modifications in unsuitable facilities.

It is not unusual for additional mounting points or cut-outs for cable runs, clearances, etc. to be needed at a stage which precludes parts replacement. Thin sections may not support the modifications required. Fibre lay-ups may not allow holes to be cut which interrupt the homogeneity of the load-carrying path.

9.3.2 Materials' characteristics

Ferrous alloys have numerous applications in which their properties of high strength, corrosion resistance and toughness are required.

Austenitic stainless steels are used for propulsion and cryogenic systems because of their excellent low-temperature toughness. Other alloy uses are in optical and precision structures where properties can be selected to match expansion criteria in a dynamic thermal environment.

Susceptibility to hydrogen embrittlement is a potential hazard for ferrous alloys, particularly where they have been treated in plating solutions. The result is similar to SCC. The corrective treatment is a severe bake-out within a limited time period—observing of course that the material properties are not affected.

Some types of stainless steel and Invar cannot be considered as nonmagnetic. This can be quite a problem when designing magnetically clean spacecraft.

Beryllium alloys are shown in Figure 9.4 to have some remarkable properties. However, they can be difficult to machine, and great care must be taken due to their toxicity. Also, they are very expensive and their use is therefore limited to specific specialist applications or to areas where weight control reaches levels of extreme concern.

Composite materials (fibre-reinforced)

A designer is now able to tailor materials to give properties in chosen directions, much as marine designers have been doing with wood and resin-bonded plywood for a long time. In the use of fibre-reinforced materials (typically with boron or carbon) advantage can be taken of the high strength offered along the fibre. Many strands of the chosen fibre are aligned parallel and held in an epoxy matrix. The resulting material exhibits high structural properties along the direction of the fibres but is limited to the properties of the matrix (glue) in other directions. However, another sheet of the composite can be aligned at an angle to the first, and in many-layer build-ups it is possible to create a material with structural properties which are tailored to the application. The failure mode of a composite is predominantly delamination—particularly in buckling—but a composite which has suffered this defect may still exhibit its full tensile properties.

Carbon–epoxy materials are finding considerable application in the fabrication (by filament winding) of struts—where the direction of loading is well defined. The structural and thermal properties combined with the light weight of the material offer significant advantages. Titanium end-fittings are usually employed in such struts. Even thermal expansion characteristics can be 'designed in' for use in precision structures such as telescopes and spectrometers.

A note of caution, particularly with carbon-fibre-based composites, is that hygroscopic absorption can add water up to 2% by weight in a normal atmosphere. Once exposed to the space environment they lose water and exhibit small dimensional changes. Methods of controlling this are being investigated and include total control of the environment in the components' lifetime, baking out prior to alignment, and even plating to seal the moisture in or to keep it out. As yet none of these methods is totally convincing.

A further caution should be addressed. Unlike metals the tension and compression bulk moduli of composite materials may differ significantly, giving a nonlinear dynamic characteristic.

Composite materials (metal matrix)

The limiting factors of an epoxy matrix (above) can be substantially overcome by employing high-strength fibres in a metallic matrix (diffusion bonding a sandwich lay-up).

It is possible to form the resulting panels, and even forging may be possible. Although this technology may represent the ultimate direction for 'designed' materials, current costs are high and NDT techniques need development. However, applications are appearing, one particular example is in lightweight mirrors [5].

In general a clear rule to be borne in mind is to choose materials (or their close equivalents) which appear in the NASA or ESA approvals lists [6, 7]. This will not only speed any approvals procedure but will provide access to the 'alert' documentation furnished by these agencies when problems occur on any related programme. When this is not possible, national defence specifications should be used for guidance. In all cases, materials traceability including treatment history from billet/raw material to finished product is one of the necessary quality assurance records to be maintained.

9.3.3 Section properties

Hollow or reduced section structural members such as tubes and 'I' beams can exhibit similar stiffness characteristics compared with solid bars. Panel deformations such as corrugations can show greatly increased stiffness and resistance to buckling compared to plane sections. It is the art of the design engineer to use material most efficiently in this way when considering all of the duties required of that section.

Honeycomb sections may be used to create panels with extremely low weight with very high stiffness. Figure 9.5 shows the general configuration of a honeycomb sandwich. A variety of materials and material combinations can be employed. For use in the space environment the designer must consider vacuum stability (out-gassing), and if thermal stability is a consideration similar materials should be used.

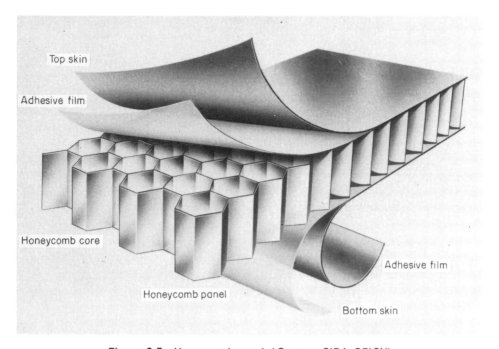

Figure 9.5 Honeycomb panel. (*Courtesy* CIBA-GEIGY)

It is not possible to tabulate the properties of such panels in the manner of materials since in essence they are designed to suit a purpose. Their suppliers produce selection charts giving variations of flexural stiffness (EI) with core thickness, skin thickness and skin modulus.

The design selection process is conducted in three stages:

1. Calculate the required panel stiffness.

2. Select the skin and honeycomb core thickness combinations which best match the loading conditions (i.e. thicker skins for concentrated loading).

3. Calculate the maximum shear stress level and then select a matching core section.

4. Consider how well the selected section can accept loads. This judgement may lead to the selection of a relatively heavier section than the uniform load capacity requires.

It is important to take care when designing the load attachment points for honeycomb panels. Each ease will tend to be unique but Figure 9.6 shows a variety of methods which can be employed. The significant consideration is that this form of panel lends itself to distributed loading, and localized loading calls for careful design with the use of inserts so that forces are spread widely into the external panels.

Two practical points are worthy of note:

1. Even with judicious employment of spare insert locations there will always be the late design change or addition after the completion of the structure. Some modification schemes exist for potting in new inserts but these should be approached with caution and thoroughly tested off-line.

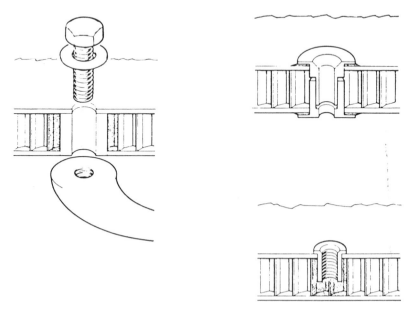

Figure 9.6 Typical attachment methods for honeycomb panels. (*Courtesy* CIBA-GEIGY)

2. The honeycomb presents a trapped volume of air. For space applications the core must have all its cell walls perforated (in a least two places for each cell) to allow the air to escape. With small holes (say 0.2 mm dia) the bulk of the air will vent rapidly, but full venting down to levels below 10^{-5} torr requires a period of days.

9.3.4 Quality assurance

The final proof of a structure's integrity involves it being tested and then inspected. The non-destructive testing of metals is a well-developed field but this is not yet fully true for composite structures.

In the field of virtually one-off spacecraft structures considerable care must be taken to ensure that integrity can be assured in the face of the variabilities due to hand laying-up, bonding and complex sections. Added to this is the problem of ensuring that the structure has not degraded during test.

At the end of the day it is the designer's judgement, confidence and instincts which will make the final arbitration.

9.4 DESIGN FOR LAUNCH

The selection of a launch vehicle plays a substantial role in setting geometric and mass limits to the design. All accelerations are imparted through the spacecraft structure to the mounting points of individual systems or units along a defined load path.

9.4.1 Load path/configuration

The accelerations applied to the base of the spacecraft are discussed in Chapter 7. With most launch vehicles standard mechanical interfaces and release equipment are well defined and set the initial criteria for the load path.

The structure of the spacecraft generally consists of a central thrust-load-bearing member (Figure 9.7), such as a cone or tube to which all systems are attached at strong points either directly or by combinations of struts, platforms and possibly shear webs.

For all these designs the paths are similar and demonstrate the need to take launch loads from the cylindrical rocket through a conical adaptor section fitted with a standard interface ring. Figure 9.8 shows such an arrangement for the Olympus communications spacecraft above an Ariane launcher; note how the mass of the nose cone has no connection with the spacecraft.

9.4.2 Multiple launches

Some launchers have the capability of launching more than one spacecraft at a time. Figure 9.9 shows an arrangement for the Ariane series which can carry two craft independently. The support arrangements for each spacecraft are quite independent but do provide standard interfaces.

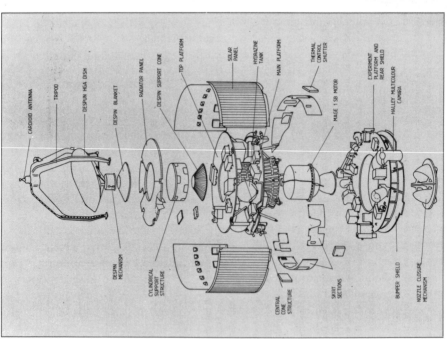

Figure 9.7 Giotto (reproduced by permission of European Space Agency)

CARDIOID ANTENNA

TRIPOD

DESPIN HGA DISH

DESPIN BLANKET

RADIATOR PANEL

DESPIN SUPPORT CONE

TOP PLATFORM

SOLAR PANEL

HYDRAZINE TANK

MAIN PLATFORM

THERMAL CONTROL SHUTTER

MAGE 1 SB MOTOR

EXPERIMENT PLATFORM AND REAR SHIELD

HALLEY MULTICOLOUR CAMERA

DESPIN MECHANISM

CYLINDRICAL SUPPORT STRUCTURE

CENTRAL CONE STRUCTURE

SKIRT SECTIONS

BUMPER SHIELD

NOZZLE CLOSURE MECHANISM

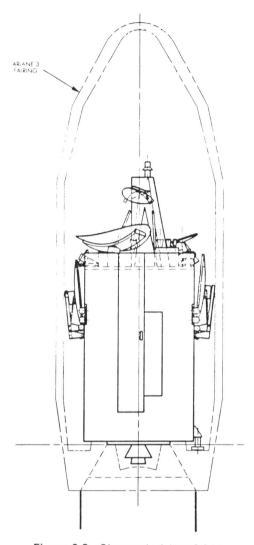

ARIANE 3
FAIRING

Figure 9.8 Olympus in Ariane 3 fairing

An arrangement on a DELTA for carrying three spacecraft which were components of a common scientific mission is shown in Figure 9.10. In this case the lower two spacecraft had to be stressed to carry the loads of the upper craft. Whilst this was quite acceptable for the common scientific mission it would be unacceptable for commercial operations, where the selection of launch partners is made late and needs to be flexible.

9.4.3 Space Transportation System (*Space Shuttle*)

The Shuttle-based system presents a variety of ways in which equipment can be launched. In some cases the spacecraft is designed solely for a STS launch (Figure 9.11). Alternatively

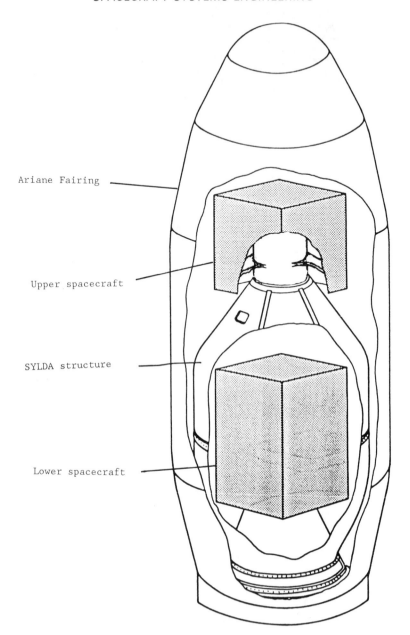

Ariane Fairing

Upper spacecraft

SYLDA structure

Lower spacecraft

Figure 9.9 Dual launch system—Ariane

it may have to be configured to be launched by either conventional rocket or the STS. This imposes a requirement that the principal load path can be in either of two orthogonal directions, resulting in a structural design which is inevitably over-designed in one direction and thus adds weight (Figure 9.12).

For other versions an adaptor is flown (Figure 9.12), which provides the payload with a standard rocket mounting—possibly together with a spin-up table if the spacecraft has

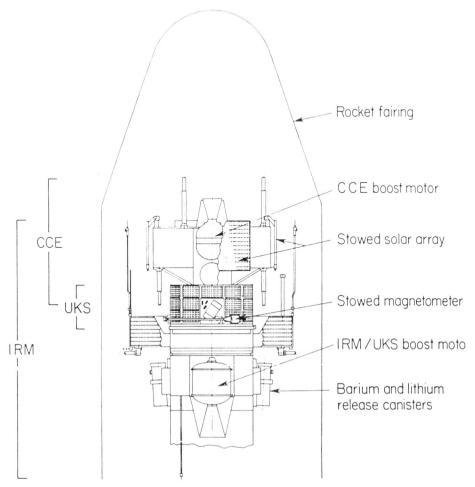

Figure 9.10 showing the stack with labels:

- Rocket fairing
- CCE boost motor
- Stowed solar array
- Stowed magnetometer
- IRM / UKS boost moto
- Barium and lithium release canisters

CCE, UKS, IRM brackets on the left.

Figure 9.10   Stack of three spacecraft forming AMPTE (active magnetosphere particle tracer explorer)

an orbit transfer stage fitted. (Typical applications are for PAM-D or the Century upper stage.)

A further arrangement is where the payload is specifically configured to fit the standard Shuttle attachment points (Figure 9.13). A typical example is that employed for UARS (see Section 9.5). A further use of these mounting points is made by the pallet system made by British Aerospace (Figure 9.14), which contains standardized mounting locations of its own for the fitting and integration of modular payloads while permitting the Shuttle to remain available for other launches.

9.4.4   General

In all cases the objective is to conduct the load as cleanly as possible into the structure and to avoid over-constraint which may result in unnecessarily large stresses and possibly

Figure 9.11   Basic structure of UARS (upper atmosphere research spacecraft). (*Courtesy* NASA/GE)

(a)

Figure 9.11 (*Continued*)

distortion of critical members. When distortion must be avoided, as in the case of attitude sensors or observational instruments, kinetic principles of mounting should be employed. These seek to limit the degrees of freedom (constrained to six generally) by means of a three-point mounting system where each mounting points has a specific flexibility. It may not be possible in all cases to achieve adequate support through the launch phase with a properly configured kinematic mount, and then it may be necessary to provide additional constraint, possibly released in orbit.

For spacecraft requiring precision alignment it may prove vital to supply an active error-correction system which can compensate for drifts in focal position, for instance as the structure deforms under residual stresses or thermal excursions.

Further aspects of the structural design will result from interactions with other requirements of the spacecraft. The interface to the launch vehicle will require perhaps four spring-loaded jacking points which allow a positive separation from the launcher and

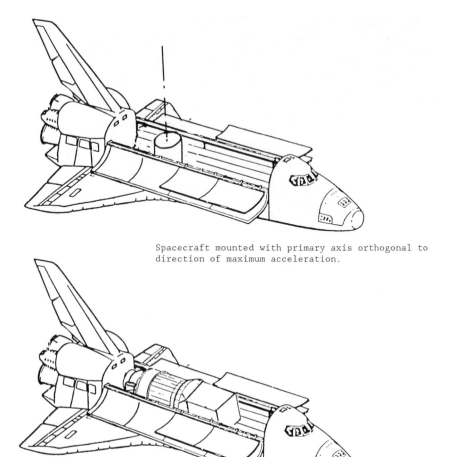

Spacecraft mounted with primary axis orthogonal to
direction of maximum acceleration.

Spacecraft (with upper stage attached) mounted with
primary axis parallel to direction of maximum acceleration.

Figure 9.12   Shuttle mountings of spacecraft which are compatible with expendable
launchers

overcome the residual grip of the umbilical connector. Heavy electrical power-convertor
units may have to be positioned for heat rejection to space, in thermally ideal areas which
are less than ideal for structural purposes.

9.4.5   Launch loads

The design of a spacecraft and its subsystems must cater for all of the loads which will
be experienced. Typical loading events are:

- testing
- transportation

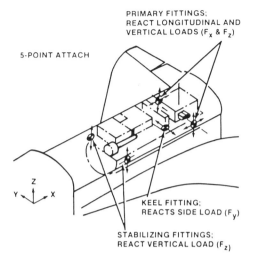

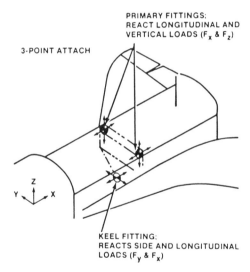

Figure 9.13 Five-point and three-point payload attachment methods for Shuttle. (*Courtesy* NASA)

- lift-off
- stage separations
- stage ignition
- stage or main engine cut-off
- maximum aerodynamic pressure
- spin-up and deployments

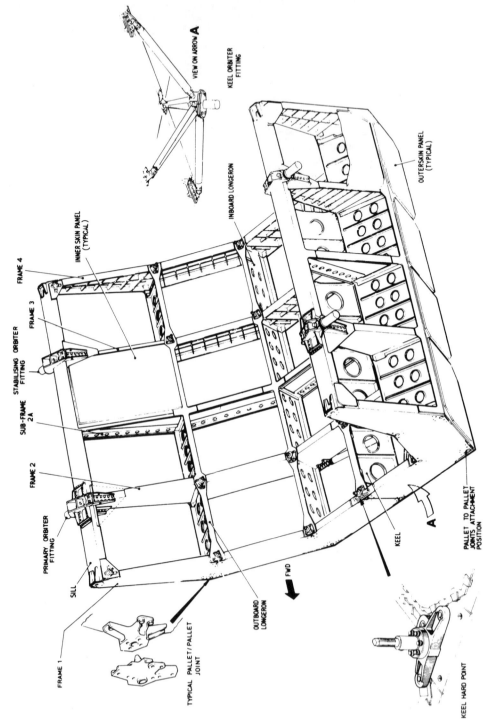

Figure 9.14 Pallet for Shuttle (reproduced by permission of British Aerospace plc)

- ACS firings

- emergency landing (STS).

Each of these will impart accelerations, shock and vibration to the structure. Careful attention to all of these factors is needed, particularly where testing in excess of launch levels is conducted. Here a damage limitation policy must be adopted and in some rare cases the test omitted in favour of analysis.

It will be found that generally the limiting factors in the structural design are set by the dynamic factors rather than by steady-state accelerations.

The assessment of load distribution within a spacecraft structure is largely an iterative process starting with generalized launcher predictions. These are used as a basis for the design of the initial spacecraft concept and thus to provide subsystem target specifications according to their location.

The primary source of loads occurs during the launch phase. All types of launcher apply different levels according to their design. For example, solid rocket engines have combustion chambers which run the length of the fuel column. They burn continuously and cannot be throttled. Liquid fuel rockets have a combustion chamber fed by separate fuel tanks. They can be throttled to reduce thrust loads at critical aerodynamic phases of the flight and to give a controlled trajectory. The tanks of liquid fuel act as a damper and attenuate vibrations from the engines.

The launcher agency will issue standard guidelines for the design and qualification of spacecraft. These apply only to the mounting interfaces at its base and are concerned with both steady and dynamic forces. They will be stated as flight limit loads, i.e levels which one would not expect to be exceeded in 99% of launches. The launcher agency will, however, require the spacecraft designer to demonstrate (by a combination of test and analysis) that the design can withstand these levels with a significant margin. Typically these factors are:

- flight limit loads   1.0

- flight acceptance   1.1

- design qualification   1.25 to 1.4.

Thus test levels for flight acceptance are set for structures and equipment which have previously (in prototype form) passed tests at the design qualification level. The implication here is that if a one-model programme is followed (i.e the prototype is actually flown) then that model must clear the higher qualification test.

The launcher agency may also specify safety factors for the use of materials (typically 1.1 at yield and 1.25 at ultimate stress for metals and up 2.0 for composites). Therefore a structure clearing design acceptance test levels plus material safety factors will have a healthy margin over actual flight loads.

9.4.6 Dynamics

It is necessary to design any structure so that its natural frequencies are significantly displaced from any forcing frequency inherent in the anticipated excitation spectrum, to avoid the build-up of excessive structural deflections during vibration.

Table 9.1   Design-goal structural natural frequency bands (Hz)

Axis	Launcher	Spacecraft	Internal equipment
	Ariane IV		
Thrust	> 31	31–60	< 80
Lateral	> 10	10–40	< 60
	Delta		
Thrust	> 35	35–65	< 80
Lateral	> 15	15–45	< 60

*For large spacecraft different levels may be specified according to the location of equipment, e.g. units mounted distant from the base of the craft will have to meet more stringent requirements.

The natural modes of vibration of a spacecraft must not match those of the launch vehicle if strong and potentially destructive coupling is to be avoided. To minimize later conflict, it is usual at the conception of a project to specify frequency bands confining major structural responses for each design area, as in the example shown in Table 9.1.

Launcher configuration plays a large role. Vehicles such as the Ariane have the advantage that the large column of liquid fuel between the engines and payload acts as a damper for vibration and some of the acoustic loads. The STS carries its liquid fuel in a massive external tank and thus presents somewhat different characteristics. However, for an initial assessment, designs which can meet the Ariane IV structural criteria will largely meet those for the Shuttle.

Even for similar launch vehicles, differences in spacecraft structural characteristics require that each configuration must be considered specifically.

Random vibrations

Random vibrations are generated by mechanical parts in movement such as turbopumps, combustion phenomena or structural elements excited by the acoustic environment. These are transmitted to the spacecraft by the launch vehicle structure.

Acoustic vibrations

Engine noise, buffeting and aerodynamic noise generate acoustic vibrations. Levels are highest at lift-off and in the transonic region. The main areas for concern are thin membrane or panel sections with limited self-damping. Acoustic vibrations propagate primarily through the atmosphere within the spacecraft at launch, but during ascent are eventually transmitted as high-frequency structure-borne noise.

Shock

The payload is subject to shocks principally during engine ignition, staging, fairing jettison and payload separation. On large spacecraft shocks imparted by attitude control thrusters can become significant.

9.5 CONFIGURATION EXAMPLES

Giotto was a comet intercept mission with parts of the spacecraft specifically designed to create a shield for protection against hypervelocity dust impacts. However, the majority of the vehicle is typical for a spin-stabilized spacecraft.

From the standard Ariane I adaptor at the base the load was carried via a corrugated aluminium alloy cone into a similar central load-bearing tube. The solid-fuel cruise engine was fitted into this volume (Figure 9.7).

Disc-shaped equipment-decks made from aluminium honeycomb were fitted to the central tube and braced by an arrangement of aluminium struts. A de-spun antenna was fitted to the upper section of the central tube. With the external wrap-around Solar cell panels removed, this configuration gives remarkably free access to the interior without any need to disturb structure or systems.

ACS fuel was carried in four symmetrically disposed tanks, arranged such that minimal disturbance to the balance of the spacecraft resulted from depletion of the fuel.

Eurostar is an example of a typical three-axis stabilized communications spacecraft and bears many features of later designs such as Olympus. Again the loads are primarily carried by a central tube. However, the nature of the mission requires that a large number of electrical units, which are substantial producers of heat, need to be mounted upon hard points on external walls to promote efficient radiation to space. In addition such units need to be grouped into systems creating the need for large areas. The solution was to employ a box shape using large aluminium alloy honeycomb panels. The load from them to the central tube is carried through four shear webs braced by carbon-fibre tubes with titanium end-fittings. This is a classical use of carbon fibre since the load direction is well defined and the construction lends itself to a well-controlled production process (Figure 9.15) [8].

The apogee engine is fitted into the interior of the central tube, and propellant tanks are grouped symmetrically about principal axes in the outer volume.

Such configurations allow good access by the complete removal of side-walls carrying whole systems. However, it may be prudent to install jury struts when such a part of the structure is removed. This point emphasizes the need to consider handling at an early stage so that appropriate lifting points are incorporated for all stages of assembly and not just for the complete spacecraft.

The solar cells are fitted to external arrays, leaving the side members of the spacecraft free for equipment mounting. Features of the design which are only used once in the mission include latches' release and hinge systems for the solar arrays. Even so they need to be strong enough to withstand launch and the shock during deployment.

Solar Maximum Mission (SMM, Figure 9.16) flew a group of large optical, X-ray and γ-ray instruments with very stringent co-alignment requirements. It also used a standard (potentially reusable) spacecraft bus (MMS—multi-mission modular spacecraft) employ-

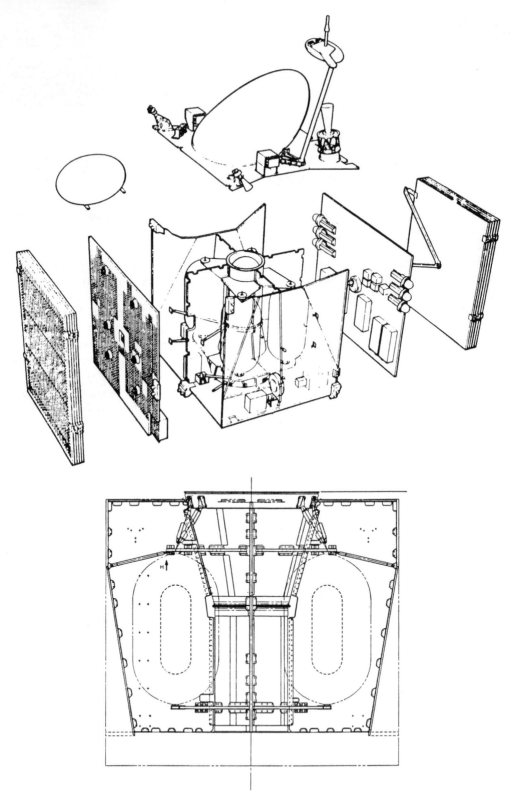

Figure 9.15 Eurostar (reproduced by permission of British Aerospace plc)

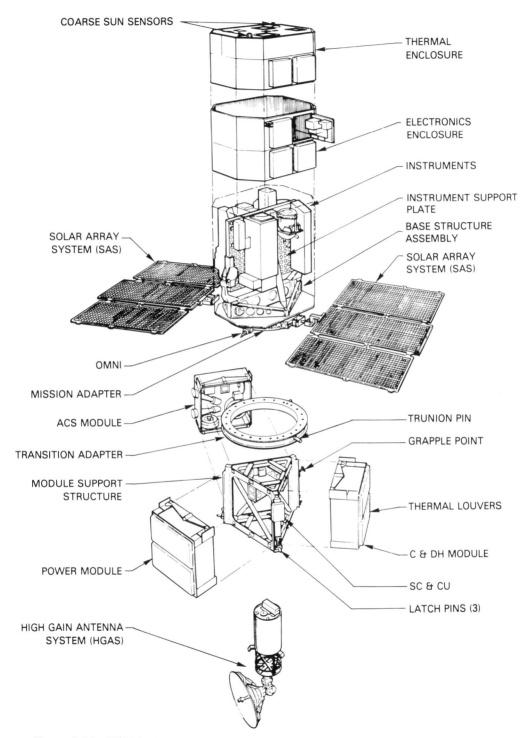

Figure 9.16   SMM (solar maximum mission) spacecraft (reproduced by permission of NASA)

ing a modular systems design. This resulted in a two-part spacecraft, observatory and bus, which could be worked on in parallel. It was designed to be launched by the Shuttle but in the event delays forced the utilization of a Delta rocket. The load interface was to the base of the bus section and carried up to the observatory via a triangular space frame. This lent itself to fitting the three system modules in a manner so that they were removed very simply with a single tool. In the event this turned out to be inspired design when later component failures occurred and a rescue mission was mounted via Shuttle to successfully replace the faulty module.

The observatory section contained a 6 × 4 ft (1.8 × 1.2 m) optical bench mounted in a relatively strain-free manner. The specifications for instrument mounting to the bench were that each should be three-point kinematic mounted, such that no strain could be passed to another and that the sub-arc minute alignment criterion could be held into orbit.

Upper Atmosphere Research Spacecraft (UARS—Figure 9.11) is a large (32 ft (9.75 m) long, 7 ton) spacecraft specifically tailored for a Shuttle launch and deployment, inter-facing in two planes with standard keel and longeron fittings within the orbiter. The launch forces are transmitted through these fittings, but it is a feature of the Shuttle environment that severe thermal excursions can exist in the payload bay due to solar aspect angle. Thus a semi-kinematic approach is necessary allowing one hard-point mount, and controlled directions of flexibility at the other locations to allow for relative expansion.

To provide services, use of an MMS bus is again made, this time fitted with a propulsion module for orbit maintenance.

A significant departure from the other spacecraft discussed is the lack of any external casing. Each of the instruments has to provide its own environmental protection and will be subject to shadowing from structure and other instruments. This creates a somewhat interrelated analysis problem!

Since the spacecraft is an observatory, instrument and sensor alignment is a driving factor in the strucutural design, and three-point kinematic mounting has been speci-fied for each unit. The primary framework employs graphite epoxy struts with tita-nium end-fittings whose low expansion characteristics aid maintenance of instrument co-alignment.

Hubble Space Telescope (Figure 9.17). This spacecraft contains a telescope capable of allowing the separation of stellar objects only 0.1 arc seconds apart. A wide thermal excursion will be experienced as it moves in and out of eclipse. Passively maintaining the 200-in (~5 m) separation of the two optical components of the telescope to 100 μin (2.5 μm) requires that only low-expansion materials could be used. Graphite epoxy was selected for the major structure, the metering truss, which is 200 in (~5 m) long and 115 in (~2.9 m) in diameter but weighs only 250 lb (114 kg). It represents a classic use of the material, a known direction of loads and the need for long-term stability and low expansion. Such are the design demands that allowance has had to be made for relaxation due to the absence of gravity in orbit. Even so the instrument had to be adjusted to be slightly off focus on the ground due to the hygroscopic nature of the material (absorption of water from the air amounting to 0.4% of the mass of the truss) resulting in a contraction of 8 000 μin (200 μm). The water will slowly outgas in the vacuum of space to restore the original dimensions, but as an insurance policy small motors fitted to the mirror cell, and structure-mounted heaters, are available to counter any out-of-focus effects which remain. This may also include relaxation effects of the structure during launch.

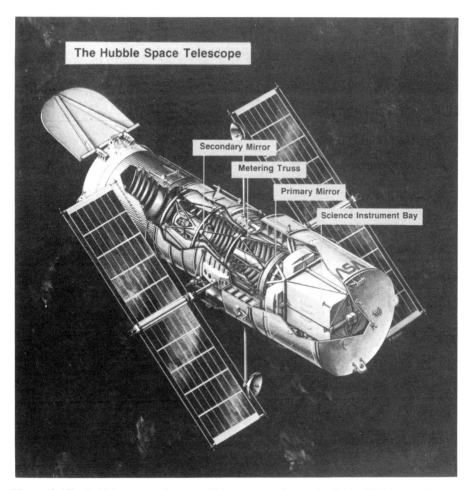

The Hubble Space Telescope

Secondary Mirror

Metering Truss

Primary Mirror

Science Instrument Bay

Figure 9.17 Hubble space telescope. This cutaway shows the Optical Telescope Assembly in the Hubble Space Telescope. The OTA includes a 94-inch (2.4 m) primary mirror, a 12.5-inch (32 cm) secondary mirror, a graphite–epoxy metering truss, and the fine guidance and control sensors (*Courtesy* Hughes Danbury Optical Systems, Inc. (subsidiary of Hughes Aircraft Company))

The 11-ton Hubble Space Telescope is Shuttle-launched, attached to the payload bay by a selection of keel and longeron mounts which minimize the constraints applied to its structure.

9.6 DESIGN VERIFICATION

An inherent part of the structural design activity is the verification of the final design, either by a full suite of tests or by modelling and analysis supported by limited testing (see Figure 9.18). The particular course of action to be followed will have to be agreed with the Launch agency. It will also be influenced by such factors as physical size (i.e can it be fitted onto a test facility?) and number of qualification models in addition to the

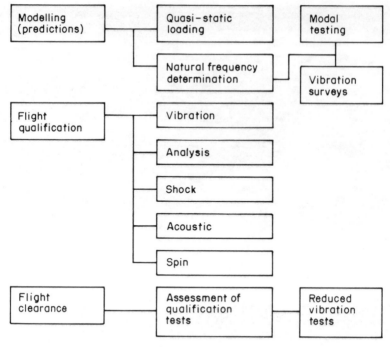

Figure 9.18 Structure verification

flight article (if only the flight model is to be built, limited exposure to testing may be a necessary policy).

9.6.1 Modelling—finite element analysis (*FEA*)

With the advent of the digital computer it is now possible to model structures mathematically in great detail, and to examine their behaviour under all possible load conditions, static or dynamic. The essence of the FEA method is to divide the structure into a large number of discrete elements, within each of which the load distributions, elastic properties and boundary conditions are known [9]. For each element appropriate parameters can be fed into the computer describing material properties, shape, degrees of freedom and connection to the next element. The computer program enables the structure to be evaluated under static loads, and it is also possible to conduct a dynamic study (i.e under fluctuating loads). In a dynamic simulation its natural frequencies can be assessed and relative phase information of deflection shapes at different locations within the structure can be indicated [10].

System design teams are required to conduct and deliver FEA analyses. A reduced model, demonstrating similar characteristics to the larger version but with a substantially reduced number of elements, is required for incorporation into the overall spacecraft simulation. This in turn is incorporated into a coupled analysis model of the launcher–spacecraft combination (see Figure 9.19), so that a full examination of the complete launch

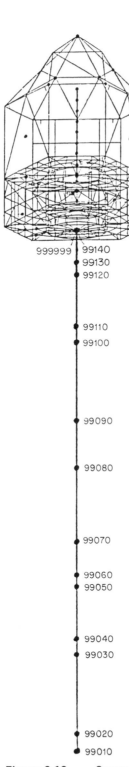

Figure 9.19 Spacecraft/
launcher coupled mathe-
matical model, for Hipparcos.
(reproduced by permission of
European Space Agency)

configuration can be made. From such an analysis may come means of accommodating individual units which cannot meet the full vibration specification. It is possible to explore trade-offs, for example with liquid fuel rockets the engine can be throttled back during passage through the phase of maximum dynamic pressure as is the case with the Space Shuttle during the transonic region, where the engines are throttled back to 65% for 25 s.

The most widespread F.E. package in use in space engineering is NASTRAN (developed initially by NASA).

9.6.2   Verification testing

The principal objective of structural testing is to qualify an assembly for launch. This is often achieved by subjecting a prototype to static and vibration loads in excess of those anticipated for flight, and is termed 'design qualification' (see Section 9.4).

The actual flight unit is subjected to near flight levels (say 110%). This is termed 'flight acceptance'.

Each launch is a unique loading case and the relevant agencies must be consulted and the test procedure agreed.

Vibration testing is carried out in each of three axes independently, and consists of:

1. *Low-level sine sweep* (usually 0.25 g). This is used to determine the natural resonant frequencies within the assembly. It is important to detect response in all three axes for each axis of excitation to assess cross-coupling. It is vital to make an assessment of the structure at this level before proceeding to higher levels. (Alternatively a low-level random excitation can employed using spectrum analysis to obtain response data.)

2. *High-level sine sweep.* This is used to provide the structural capability of the design but it is unrepresentative of flight conditions and is often deleted in favour of the following:

3. *Random vibration.* The design is subjected to a representative spectrum of random vibration such as will be encountered at launch. This has the advantage of subjecting the structure to a wide band of frequencies simultaneously such that no resonance can build up. The outputs of the sensors can be spectrum analysed and compared to the input.

4. *A repeat of the low level survey* as a quality tool to show that no frequency shifts occurred due to exposure to the loads.

Typical input specifications are given by Hecks [8]. These are derived from an assessment of a coupled launcher and generalized spacecraft analysis. The envelopes so depicted are drawn to include all of the maximum conditions which will be transmitted to the spacecraft. *It is therefore a worst case.* It may well be that the low-level survey shows a dangerous amplification at a frequency which is forecast as unlikely to be dominant in the final coupled analysis. In this case consideration can be given to 'notching' the input in a narrow band centred upon the critical frequency (see Figure 9.20).

At all stages it is necessary to compare the output responses to those predicted by the

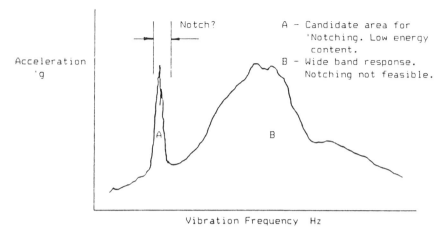

Figure 9.20 Typical structural frequency response

FEA in order to validate the model. It is quite certain that a vibration input constrained to one axis does not completely replicate the true condition of launch and in fact may cause an over-test in some locations and an under-test in others. Recent work has concentrated on the development of hydraulic multi-axis vibrator facilities to produce a more realistic test.

Following prototype tests the spacecraft designers can give to the equipment suppliers a preliminary vibration regime for their particular location in the spacecraft. If necessary this will be adjusted when a final coupled-loads analysis is conducted for the mature spacecraft configuration and the launch vehicle.

9.6.3   Large structures—modal testing

The foregoing assumed that the spacecraft or subsystem was excited as a whole by a single vibrator. This is not always possible, and then a method relying more heavily upon the modelling process must be adopted. The principle adopted is to FE model the structure as a series of sub-units with definable interactions with each other and to survey each, locally, for natural frequencies, mode shapes and damping [9]. A test is then formulated which employs small vibrators suitably positioned within the structure. The response is correlated with the FEA prediction as a process to validate the model [10]. This method requires considerable experience in its interpretation since no flight load test is performed.

Shock testing

This test is now usually conducted as an adjunct to the vibration test. A simulated shock spectrum is fed into the vibrator to excite the spacecraft. For structures too large for this to be achieved it is possible to simulate local shocks due to releases or deployments, by the utilization of small exciters in appropriate locations.

Quasi-static load testing

Realistically, static load testing can be done only on a very large centrifuge. Practically, such tests are confined to relatively small units to avoid gravity gradients due to the geometry of the test machine. The method usually adopted is to attach loads (direct weights or hydraulically), to appropriate nodes of the structure and to assess the resultant deflections against prediction. The level to which this is done depends upon the model philosophy chosen for the programme. On a full model programme a structural test article can be overloaded significantly to assess safety factors, but on a Protoflight programme (one model—and it flies) the structure may be under-tested provided that it closely matches the predicted (FEA modelled) behaviour.

Spin testing

Again this requires specialist equipment in the form of a robust spin table and a strong safety enclosure. Essentially, this test is applied to any craft which will undergo spinning in operation, not only to test structural integrity but to allow the dynamic balance of the spacecraft to be measured.

9.7   THE FUTURE FOR SPACE STRUCTURES

As confidence grows in the accuracy of structural analysis and loads prediction there will be a continued trend towards verification of mathematical models by minimal, low-level testing. This will allow a reduction in safety margins since the over-test of design qualification can be removed.

Orbiting large open structures such as large antennas or the space station main trusses are too big for launching in their assembled form to be contemplated. Thus the designer is faced with deployment or perhaps assembly after achieving orbit. (Figure 9.21).

Potentially the task of designing for launch can be easier since the structure may be stowed compactly within the launch vehicle, and an extremely low-mass design may then proceed for withstanding only the zero-*g* environment. Careful consideration needs to be given for loads created in hinges or members during automatic or remote deployment where sections can build up momentum as they are swung into place and are then suddenly stopped when positioned. A similar argument prevails for attitude control or orbit maintenance manoeuvres by the space station. Not only are these likely to excite very low-frequency cross-coupled oscillations within the structure but the designer is faced with a continuously changing situation as mass is added, moved or subtracted from the configuration. This is potentially a fertile area for the application of distributed thrusters and an 'expert system' or 'intelligent' attitude control system to phase the individual firings so as to counter undesirable deflections.

It can be expected that new families of wire-tensioned structures will emerge which can only exist in a zero-*g* environment.

Much consideration is being given to building sections in orbit directly, from material coiled flat on a roll within a machine to be located within the Shuttle cargo bay. This would roll the stock into a structural section as it emerged. Here, if the concept can be

Figure 9.21 Space station configuration. (*Courtesy* NASA)

successfully developed, the designer will face the difficult problem of bridge designers on Earth—that of ensuring structural safety through all stages, i.e., in this case, before momentum control systems have been installed and the structure is only partly built.

Finally, free thinkers contemplate the use of non-terrestrial materials, such as from the Moon or even asteroids, which can be placed in orbit with less use of energy than from Earth and do not necessarily require demanding material properties.

REFERENCES

[1] Crane, F.A.A. and Charles, J.A. (1984) *Selection and Use of Engineering Materials.* Butterworths.

[2] Ashby, M.F. and Jones, D.R.H. (1980) *Engineering Materials—An Introduction to their Properties and Applications.*

[3] Gordon, J.E. (1978) *Structures, or Why Things Don't Fall Down.* Pelican, London.

[4] Kinnersley, M.A., Stark, J., and Swingard, B. (1989) Development of, and initial results from a high fluence, high velocity atomic oxygen source. 4*th European Symposium on Spacecraft Materials in the Space Environment.* CERT/CNES/ESA, 701.

[5] Mohn W.R. and Gegel G.A. (1986) Dimensionally stable metal matrix composites and optics applications. *Advanced Composites Conference Proceedings,* ASM International and Engineering Society of Detroit.

[6] *NASA Materials Guidelines.* N-75-24848 (SP-3094).

[7] *ESA Materials Guidelines.* ESA PSS-01-701.

[8] Hecks, K. (1987) Mechanical design of the Eurostar platform. *J. British Interplanetary Society,* **40**, 130–9.

[9] Haigh, M.J. (1985) *An Introduction to Computor-aided Design and Manufacture.* Blackwell Scientific, Oxford.

[10] Robinson, J. (1981) *Understanding Finite Element Stress Analysis.* Robinson and Associates.

[11] Bruel and Kjaer, *Modal Analysis of Large Structures—Multiple Exciter Systems.*

10 *ATTITUDE CONTROL*

Peter W. Fortescue

Department of Aeronautics and Astronautics, University of Southampton

10.1 INTRODUCTION

Although the prime purpose of the attitude control system (ACS) is to orientate the main structure of the spacecraft correctly, and to the required accuracy, it is worthwhile considering it also as a momentum management system. Angular momentum is a commodity which can be acquired and disposed of, or stored.

The ACS designer must assess which is his best policy and provide the hardware to achieve it. He will need to assess the momentum implied by the pointing requirements and to specify torquers, storage etc. to provide it. This view of an ACS will be helpful in understanding how it operates.

When one looks at the types of ACS in existing spacecraft one finds that a profusion of different methods are employed to meet this seemingly simple objective. The structures may be spinning or not, or may be doing so only during certain phases of the mission. The ACS may or may not use momentum bias and/or momentum storage; it may use a variety of torquing methods, in combination or on their own. The number of options open to the designer is large. What then influences him to select one solution as opposed to another?

The decisions that he takes are not based solely upon the considerations presented here. He will be influenced by the experience and the history of the company in which he works. A feature of attitude control is that different technical solutions may have very similar performance, and he will choose well-trodden paths, the type of solution of which his company has experience. And rightly so.

There are complete books on the subject of attitude dynamics and control [1–3], and this solitary chapter cannot present a comprehensive account of the subject. The approach adopted is to look at the fundamentals, to progress from the system level downwards, and to identify the options referred to. Detail design is not covered.

Spacecraft Systems Engineering. Edited by P.W. Fortescue and J.P.W. Stark
© 1991 John Wiley & Sons Ltd

10.2 ACS OVERVIEW

10.2.1 The design objective

The orientation required of the spacecraft's structure will be determined by the mission. The structure will be seen as the mounting base for the payload(s), and for several 'housekeeping' subsystems which have objects which must be pointed in specific directions. Among the latter will be solar arrays to be pointed at the Sun, thermal radiators to be pointed at deep space, communication transmitters to be pointed at their receiver and, prior to firing, thrusters to be pointed in the correct direction. Add to this the pointing requirement of the payload and it will be seen that there is an essentially three-dimensional problem to solve, at the end of which will emerge a proposed layout for the spacecraft with locations for the objects to be pointed, and a specification for the orientation of their mounting base, the main structure.

The required orientation will frequently be related to an Earth-based frame of reference. For example 'one face of the spacecraft must point down the local vertical'. The ACS designer must then analyse the orbit and the mission in order to assess the motion which is required of the structure, and the disturbance torques, etc., to which it will be subjected. For this objective the structure requires an angular rate equal to $\dot{\theta}$, as given in equation (4.13).

There will often be several alternative configurations which will meet the overall objective for a given payload, and these may pose very different problems. The Intelsat series of satellites illustrates this point. Their payloads are all for communications purposes and have antennas to be pointed at locations on the Earth. But their configurations include both dual-spin and three-axis stabilized types.

In general pointing mechanisms are to be avoided. However, they are often needed in order to enable an object to remain pointing in one direction whilst the main structure changes its orientation, or vice versa. For example, a solar array needs to remain pointing at the Sun whilst the main structure turns to align with the local vertical.

The required accuracy of orientation will be set by the payload. The accuracy with which its direction can be controlled will be less than that to which it can be measured, and this is in line with the requirement. For example, spacecraft astronomers will need to know where a telescope's axis is pointing to great accuracy (typically arcseconds), but the control of its direction may be less accurate, related to its field of view.

A full accuracy specification for both measurement and control of the main structure's attitude may then be determined from the various pointing requirements.

10.2.2. Mission-related system considerations

The ACS designer will need to know the history of the rotational motion, the angular velocity ω which is required of the spacecraft, and of any parts of it which can move independently on bearings. The angular momentum $\mathbf{H}_c$ and the torque $\mathbf{T}$ needed to produce it may then be calculated from the Newtonian law (from equation (3.25)):

$$d/dt(\mathbf{H}_c) = \mathbf{T} \qquad\qquad (10.1)$$

where $\mathbf{H}_c$ is the angular momentum referred to the centre to mass C, detailed in equations (3.30)–(3.33) and $\mathbf{T}$ is the torque.

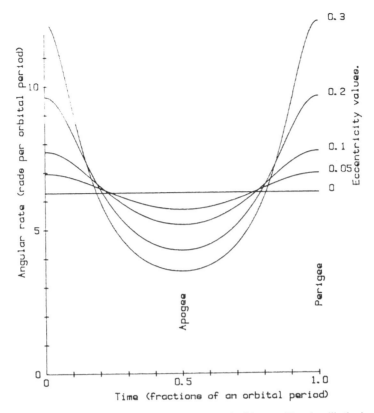

Figure 10.1   Angular rate of the local vertical for satellites in elliptical orbits

A circular orbit provides an interesting example. The local vertical will rotate at a constant rate equal to 1 revolution per orbit about the normal to the orbit plane. If one face of a structure with no momentum bias is required to point down the vertical, then no torque is needed to maintain this condition provided that a principal axis is aligned with the normal to the orbit plane (cf. page 50). If the normal is not a principal axis, then equation (10.1) will indicate that a cyclic torque is needed.

From an elliptical orbit, on the other hand, the local vertical will not have a constant angular velocity. An assessment over one orbit indicates that the angular rate will vary as shown in Figure 10.1. It has a cyclic variation about a mean level of one revolution per orbit.

If one face of the structure is required to point down the vertical, and a principal axis is aligned with the normal to the orbit plane, then no torque is needed to maintain the mean angular rate, constant angular momentum. But torques *will* be needed to produce the fluctuations in momentum about the mean value.

If the structure is not rotating, frequently the case for astronomical missions, then again no torque is needed to maintain this condition once it has been achieved. This will be approximately true of solar arrays too, since their angular velocity is likely to be virtually zero as they point towards the Sun, even if the main structure is rotating.

The torque requirements for each part of the spacecraft may be assessed as above, on the assumption that it is rigid, and these will be the *minimum* torques needed. Extra will be required in order to combat the uncontrolled (disturbance) torques such as that due to solar radiation pressure, movement of fuel, mechanisms etc, and so these must be evaluated too (see Section 10.4). The assessment must cover *all* phases of the mission, and all three axes of the spacecraft. It may be expected that the largest torques will be needed during the early stages, between final separation and being on station.

Momentum storage may be used when there is a requirement for fluctuating torque and momentum. A momentum storage device is basically a wheel fitted with a torque motor (see Section 10.4). In operation the motor drives the spacecraft structure in order to give it the angular motion which it requires for its mission, and consequently the momentum H_c which it needs. This momentum is extracted from the store. If the mission calls for a fluctuating momentum which over a period returns to its original value, then the stored momentum would also return to its original value and the manoeuvre would be completed without any fuel being expended by gas jets. The elliptical orbit illustrates this, Figure 10.1 indicating that the momentum will return to its original value after one orbit. Re-pointing a telescope from one fixed direction to another is another example.

In practice the torque motor will have to contend with the disturbance torques too. These will increase the momentum of the spacecraft as a whole as indicated by equation (10.1). The torque motor will automatically place this momentum into the store as it carries out the process of controlling the structure. Eventually the store will become full—the wheel will reach its maximum permitted speed. The slowing-down process is called 'momentum dumping'. It can only be achieved by applying an *external* torque to the spacecraft (see page 43).

If the designer chooses to include a momentum storage system—it is optional—then his assessment of the momentum fluctuations will indicate the size of store he will need. For the elliptical orbit, Figure 10.1 indicates the storage required. It must accommodate at least half the difference between maximum and minimum in order that dumping is not needed in every orbit.

10.2.3 Momentum bias

The level of momentum which is involved in a storage system will be quite small, and much smaller than will bring any significant benefit from gyroscopic rigidity. This will come from momentum bias, which makes the direction of one axis of the spacecraft highly resistant to change.

For example, if the mission requires that one axis of the spacecraft shall always lie in the direction of the normal to the orbit plane, or at right angles to the Sun vector, then the designer might include momentum bias in that direction. The magnitude of the bias, H_b, whilst not critical, is likely to be an order of magnitude greater than the storage system will handle. A torque T at right angles to it will cause the axis to precess at a rate T/H_b (see equation (3.26)). The attitude response *about* the bias direction will not be altered.

Momentum bias devices may be used for momentum storage too. If more than one bias device is used, and this may well be done in order to enhance the reliability, then it must be remembered that their momenta add vectorially to produce only *one* gyroscopically rigid axis.

10.2.4 The ACS block diagram

The block diagram in Figure 10.2 shows the major components of a general ACS system. The links between components identify major interactions, with arrows indicating that there is a cause–effect relationship; it is convenient to think of them as channels along which information flows. For example the main structure of the spacecraft is subjected to time-varying torques from torquers, and will respond with attitude motion which will be detected by the sensors. Outputs from these will be sent to computers, on-board and at the ground control station, and their information will be used to determine the torques which should be applied to the structure.

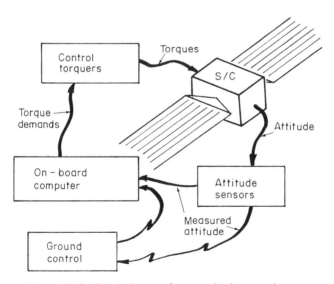

Figure 10.2   Block diagram for an attitude control system

10.3 THE SPACECRAFT ATTITUDE RESPONSE

In the context of the ACS block diagram the spacecraft may be seen as an object which changes its orientation in response to a torque, which is the simple message conveyed in Figure 10.2. The way in which it responds will depend upon whether it has momentum bias or not, and the response equations are contained in Sections 3.3 and 3.4.

Whilst the decision to include bias and the method of doing so will be decided at system level (cf. Section 10.2), it is as well to recall the main types of structure and their characteristics that the ACS designer will have to deal with. For each type he will be concerned with the following:

- the equilibrium state,

- the response to a steady torque,

- the stability, and the existence of any oscillatory modes.

10.3.1   The three-axis-stabilized spacecraft, with no momentum bias

The dynamic equations for this craft, treated as a rigid body, are covered on pages 48–9, using principal axes for the analysis. With small angular velocities the responses about these axes are largely uncoupled and may be approximated to

$$I_{xx}\dot{\omega}_x = T_x; \quad I_{yy}\dot{\omega}_y = T_y; \quad I_{zz}\dot{\omega}_z = T_z \qquad (10.2)$$

Any of the torque components on its own will produce an acceleration about its own axis. Combinations, however, will not do so; they will produce a cross-coupled response except when moments of inertia are equal. Cross-couplings will increase as the angular velocity increases.

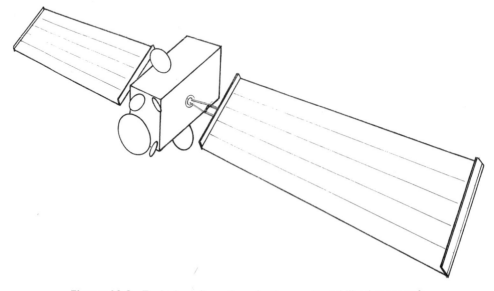

Figure 10.3   Typical configuration of a three-axis-stabilized spacecraft

This type of craft usually has flexible solar arrays attached to the main structure as shown in Figure 10.3, with lightly damped flexure modes extending down to low frequencies. Typically up to about 20 of these modes may be mathematically modelled in order to identify adverse reactions with the ACS.

10.3.2 The spinning spacecraft

The dynamics of a spinner are covered on pages 49–53. It is assumed here that the vehicle is spinning about its z-axis with an angular rate S, and that its mass distribution is axially symmetric so that $I_{xx} = I_{yy}$. The response to a torque will be of interest in one or both of two sets of axes. A typical 'spinner' is shown in Figure 10.4.

The response in axes fixed in the structure will be of interest when the torque components are in these axes or when the attitude sensors measure components in these axes. The equations, developed from equation (3.37), are

$$I_{xx}\dot{\omega}_x + S\omega_y(I_{zz} - I_{xx}) = T_x$$

$$I_{xx}\dot{\omega}_y - S\omega_x(I_{zz} - I_{xx}) = T_y \qquad (10.3)$$

$$I_{zz}\dot{S} = T_z$$

Constant components of torque T_x, T_y will produce constant components of angular velocity ω_x, ω_y given by

$$\omega_x = -T_y/S(I_{zz} - I_{xx}); \quad \omega_y = T_x/S(I_{zz} - I_{xx}) \qquad (10.4)$$

When these are combined with the spin motion the result is a coning rotation at the spin frequency S when viewed from outside the spacecraft.

One of the major reasons for spinning a spacecraft is to counter the effect upon the trajectory which is caused by a thrust offset when a high-thrust motor is being used, such

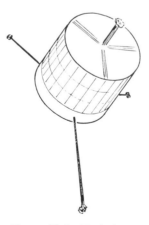

Figure 10.4 Typical configuration of a spinning spacecraft

as during orbit-changing manoeuvres or during the later stages of expendable launch vehicles. Equations (10.4) indicate that the torque due to the offset will then cause coning and spinning which is acceptable, with an average thrust in the correct direction.

The z-component of torque T_z produces an acceleration about that axis which is uncoupled from the other two axes.

The nutation mode (pages 52–3), observed in the structure axes, has a frequency

$$\omega_{nut} = S((I_{zz}/I_{xx}) - 1) \tag{10.5}$$

and needs stabilizing (see Section 10.3.4).

The response in non-spinning axes will be of interest when the spin axis has to be redirected and when the reason for spinning the structure is to provide gyroscopic rigidity via momentum bias. One of the axes is aligned with the spin axis (z-axis); the x- and y-axes do not rotate with the structure. The response equations become (see equation (3.44))

$$I_{xx}\dot{\Omega}_x + I_{zz}S\Omega_y = T_x$$

$$I_{xx}\dot{\Omega}_y - I_{zz}S\Omega_x = T_y \tag{10.6}$$

$$I_{zz}\dot{S} = T_z$$

Constant torque components T_x, T_y produce precessional angular rate components:

$$\Omega_x = - T_y/I_{zz}S; \qquad \Omega_y = T_x/I_{zz}S \tag{10.7}$$

These are rotations which will be observed from outside the spacecraft, and they demonstrate rigidity in that as the bias ($I_{zz}S$) increases, so the response to a given torque decreases. The nutation mode, observed in these non-spinning axes, has frequency

$$\omega_{nut} = S(I_{zz}/I_{xx}) \tag{10.8}$$

and will need stabilizing (see Section 10.3.4).

10.3.3   Hybrid spacecraft

The dynamics of these craft are covered on pages 53–4. In the dual-spin or partially de-spun versions momentum bias is provided by rotating a piece of structure on a bearing attached to the non-spinning part; it will be assumed that the spinning part has an axially symmetric mass distribution (see Figure 10.5).

In a three-axis-stabilized version with momentum bias the bias is provided by a momentum wheel (see page 244).

The response equations for the hybrid craft with their bias H_z along the z-axis are equations (3.54). For low body rates they approximate to

$$I_{xx}\dot{\omega}_x + H_z\omega_y = T_x$$

$$I_{yy}\dot{\omega}_y - H_z\omega_x = T_y \tag{10.9}$$

$$I_{zz}\dot{\omega}_z + H_z = T_z$$

Constant torque components T_x, T_y produce precessional angular rates:

$$\omega_x = - T_y/H_z; \qquad \omega_y = T_x/H_z \tag{10.10}$$

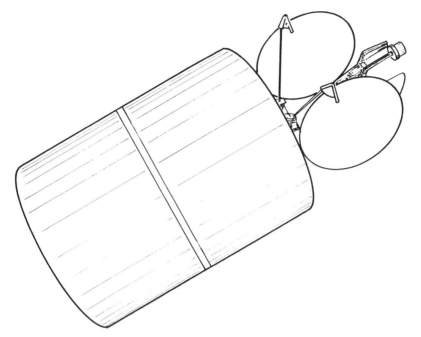

Figure 10.5 Typical configuration of a dual-spin or partially de-spun hybrid spacecraft

The torque component T_z about the bias direction produces an acceleration response about that direction, which does not cross-couple into the other axes. In the case of momentum wheels the total bias may be the vector sum of the biases from two or more wheels in different directions; nevertheless it is the component of torque in the total bias direction which produces the acceleration response described above.

The nutation mode has a frequency

$$\omega_{\text{nut}} = H_z/\sqrt{(I_{xx}I_{yy})} \qquad (10.11)$$

and needs to be stabilized (see Section 10.3.4).

10.3.4 Oscillatory modes

A characteristic of the space environment is that oscillatory modes have very little damping. The ACS has to avoid undue excitation of these and must include means of damping them. Damping may be enhanced by means of energy dissipation or by active control techniques.

The energy dissipation method is based upon the fact that the energy present in an oscillatory mode is exchanged between kinetic and potential types during each cycle. A constant total amount corresponds to a constant amplitude of oscillation which is proportional to the square root of the energy. By using the oscillation to excite an energy dissipator the amplitude will decrease and the mode becomes damped. Active damping entails sensing the oscillation and applying a suppressing torque in the correct phase.

Nutation damping may be implemented either way. A passive damper may consist simply of a ball which is free to move in a tube containing viscous fluid. Relative movement leads to energy being dissipated in the fluid. The damper has to be mounted at a location where it will be shaken by the nutation. For the long-term stability of the dual-spin spacecraft it should be mounted in the non-spinning part.

Active damping may be achieved by applying a torque component proportional to the angular velocity component along the same axis, i.e. T_x proportional to ω_x. The axes may be the spinning (body) axes or non-spinning ones. For the hybrid spacecraft equations (10.9) become

$$I_{xx}\dot{\omega}_x + H_z\omega_y = T_x = -K_x\omega_x$$
$$I_{yy}\dot{\omega}_y - H_z\omega_x = T_y = -K_y\omega_y \qquad (10.12)$$

where K_x, K_y are the feedback constants expessing the torque per unit angular rate.

The characteristic equation, in terms of the Laplace operator s, is

$$I_{xx}I_{yy}s^2 + (I_{xx}K_y + I_{yy}K_x)s + (K_xK_y + H_z^2) = 0 \qquad (10.13)$$

This represents damped harmonic motion when K_x and/or K_y are non-zero and positive.

In practice there will be other oscillatory modes present, and excited by the same torquers. Then the suppression of one mode may lead to the excitation of another. The choice of algorithms to provide damping for multiple modes is a specialist task which is beyond the compass of this book.

The main oscillatory modes are follows:

- *Structural flexibility*, particularly slender appendages such as solar arrays, antennas, tethers etc. These will have an infinite number of modes and perhaps the lowest twenty frequencies will be considered during the design stage. If an array with mass m is represented as a uniform cantilever of length l, with flexural stiffness (EI), then its modal frequencies will be given by

$$\omega = \lambda^2\sqrt{[(EI)l/m]} \text{ rad/s} \qquad (10.14)$$

  where λ takes values given by the solutions of

$$\cos \lambda l \cosh \lambda l = -1 \qquad (10.15)$$

- *Nutation*, due to momentum bias.

- *Libration*, the pendulum effect of free motion under gravity-gradient torque (see page 242). This mode is important in tethered configurations.

- *Fuel slosh*. The movement of fuel in its tanks may set up an oscillatory modal effect.

The modes of a spacecraft may be separated into the so-called rigid-body ones, and those associated with mass movement and flexure. The mathematical modelling covered in this chapter is restricted to the rigid body and thus will cover nutation and libration but not the other modes. A more comprehensive coverage, is given by Wertz [1] and Hughes [3].

10.3.5 Summary

The ACS is a feedback system in which the spacecraft is the plant, shown as a block in a block diagram (Figure 10.2) in which torque affects attitude. The dynamic equations linking these are presented in Chapter 3, and above they are shown when the torque components are about the principal axes.

Whilst the torques will excite oscillatory modes, and these must ultimately be stable, their prime purpose is to control the orientation of the axes of the spacecraft. Each component of torque will have an effect, and by resolving into principal axes their effects are largely uncoupled from each other. These simple responses to each separate component of torque are useful to preserve as the dominant responses of the block. There are just two types—the precessional and the non-precessional:

- *The non-precessional response* obeys, for example,

$$I_{xx}\dot{\omega}_x = I_{xx}\ddot{\phi} = T_x \tag{10.16}$$

The rotational response is an angular acceleration about the same principal axis as the torque. This applies to each axis when there is no momentum bias, but applies only to motion about the bias axis when it is present.

- *The precessional response*, approximated by omitting the nutation mode, leaves the steady-state response, which is

$$\Omega_y = \dot{\theta} = T_x/H_z \quad \text{and} \quad \Omega_x = \dot{\phi} = -T_y/H_z \tag{10.17}$$

The response equations with nutation included, in state-space form, are

$$\mathrm{d}/\mathrm{d}t \begin{bmatrix} \Omega_x \\ \Omega_y \\ \phi \\ \theta \end{bmatrix} = \begin{bmatrix} 0 & -H_z/I_{xx} & 0 & 0 \\ H_z/I_{yy} & 0 & 0 & 0 \\ 1 & 0 & 0 & 0 \\ 0 & 1 & 0 & 0 \end{bmatrix} \begin{bmatrix} \Omega_x \\ \Omega_y \\ \phi \\ \theta \end{bmatrix} + \begin{bmatrix} 1/I_{xx} & 0 \\ 0 & 1/I_{yy} \\ 0 & 0 \\ 0 & 0 \end{bmatrix} \begin{bmatrix} T_x \\ T_y \end{bmatrix} \tag{10.18}$$

where $\dot{\theta} \approx \Omega_y$, and $\dot{\phi} \approx \Omega_x$.

10.4 TORQUES AND TORQUERS

The torques, moments of forces about the centre of mass, and couples, must be identified as being *external* or *internal* to the spacecraft. The former will affect the total momentum, whereas the latter will affect only its distribution. The case has already been made (cf. page 43) that it is necessary to include controllable external torquers whereas internal ones, with their momentum storage facility, are optional.

The main sources of torques, occurring either naturally or as disturbances, are introduced below and summarized in Table 10.1. Some of these phenomena may be used to provide controllable torquers. The magnitude of torques in space is small when compared with terrestrial standards. Even very small ones become significant when there is no friction to oppose them and when the orientation has to be very accurate.

Table 10.1 Disturbance torques

EXTERNAL TORQUES Source	Height range over which it is potentially dominant
Aerodynamic	< about 500 km
Magnetic	500–35 000 km
Gravity gradient	500–35 000 km
Solar radiation	> synchronous
Thrust misalignment	all heights
INTERNAL TORQUES Source	
Mechanisms	
Fuel movement	
Astronaut movement	
Flexible appendages	
General mass movement	

10.4.1 Thrusters (*external type*)

Thrusters are potentially the largest source of torque on the spacecraft. Being external they affect the total momentum. Ideally the thrust vector passes through the centre of mass, but inevitably there is a tolerance on this and consequently a disturbance torque arises.

The main means of countering the effects of this torque when large thrust levels are present are either to spin the vehicle about the intended thrust direction or to provide means of controlling the achieved thrust direction. At lift-off, for example, the latter method must clearly be used. This involves mounting some of the thrusters in gimbals, or using secondary fuel injection into the rocket nozzle. At other times the spacecraft will usually be made to spin about the axis of the thruster in order to average out the offset effects.

Thrusters are in common use in attitude-control systems for providing controllable external torquing, and hence of controlling the total momentum of the spacecraft. Thrusters for this purpose will be mounted in clusters on the surface of the vehicle, pointing in different directions in order to provide three components of torque. They have a number of advantages and disadvantages compared with their main rival, the magnetic torquer.

Their main advantage is that their torque level is independent of altitude and there is potentially no limit to its magnitude. However, the magnitude is not controllable when installed, but only the switch-on time. This torquing system integrates well with the station-keeping requirement for thrusters since a common fuel and control system can be used.

Thrusters have a number of disadvantages. Their restriction to an on–off type of control leads to a limit-cycle occurring, as indicated in Section 10.6. If they are used to achieve accurate pointing, then the torquers will need to provide small and consistent impulses, and their minimum switch-on time of several milliseconds leads to a low maximum torque being needed. Typically a minimum thrust impulse of order 10^{-4} N s may be required, with a thrust level as low as 10^{-2} N. This may be relaxed when they are used just for momentum dumping. A variety of thruster systems may be used, ranging from cold gas to electric propulsion, as indicated in Sections 6.3 and 6.4.

The fact that thrusters use fuel is another of their disadvantages. Although a large amount of fuel is not normally needed for attitude control it will eventually be exhausted, and a number of spacecraft have reached the end of their useful life because of this.

10.4.2 Magnetic torque (*external type*)

The magnetic field generated by a spacecraft interacts with the local field from the Earth and thereby exerts an external couple on the vehicle. This is a similar effect to that of a compass needle which attempts to align itself with the local direction of the field. If the spacecraft's magnetism is represented as a dipole whose magnetic moment is **m**, then it reacts with the local flux density **B** to produce a torque **T** given by

$$\mathbf{T} = \mathbf{m} \times \mathbf{B} \tag{10.19}$$

Care must be taken that electric currents and spurious magnetic effects do not cause a significant torque (see Chapter 16).

Electromagnets may be used to provide a controllable external torque. Their strength can be controlled by means of the current i. Their reaction with a local field **B** leads to the couple:

$$\mathbf{T} = niA(\hat{\mathbf{c}} \times \mathbf{B}) \tag{10.20}$$

where n is the number of turns,

        A is the cross-sectional area of the coil,

and $\hat{\mathbf{c}}$ is the unit vector in the direction of the coil's axis.

The magnet may take the form of wire wrapped round the spacecraft (with assembly problems), or may be a purpose-built rod which may be bolted to the structure. Three orthogonal magnets enable the direction and magnitude of the dipole to be controlled.

Magnetic torquers are in common use in satellites orbiting at altitudes up to geostationary altitude, but they become less popular at the higher altitudes since the strength of the Earth's field reduces with height (see Figure 2.13).

The field's strength and direction also vary with the position of the spacecraft in its orbit in general, and when using magnetic torquers it is common practice to carry a magnetometer to measure the local field. A feature of magnetic torquers is that they cannot produce a torque component about the local field direction. In a polar orbit any required torque direction can always be achieved at some point in the orbit since the field direction changes round the orbit. In an equatorial orbit the field lines always lie horizontally north–south; but even so a rotation about this direction can be achieved by rotating the torque vector about the field line. In the resulting coning motion the required reorientation will occur.

An advantage of magnetic torquers is that they require no fuel and so have virtually unlimited life. They do of course require electrical power. But there is no exhaust pollutant and by providing a couple they are not sensitive to movement of the centre of mass.

10.4.3 Gravity-gradient torque (*external type*)

This source of torque occurs because in a gravitational field which gets weaker with increase of height a body will only be in stable equilibrium if its axis of minimum inertia is aligned with the local vertical.

The gravitational force dF on an increment of mass dm is

$$dF = \frac{\mu \, dm}{r^2} \tag{10.21}$$

where μ is the Earth's gravitational constant, $= 0.3986 \times 10^{15} \text{ m}^3/\text{s}^2$,
         r is the distance from Earth's centre.

By summing the moments about the cenre of mass C the torque components may be shown to be

$$T_x = (3\mu/2r^3)(I_{zz} - I_{yy}) \sin 2\phi \cos^2 \theta$$

$$T_y = (3\mu/2r^3)(I_{zz} - I_{xx}) \sin 2\theta \cos \phi \tag{10.22}$$

$$T_z = (3\mu/2r^3)(I_{xx} - I_{yy}) \sin 2\theta \sin \phi$$

where θ, ϕ refer to the pitch and roll angles, using the aircraft axis convention (see Figure 3.A2).

An oscillatory 'libration' mode will occur if these torques govern the motion about the equilibrium state. For small oscillations of an axisymmetric spacecraft ($I_{yy} = I_{xx}$) the motion is like a conical pendulum, whose frequency is

$$\omega_{\text{lib}} = \sqrt{[(3\mu/r^3)(1 - I_{zz}/I_{xx})]} \text{ rad/s} \tag{10.23}$$

Gravity-gradient torque provides a passive self-aligning torque which has been used (e.g. Seasat), but the libration does need damping to be incorporated. The torque levels will be low unless a long thin configuration is used, or in the case of tethered satellites.

10.4.4 Aerodynamic torques (*external type*)

Aerodynamic torques are dominated by the drag force, which is dependent on frontal area A. Their total moment about the centre of mass C may be assessed by considering the projection in the direction of travel.

The torque contribution about the z-axis is

$$T_z = \tfrac{1}{2}\rho C_D \int yV^2 \, dA \tag{10.24}$$

where ρ is the atmospheric density (see page 22),
         C_D is the drag coefficient, normally taken to be about 2.2.

For zero torque spacecraft designers will of course aim to locate the centre of mass close to the centre of area, but tolerances, shifts of the centre of mass and thermal distortion will affect the balance.

The torque is height-dependent and is not an important effect above about 600 km, depending on the spacecraft configuration.

10.4.5   Solar radiation pressure (*external type*)

Solar radiation produces a force on a surface which depends upon its distance from the Sun; it is independent of the height above the Earth. Large flat surfaces with a significant moment arm about the centre of mass, such as solar arrays, may produce a significant torque.

The pressure depends upon the momentum change brought about by the absorption, reflection or diffusion of the radiation. If $\hat{s}$ is the unit vector from the spacecraft to the sun and dS represents an increment of surface area, directed along its normal, then the force due to radiation pressure will be

$$-P \cos i[(1 - C_s) \, dS \, \hat{s} + 2(C_s \cos i + C_d/3) \, dS] \qquad (10.25)$$

where P is the mean momentum flux, $= 4.4 \times 10^{-6}$ N/m^2 at Earth,
          i is the angle of incidence of the radiation,
and C_s, C_d are the coefficients of specular and diffuse reflection.

The torque may be obtained by integrating the moment over the spacecraft's surface area.

Both aerodynamic and solar radiation torques are in principle usable for counteracting momentum build-up but usually the spacecraft is designed such that the forces balance out to give zero torque.

10.4.6   Mass movement (*internal type*)

The movements of masses within a spacecraft exert torques upon the main structure and are liable to alter the location of the centre of mass and also the inertial properties. They are classified as internal torques and do not affect the total momentum.

The centre of mass C has been identified as a key reference point for establishing the dynamic behaviour (Chapter 3). Moving the point affects the balance of the vehicle in dynamic ways. It also affects the torques due to forces on the vehicle, but not the couple of the magnetic torquer. In principle the centre of mass location could be controlled in order to balance out the disturbance torques.

A major source of mass movement is that of the fuel. The tanks are normally located in such a way that as their contents are used up the centre of mass does not shift. Fuel movement within the tanks causes a different sort of problem in that it moves in a dynamic way in response to the motion of the spacecraft—fuel slosh—affecting its modal characteristics.

Mass movements from one position to another, such as the erection of solar arrays and other appendages, and movement of astronauts, etc. have an effect upon attitude which is best assessed by using the fact that angular momentum is conserved.

10.4.7   Momentum storage torquers (*internal type*)

Torquers associated with momentum storage such as reaction wheels (RWs) and momentum wheels (MWs) are essentially internal torquers, suitable for attitude control but not for controlling the total momentum.

These devices are purpose-built precision-engineered wheels which rotate about a fixed axis, with a built-in torque motor which accelerates the wheel in either direction. The reaction wheel has a nominally zero speed, whereas the momentum wheel has a high nominal speed which is typically of order 10 000 r.p.m., which provides momentum bias. Both types provide momentum storage, and need to be used in conjunction with external torquers, as described on page 232.

The principle of momentum wheels has been extended by the development of more advanced forms, such as control moment gyroscopes (CMGs). By mounting the wheel in gimbals fitted with torque motors all three components of torque may be developed from a single wheel. There is potential for incorporating attitude-sensing with momentum-storage and momentum bias in sophisticated devices of this type.

10.5   ATTITUDE MEASUREMENT

10.5.1   Attitude: its meaning and measurement

The meaning of 'attitude' or 'orientation' usually presents no conceptual difficulties. There must be some datum frame of reference, and once this has been chosen then the attitude of a spacecraft refers to its angular departure from this datum. A right-handed set of axes is normally used in order to define a frame of reference, and if both a datum set and a set of spacecraft axes are chosen, then the attitude may be defined in a way that may be quantified.

Specifying attitude may be done in a number of ways such as Euler angles, direction cosines, quaternions, etc. [1]. Three pieces of information are needed. A common way is to use the three Euler angles which are defined in the same way as is standard practice for aircraft. These are the angles of yaw ψ, pitch θ, and roll ϕ, as measures of the rotations about the z-, y-, and x-axes respectively, in that sequence, which are needed to bring the datum axes into alignment with those of the spacecraft. Figure 3.A2 illustrates these rotations.

For a spacecraft in circular orbit, whose z-axis is nominally down the local vertical and whose x-axis is nominally in the direction of travel, the aircraft's standard is frequently used. For other applications a star-fixed set would be better. There is, however, no universally accepted standard for specifying a spacecraft's attitude.

A potential problem when using Euler angles as above is that there is a singularity when the pitch angle θ is 90°. Whilst a set of angles (ψ, 90°, ϕ) may be chosen to specify any such attitude, it is not a unique set. For example if the aircraft's set is used for Shuttle, namely x is horizontal, z is vertically down and y completes the right-handed set, then when it is on the launch-pad its pitch attitude is 90° and its yaw and roll angles cannot be uniquely specified. Such a problem may be overcome by choosing a more suitable datum set.

It is worth noting that angles, and consequently attitude, are not vector quantities. The combination (ψ, θ, ϕ) should not be thought of as three components of a vector. On the other hand the rates of change of $\dot{\psi}$, $\dot{\theta}$, $\dot{\phi}$ can be interpreted as vector quantities, whose directions are along the (non-orthogonal) axes about which the rotations take place (see Figure 3.A2). Resolving $\dot{\psi}$, $\dot{\theta}$, $\dot{\phi}$ along spacecraft axes enables the components of the spacecraft's angular velocity $\boldsymbol{\omega}$ relative to the datum axes to be expressed as

$$\omega_x = \dot{\phi} - \dot{\psi} \sin\theta$$
$$\omega_y = \dot{\theta} \cos\phi + \dot{\psi} \cos\theta \sin\phi \tag{10.26}$$
$$\omega_z = \dot{\psi} \cos\theta \cos\phi - \dot{\theta} \sin\phi$$

The inverse relationship is

$$\dot{\psi} = (\omega_y \sin\phi + \omega_z \cos\phi)/\cos\theta$$
$$\dot{\theta} = \omega_y \cos\phi - \omega_z \sin\phi \tag{10.27}$$
$$\dot{\phi} = \omega_x + (\omega_y \sin\phi + \omega_z \cos\phi) \tan\theta$$

When the angles are small, then $\dot{\psi} \approx \omega_z$, $\dot{\theta} \approx \omega_y$, and $\dot{\phi} \approx \omega_x$.

Equations (10.27) indicate how, by integration, the attitude in the form of the Euler angles (ψ, θ, ϕ) may be obtained from measured components of angular velocity. The singularity at $\theta = 90°$ shows up in the form of $\tan\theta$ and will lead to problems with the integration as θ approaches this value.

10.5.2 Measurement system fundamentals

Fundamentally, the measurement of attitude requires the determination of *three* pieces of information which relate the spacecraft axes to some datum set, whether they are in the form of Euler angles or in other forms. The measurement subsystem must include sufficient sensors to enable the information to be extracted with the necessary accuracy, and with reasonable simplicity. This must be done at all phases of the mission.

There are *two categories of sensor*, and they are commonly used to complement each other in a measurement system. *The reference sensor* gives a definite 'fix' by measuring the direction of an object such as the Sun or a star, etc., but there are normally periods of eclipse during which its information is not available.

In many cases there is a requirement for an axis of the spacecraft or its equipment to point towards a detectable object, for example the solar array towards the Sun. Then the reference sensor may be used in a nulling role, providing direct information about the pointing *error*. It will need an accurate calibration of its null point and only a narrow field of view. It will, however, need to be backed up by sensors with wide fields of view which can be used during the initial 'capture' stage of orientating the spacecraft correctly.

Inertial sensors measure continuously, but they measure only *changes* in attitude, effectively relative to a gyroscope rotor. They therefore need a fix—a calibration from reference sensors. In between fixes their errors progressively increase due to random drifts.

A measurement system may be formed by using reference and inertial sensors to complement each other. In a simple combination the reference sensor will calibrate the inertial sensor at discrete times and the latter will then effectively 'remember' the reference object's direction until the next calibration. This allows a period in eclipse to be covered. The

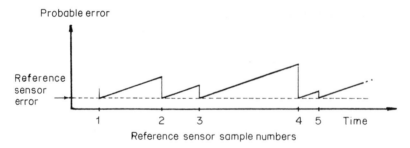

Figure 10.6   Error history for a mixed reference/inertial sensor system

accuracy of the system will fluctuate, being that of the reference sensor at the calibration instant, and steadily degrading until the next calibration, as shown in Figure 10.6.

It is clear that the achievement of good system accuracy calls for good accuracy from the reference sensors, and a low degradation (drift) rate from the inertial sensors. In practice it is likely that the mixing will take place in a Kalman filter, whose design to minimize the errors is a specialist topic.

Complete attitude information requires *three* pieces of information as explained above. Most reference sensors are based upon detecting the direction of a vector, and any one sample of their outputs is incapable of providing all three pieces. A sun sensor cannot detect any rotation of spacecraft about the Sun vector for example. Two vector directions, ideally orthogonal are needed for complete attitude information.

In some cases the two vector directions can be provided by just one sensor. This happens when the vector direction changes between samples so that the missing information is provided by later samples. A star mapper is an example. It scans the heavens and so one sensor uses different star directions for complete attitude determination. The time lapse between viewing two stars in orthogonal directions needs to be short enough to avoid significant build-up of errors. An Earth sensor (detecting the local vertical) can be used in a similar fashion. The local vertical rotates in space through 90° in a quarter of a circular orbit, a lapse of about 23 minutes in LEO and 6 hours in geostationary orbit. The Sun vector's 90° time is three months!

The accuracy required will normally be set by the payload, its pointing direction and the required measurement accuracy. There will be an ultimate accuracy associated with the object used by the sensor; stars provide the most accurate sources, with the Sun and Earth being progressively less accurate by virtue of the angle which they subtend at the sensor, and the fuzziness of their horizons. A rough guide to accuracies is shown in Table 10.2.

The extent to which the ultimate accuracy is realized in a given instrument depends upon its design, and this will be related to its intended use. Each sensor will have a limited field of view, beyond which it gives no information, and in getting a spacecraft to its intended attitude it will normally be necessary to include very wide-angle low-accuracy sensors for use when steering the craft towards the state in which the accurate sensor has its objective in its field of view.

Each phase of the mission must be addressed when the list of sensors is being compiled. In the early stages the visibility conditions will be quite different from those when it is on station, partly because the orbit is different, but also because of stowed arrays etc. The

Table 10.2 Potential accuracies of reference sensors

Reference object	Potential accuracy
Stars	1 arcsecond
Sun	1 arcminute
Earth (Horizon)	6 arcminute
RF beacon	1 arcminute
Magnetometer	30 arcminutes

This table gives only a guideline.

sensor list must also cover the possibility that the spacecraft may need to be recaptured following a failure of some sort.

The datum axes which are used when defining the attitude of the spacecraft are normally related to its payload. A telescope is likely to need star-based datum axes for specifying its attitude, for example. For the control system it is the error measured from the intended attitude which is important, and in many systems sensors will be chosen to measure this directly. For example the pointing of the solar arrays towards the Sun will use a Sun sensor. Earth-facing hardware will possibly use the local vertical as a datum, or maybe beacons on the ground which have been set up for the purpose. The error in measuring the pointing direction of the payload will be reduced if the sensor uses a related pointing object.

Errors in measuring the payload's pointing direction will also depend upon the separation between it and the sensor. If each is mounted on the same base, then the mounting tolerances of both instruments will be sources of error, and so will distortions of the base due to thermal or other effects. Error paths of this type should be kept to a minimum and must be carefully assessed when the location of sensors measuring payload pointing directions is being considered. For extreme accuracy the payload pointing will be calibrated in space.

10.5.3 Types of reference sensor

There are numerous different forms of sensor. Only a brief review of the selection on offer is given here.

Sun sensors

The Sun subtends an angle of about 32 arcminutes at Earth, and provides a well-defined vector, which is unambiguous because of the intensity of the radiation. Sensors range from mere presence detectors which determine whether the Sun is in a specified field of view, to instruments which measure its direction to an accuracy of better than one arcminute.

The basic essentials for the sensor are a detector element, which may be photocells, and an optical system which defines the direction of the Sun's rays when the cells are

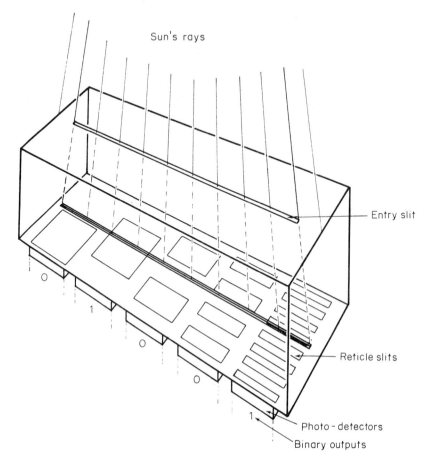

Figure 10.7 Sun sensor with digital output

illuminated. For example, by exposing an entry slit to the Sun a plane of light is obtained in the instrument. Reticle slits mask detectors as in Figure 10.7 such that their sequence of binary outputs indicates the plane in which the Sun lies. Two such instruments give the two pieces of information which define the Sun's vector direction. With a spinning mounting base a single reticle slit may be used in order to detect when a fan beam passes across the Sun; by using two angled beams and noting the time lapse between the passage of the Sun across them the direction of the Sun vector can be determined. This instrument will also measure the spin rate of the mounting base.

Other types of optical/detector combinations are available.

Earth sensor

The Earth, radius R, subtends the angle $2 \arcsin(1/(1 + h/R))$, at a satellite at a height h. At 500 km it is about 135°, falling to 17.5° at geostationary altitude. Sensing the direction of the local vertical entails bisecting the directions to the horizons at the ends of a diameter of its disc, and horizon sensors provide the means of doing this.

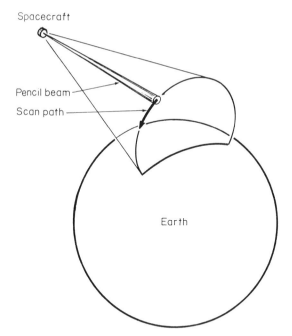

Figure 10.8 Scanning technique for an Earth horizon sensor

A horizon sensor typically uses an infra-red detector together with a fan or pencil beam in order to sense the abrupt change in intensity as the beam sweeps from deep space across the horizon. The beam scans the horizon either by means of an internal rotating mechanism or by using the rotating base of a spinning spacecraft structure.

Figure 10.8 shows a schematic diagram in which a pencil beam traces out a cone which intersects the Earth's surface, generating a pulse during that part of the revolution. This is just one type of scan. If the pulses from two such cones on opposite sides of the spacecraft are of equal duration, then the local vertical lies in the plane bisecting the cone angles, and this can form the basis of a nulling control system for roll motion. Pitch sensing uses the fact that the local vertical lies in the plane defined by the bisector or the pulse from each conical scan.

For this system one cone angle will be suitable for only a limited range of altitudes. In some sensors the angle may be adjusted to accommodate the changes in altitude as the spacecraft moves from LEO to geostationary height.

The accuracy of Earth sensors is typically about 5–10 arcminutes.

Star sensors

The stars are the most accurate reference sources in common use for measuring attitude. Accuracies down to 1 arcsecond or better may be obtained. But the star sensor is large, heavy, and expensive, and there is a need to identify the particular star(s) in its field of view.

The *star tracker* is a form of sensor which maintains one or two particular stars in its field of view, which is typically less than 1°, either by mounting the optical system on gimbals or by using electronic tracking techniques. For a spacecraft which provides a spinning base a *star scanner* may be used. This detects a pattern of stars passing across its field of view and the spacecraft attitude is determined by recognized the pattern.

Radio frequency beacons

Direction-finding techniques may be used to detect the direction of an RF source, with an accuracy of order 1 arcminute. There are several techniques by which this can be done. It can be a useful sensor when the spacecraft is using Earth-facing antennas.

Magnetometers

The magnetometer is a robust instrument but with an accuracy which is limited to about 0.5°. It measures the direction and possibly strength of the local magnetic field. But the field is not well mapped and has abnormalities which make the sensor of limited use for attitude sensing. It is used in conjunction with magnetorquers as described on page 241.

10.5.4   Inertial sensors

Gyroscopes form the basis of the inertial sensing system for attitude. The conventional, wheel, type of gyro has a rotor mounted in an environment which is very carefully controlled. In the rate- and rate-integrating types the wheel is torqued so that it follows the motion of the spacecraft. The torque is a measure of the angular rate about the instrument's sensitive axis (see equation (3.26)).

The laser gyro is an alternative type which has no moving parts. The principle is that the time taken for light to travel the length of a fibre-optic coil will change if the coil is rotating about its axis. The difference in time is a measure of the angular rate.

A set of three orthogonal rate-gyros will measure the components $(\omega_x, \omega_y, \omega_z)$ of the spacecraft's angular velocity. The output of a rate-integrating gyro (RIG) is the integral of the angular velocity component, such as $\int \omega_x \, dt$ etc. Only when the direction of the RIG axis remains fixed in space does its output represent the angular displacement about the axis. In more general motion equations (10.27) apply.

In high-quality sensors, drift rates of less than 0.01 degree/h are obtainable.

10.6   ACS COMPUTATION

10.6.1   The computer

The development of digital computers for use in spacecraft has proceeded rapidly, and is still doing so. They must perform reliably in the radiation environment of space, and a number of space-qualified ones now exist. Further development is providing more power and speed, the use of floating-point arithmetic, and the capability of being programmed in higher-level languages. These on-board computers (OBCs) link with ground-control computers, which will normally host their software development tools (see Chapter 15). The availability of powerful computers means that spacecraft will be given greater autonomy, and many of the sophisticated control techniques which find applications in ground-based systems may be used on spacecraft.

Robustness is a requirement for ACS and other on-board systems. For example the ACS must potentially operate with large flexible structures such as solar arrays, whose natural frequencies cannot be established accurately before launch. Fixed algorithms will tolerate limited variation from their expected value whereas adaptive control, which requires computer support, may allow greater tolerance. This form of control may also compensate for parameter changes which take place over the lifetime of the spacecraft, and with some forms of hardware failure.

Computer power will also benefit the attitude measurement subsystem. The mixing of sensor outputs to achieve maximum accuracy via the Kalman type of filter requires computer modelling. In addition they can provide the substantial data back-up which is needed when star mappers are used.

10.6.2 A simple control example

The sophisticated techniques above are specialist topics which are beyond the scope of this book. However, control of the rotation of a rigid body about one axis can be achieved with a simple PID algorithm (proportional, integral, and differential). This is the basis of many such systems and is used in the following example. It is important that cross-couplings between the motions about the three axes is not severe (see Chapter 3), and any large reorientations may be implemented sequentially to avoid them.

The following example illustrates the type of motion which may be expected when a simple three-term PID controller is used to control the roll attitude of a spacecraft.

When no momentum bias is present the roll error ϕ will respond to a roll torque T_x as follows (see equation (10.16)):

$$I_{xx}\ddot{\phi} = T_x \tag{10.28}$$

A PID controller generates a demanded torque signal based upon measured values of roll error and of roll rate, as in

$$T_{xD} = -K_p\phi_m - K_i \int \phi_m \, dt - K_d(\dot{\phi})_m \tag{10.29}$$

This can be made to represent a stable system by appropriate choice of the constants K_p, K_i, and K_d, due allowance being made for any delay in implementing the torque. The presence of any flexure modes, fuel movement, etc., will also impose constraints on the choice of these constants.

Pointing errors will result as a consequence of zero errors from the roll-error sensor and drift in the integration process, but constant disturbance torques and a constant error from the roll-rate sensor will be automatically compensated for.

With an on–off control torquer such as gas jets the outputs from equation (10.29) will be interpreted as a switching signal e rather than a demand for torque, as shown in Figure 10.9. Torque impulses are generated whenever e exceeds a threshold set by the accuracy required, and the spacecraft coasts between the thresholds on each side of the nominal attitude required. The period T is

$$T = 8\phi_d/\Delta \text{ seconds} \tag{10.30}$$

and typically has a value in excess of 10 s. The fuel consumption is

$$\Delta^2 I_{xx}/4\phi_d l I_{sp} g_0 \text{ kg/s} \tag{10.31}$$

where $\pm \phi_d$ is the permitted pointing error (rad),
     Δ ($=$ torque-impulse/I_{xx},) is the change in angular rate caused by a torque impulse (rad/s),
     I_{sp} is the specific impulse (s),
and l is the moment arm of the thrusters (m).

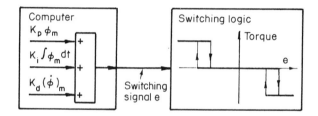

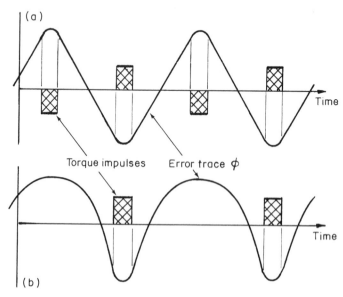

Figure 10.9 Limit cycle for an on–off control system, with no momentum bias: (a) no disturbance torque—hard limit cylce; (b) with disturbance torque—soft limit cycle

A constant disturbance torque will bias the cycle towards one side, and if it is strong enough ($> \Delta^2 I_{xx}/8\phi_d$) it will lead to the one-sided cycle shown in Figure 10.9. This roughly halves the fuel consumption, with little change in the period.

This form of control would suit the momentum-dumping operation in a momentum-storage system, with the gas jets taking over the pointing role during the dumping process.

With momentum bias present the limit cycle takes a different form, the drift between the thresholds being due to disturbance torques, with nutation motion superimposed. The philosophy of this control is similar to that of stationkeeping (see Section 5.2).

REFERENCES

[1]   Wertz, J.R. (1978) *Spacecraft Attitude Determination and Control*, Reidel,
[2]   Kaplan, M.H. (1976) *Modern Spacecraft Dynamics and Control*, Wiley,
[3]   Hughes, P.C. (1985) *Spacecraft Attitude Dynamics*, Wiley,

11 ELECTRICAL POWER SYSTEMS

John P.W. Stark

Department of Aeronautics and Astronautics, University of Southampton

11.1 INTRODUCTION

Provision of electrical power for space vehicles is, perhaps, the most fundamental requirement for the satellite payload. Power-system failure necessarily results in the loss of a space mission, and it is interesting to note that many of the early satellite systems failed due to such loss. The demand for power has increased and is characterized by enhanced spacecraft operational complexity and sophistication. This may be seen from Figure 11.1. The earliest spacecraft, such as Vanguard 1, typically required a power raising capability of only ~1 W, whereas current communications satellites typically require three orders of magnitude greater than this. Evolving trends suggest that a further two orders of magnitude will still be needed, and even more if developments such as the satellite power systems (SPS) concept come to fruition [1], wherein some 5 GW of power will be generated.

The best methods of raising power can be broadly related to power level and mission time as shown in Figure 11.2. It is apparent that photovoltaic (solar cells) or radioisotope generators (RTG) are appropriate for the power requirements typical of present generation spacecraft, namely a few kilowatts for missions of several years. For shorter periods fuel cells show advantage, whereas for periods of less than a few days batteries come to the fore. It is not surprising, therefore, to discover that batteries are used in launch vehicles to provide the primary energy source, fuel cells are used in Shuttle and both photovoltaic devices and radioisotope generators are used for general spacecraft operation, dependent upon the mission. It should be noted that nuclear sources of power such as RTG and nuclear dynamic systems are used for military applications, but are not generally acceptable for civilian vehicles in Earth orbit. Before the individual elements of a spacecraft power system are considered, the overall power system configuration will be described briefly.

Spacecraft Systems Engineering. Edited by P.W. Fortescue and J.P.W. Stark
© 1991 John Wiley & Sons Ltd

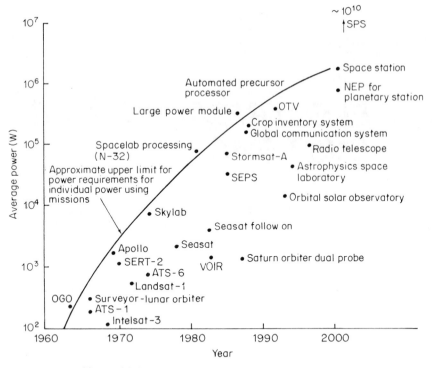

Figure 11.1 Evolution of mission power demand [4]

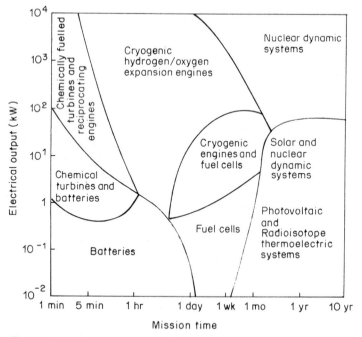

Figure 11.2 Power outputs: mission duration relationship between energy source and appropriate operational scenario [2] (Reproduced with permission from Angrist, S.W. (1982) *Direct Energy Conversion*, 4th edn. Allyn and Bacon, New York.)

11.2 POWER SYSTEM ELEMENTS

In general a spacecraft power system consists of three main elements: primary and secondary energy sources, and a power control/distribution network. These are shown schematically in Figure 11.3.

The primary energy source converts a fuel into electrical power. On early space flights and on launch vehicles batteries have provided this. Strictly these systems do not have a fuel element, in that a battery is a device which stores energy rather than performing a direct energy conversion process.

The majority of present-day spacecraft use a solar array as the primary energy source. The 'fuel' in this case is solar radiant energy, which is converted via the photovoltaic effect (see below) into electrical energy. On manned missions fuel cells have been used most frequently as the primary source. These electrochemical devices perform a controlled chemical reaction in such a way that electrical energy may be derived rather than heat energy. The fuels used for space operation are hydrogen and oxygen, yielding water as the reaction product; this may then be drunk by astronauts. Nuclear systems utilize either a radioactive decay process (RTG) or a nuclear fission process as the energy source. RTG make use of the thermoelectric effect, whereas fission reactors operate in a manner similar to terrestial nuclear power plants [3].

The secondary energy source is required to store energy and subsequently deliver electrical power to the satellite system and its payload when the primary system's energy is not available. The most usual situation when this condition arises is during an eclipse period when the primary system is a solar array. The eclipse's duration depends on the spacecraft orbit; typically for LEO a 35 minute eclipse occurs in each orbit for low-inclination satellites; in GEO, eclipses occur only during equinoctial periods, with a maximum duration of 1.2 h in a 24-hour period. For such short times batteries demonstrate the highest efficiency. However, for systems which require high power levels, typically 100 kW, a solar array/regenerative fuel-cell combination has improved characteristics over a solar array/battery combination [4]. Regenerative fuel cells operate in a closed fuel cycle: H_2/O_2 fuel is consumed to form water on the 'discharge' cycle and electrolysis of water is performed during the 'charge' cycle, with power for this being derived from the solar array. Whilst the net efficiency is low, only 50–60% compared with nearly 90% for a battery, it is possible by judicious sizing of the fuel (H_2/O_2) component to reduce the size of the solar array required for primary power raising. For LEO operations where aerodynamic drag is significant, the reduction in array area reduces the

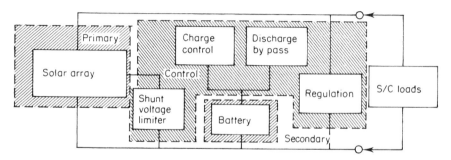

Figure 11.3 Schematic of typical spacecraft power system block elements

mass of propellant required for orbit control, leading to a lower wet mass of the system at launch compared with the conventional array/battery configuration.

The power control and distribution network is required to deliver appropriate voltage–current levels to all spacecraft loads when required. Several salient features should be noted.

The primary power source always degrades during the mission. Thus at its start an excess of power will be generated, and it is necessary to provide an ancillary load to dissipate it. The simplest way to do this is to use a resistive load, generally external to the main spacecraft structure in order to simplify the thermal design.

Both primary and secondary power system characteristics will change, leading to a requirement for voltage, and/or current regulation. The changes arise due to both de-gradation effects—such as cell failures, and also illumination variations caused by chang-ing solar array aspect angles with respect to the Sun. The customary approach is to use a voltage shunt regulator across the array. A variety of types have been used, perhaps the simplest being either a pulse-width modulation scheme [5], or sequential switching shunt regulation, known as S^3R [6].

Charge control of a battery system is particularly important to maintain the lifetime and reliability of battery units. It generally necessitates both current and voltage control. A variety of techniques may be used to sample the state of a battery and these will be discussed in Section 11.6. Discharge control is also required in order to limit current output.

A spacecraft's various electrical loads will require specific voltage inputs for satisfactory operation, and the electrical 'bus' may therefore be required to provide a variety of voltages. Generally within Europe the trend has been to have a regulated d.c. power bus, typically at 28 V or 50 V. As an example the Exosat bus provides 28 ± 0.3 V with 0.2 V peak to peak ripple whereas ECS uses a 50 ± 1 V power line. In contrast US spacecraft generally use unregulated buses; for example the NASA standard for unmanned space-craft provides a voltage in the range 21–35 V d.c. Present spacecraft are being designed with higher bus voltages (~ 150 V) to reduce resistive losses and harness mass. For both regulated and unregulated systems d.c.–d.c. converters are required to provide the variety of voltages needed; such conversion frequently takes place at an equipment level rather than centrally.

An a.c. bus is sometimes used to augment the d.c. one. The hybrid system can provide mass savings due to both the simplicity of conversion from a.c. to a variety of d.c. levels and also since it is possible to run the power distribution harness at higher voltages if a.c. supply is used. Indeed it has been noted [7] that equivalent wiring cross-section is 6 times smaller for a three-phase a.c. distribution network than for a d.c. one. A.c. distribution is mainly applicable to high-power spacecraft, and where a large number of d.c. voltage levels are required at equipment level. A.c. buses generally have a square-wave voltage waveform, an example being Hipparcos. A notable exception to this is on Shuttle Space-lab, where the a.c. power bus is more sophisticated, providing a three-phase sine-wave voltage form at 400 Hz.

11.3 SOLAR ARRAYS

A solar array is an assembly of many thousand individual solar cells, connected in a suitable way to provide d.c. power levels from a few watts to tens of kilowatts. For a

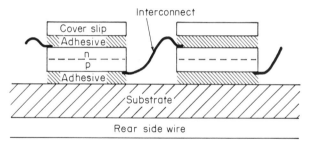

Figure 11.4 Schematic of a typical solar cell assembly

detailed description of both terrestrial and space solar arrays Rauschenbach [8] is recommended.

Each solar cell assembly has a semiconductor p–n junction as shown schematically in Figure 11.4. For spacecraft applications the base material typically has a resistivity of between 10^{-3} and $10^2\,\Omega\,cm$. Using silicon for example, it might be doped with boron to form the p-type material (electron deficient), and with phosphorous for the n-type (electron excess).

With no illumination the junction achieves an equilibrium state in which no current flows. But when it is illuminated with suitable radiation, photons with sufficient energy will create electron–hole pairs and the radiation is converted to a potential across the cell, with usable electrical power. The incident photon energy needed for this must exceed a *band gap* which depends upon the material, as shown in Table 11.1. Photons with excess energy dissipate it as heat within the cell, leading to reduced efficiency.

Characteristic voltage/current curves for cells are shown in Figure 11.5. Typically open circuit voltages for silicon cells lie between 0.5 and 0.6 V under solar illumination. The plot of power against voltage (Figure 11.6) has a clear maximum, with a particularly rapid fall once the optimum voltage is exceeded. On some spacecraft maximum-power-point tracking is used to operate the array most efficiently. Increase in cell temperature results in decreasing open-circuit voltage with only a modest increase in short-circuit current. The theoretical maximum efficiency of both silicon (Si) and gallium arsenide (GaAs) cells is shown as a function of temperature in Figure 11.7. It shows the particular sensitivity of Si to temperature and also the improved performance of GaAs at high temperatures. This means that it is theoretically possible to use a focusing optical arrangement for GaAs cells such that these cells may be illuminated by an intensity greater than the nominal radiation intensity of $1.4\,kW/m^2$ in Earth orbit. Fewer cells would then be required to

Table 11.1 Properties of semiconductor materials

Material	Band gap (eV)	Maximum wavelength (μm)
Si	1.12	1.12
CdS	1.2	1.03
GaAs	1.35	0.92
GaP	2.24	0.554
CdTe	2.1	0.59

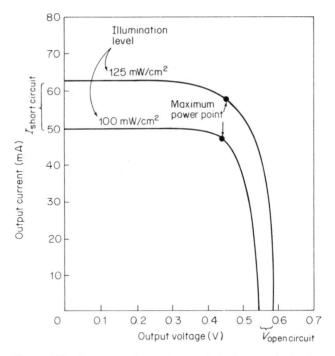

Figure 11.5 Current–voltage characteristic for a typical solar cell. The short-circuit current is dependent upon both the illumination level and the size (area) of the cell (Reproduced with permission from Angrist, S.W. (1982) *Direct Energy Conversion*, 4th edn, Allyn and Bacon, New York.)

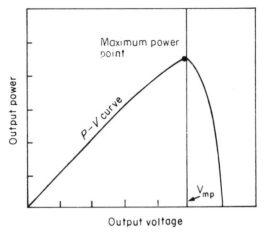

Figure 11.6 Power–voltage characteristic for a typical solar cell (from *Solar Cell Array Design Handbook* by H.S. Rauschenback, copyright © 1980 by Van Nostrand Reinhold. All rights reserved)

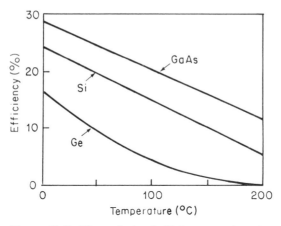

Figure 11.7 Theoretical cell efficiency as a function of temperature for three semiconductor materials (from *Solar Cell Array Design Handbook* by H.S. Rauschen-back, copyright © 1980 by Van Nostrand Reinhold. All rights reserved)

provide a given power level, which could result in array cost reductions since the cell cost is a large proportion of the total cost. Concentration ratios as high as 100:1 for GaAs cells have been investigated in such studies, but at present the benefits of these systems are not conclusive.

The n-type contact on the upper surface of the cell is in the form of a multiple-finger arrangement. These fingers are required for efficient current collection, whilst maintaining good optical transparency (typically $\sim 60\%$). They are connected at a bar, along one edge of the cell. Frequently titanium/silver (Ti/Ag) is used for this.

Radiation damage is a problem with solar cells. In general Si cells having higher base resistivity ($\sim 10\,\Omega$ cm) are the most tolerant of radiation. Furthermore cells have the n-type material uppermost, since on early space flights it was discovered that cells having the p-type material as the upper region rapidly suffered from radiation damage. Thin cells suffer less than thicker ones, but at present they have a lower conversion efficiency. GaAs cells are more radiation tolerant than Si and for this reason there is considerable interest and effort in their development.

The cover glass provides environmental and radiation protection. For design purposes the particle fluence of a spacecraft's radiation environment may be expressed as an equivalent fluence of monoenergetic 1 MeV electrons (see Figure 11.8). Degradation of cell output to this irradiation is generally available from manufacturers' data; Figure 11.9 shows typical degradation curves for cells with a variety of thicknesses.

The effectiveness of the cover glass depends on its density and thickness. Suitable glass microsheet is commercially available in several thicknesses from 50 μm to 500 μm [9]. Their absorption of radiation follows approximately an exponential law, so that the intensity of radiation after traversing a depth x into the glass is

$$I \sim I_0 \exp^{-k\rho x}$$

where I_0 is the initial radiation fluence at $x = 0$, ρ is the density of material and k is an energy-dependent absorption coefficient. For fused silica Figure 11.10 shows the effect of

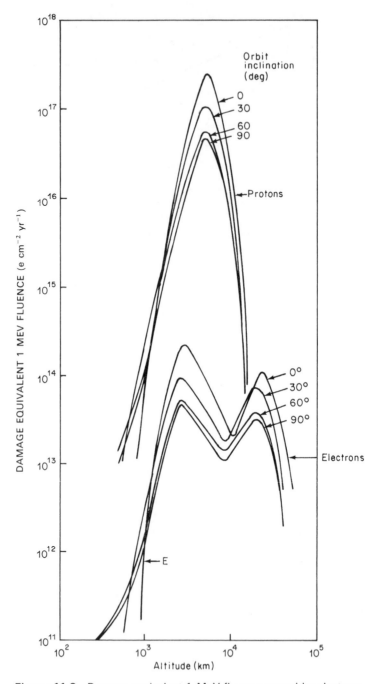

Figure 11.8 Damage equivalent 1 MeV fluence caused by electrons and protons due to trapped particles, to silicon cells protected by 150 μm fused silica covers and infinitely thick rear shielding [8] (from *Solar Cell Array Design Handbook* by H.S. Rauschenback, copyright © 1980 by Van Nostrand Reinhold. All rights reserved)

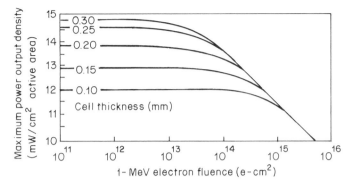

Figure 11.9 Effects of thickness and fluence on conventional non-p$^+$ silicon solar cell performance [8] (from *Solar Cell Array Design Handbook* by H.S. Rauschenback, copyright © 1980 by Van Nostrand Reinhold. All rights reserved)

changing glass thickness. It should be noted that in order to evaluate the total radiation fluence absorbed by the solar cell both front-side and rear-side fluences must be calculated. On rigid panel arrays mounted on honeycomb structure, and on solar arrays bonded to the spacecraft wall (e.g. spinning spacecraft) the rear side is effectively screened from radiation, whilst on lightweight arrays this is not so valid.

Additional features required of the cover glass are that it provides good optical coupling between free space and glass and also between glass and adhesive, and that it provides suitable wavelength selection, limiting the u.v. flux to the adhesive layer and the cell. These features are achieved using an anti-reflection coating, such as magnesium fluoride on the upper surface, perhaps with an additional indium oxide conductive coating as described in Chapter 2; a u.v. filter coating may be applied to the underside of the cell, to reflect u.v. radiation. For cover glass with cerium doping additional u.v. filtering is unnecessary.

For *efficient cell operation* and insensitivity to radiation a shallow junction depth (typically less than 10 μm) is required. Various Si cell configurations have been investigated to improve conversion efficiency. These include the following.

- back-surface reflectors (BSR) to reflect unabsorbed radiation from the rear side of the p-region back through the cell. This reduces cell heating.

- the introduction of a p$^+$-region or back-surface field (BSF) at the rear of the p-regions. This exhibits higher output due to enhanced carrier collection efficiency, but the improvement is lost under high fluence damage.

- the use of a textured front surface of the cell reduces reflection from the cell surface.

Solar arrays using Si cells are made from individual cells which at present are generally rectangular, 2 cm × 4 cm, having a conversion efficiency of ∼ 12–14%. To improve packing efficiency, much work on high-efficiency 5 cm × 5 cm cells is under way. The cells have a thickness of between 50 and 250 μm, ∼ 200 μm being used for the majority of arrays. Reliable production of ultra-thin cells (< 100 μm) has not yet been demonstrated.

Interconnections between cells represent a major array failure hazard. This arises due to the thermal cycling inherent upon entry/departure from sunlight to eclipse. Since the materials used for cell and substrate are different, differential expansion takes place during

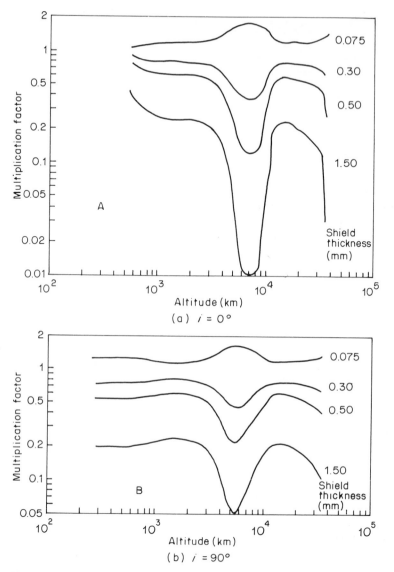

Figure 11.10 Multiplicative factors to be applied to damage fluence on a solar cell as a function of coverslip (shield) thickness, and operational orbit height [8]

the rapid temperature change ($\sim 100°C$ in a few minutes). Thermal stress-relieving loops are required to reduce such failure mechanisms as interconnect lift-off and fracture.

A variety of *substrate materials* have been used and proven in space. Frequently Kapton with fibre reinforcement, $\sim 100~\mu m$ thick, forms the immediate interface with the cell, which may be mounted on a honeycomb panel for rigidity (for example TDRS solar array). Increasingly, flexible cell blankets are used to reduce mass. The space telescope solar array is a typical example, with glass fibre reinforcing the Kapton. For the largest

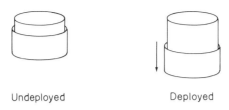

Undeployed   Deployed

Figure 11.11

arrays flexible substrate materials offer mass savings. But it is interesting to note that recent work [10] has shown that for power levels up to 6 kW, mass savings occur when using advanced rigid arrays, wherein Kapton is reinforced with carbon fibre.

As noted earlier individual cells produce power at a voltage of ~ 0.5 V, and it is necessary to connect many cells in series. Reliability is then achieved by additional parallel coupling at each cell; typically three or four cells form a parallel combination. This series–parallel arrangement is called a solar cell string. Further protection is afforded using shunt diodes which provide current bypass paths should individual cells become shadowed. Shadowing can cause cell failures since if a cell is unable to generate power because of loss of illumination, then the entire string voltage may appear across the cell.

Now consider the *system level interactions* between the solar array design and the vehicle itself. The relatively low conversion efficiency of an array results in the need for large areas of solar cells to intercept sufficient solar radiation for the power demand.

For a spinning satellite—using either dual spin or simple spin ACS the 'drum' size evidently limits the power that can be generated. The drum itself is limited by the launch volume. A possible solution is to introduce an additional mechanism such as the 'drop skirt' used on Intelsat VI. In this a larger-diameter hollow cylinder is deployed in the manner shown in Figure 11.11 to expose additional cell area. Typically within the drop skirt thrusters are located which may impact on the skirt material causing both an unwanted disturbance and a loss of propulsive efficiency, together with the possibility of contamination.

Other aspects to note for the spinning satellite solar array are coupled with the thermal environment. Assuming a typical spin rate of ~ 50 r.p.m., the average temperature of the array can be maintained at a lower value than for the three-axis configuration. This leads to an increased efficiency of the individual cells, and thus a decrease in the required active cell area. However, since not all the array is instantaneously illuminated, a factor of $\sim \pi$ times the number of cells is required for the same collected power. Since also on the spinner the array is mounted essentially on the body of the spacecraft the temperature excursions noted on the array (between sunlit and eclipse phases) are reduced, having a significant impact on reducing the thermal shock characteristic of a three-axis deployed solar array.

Considering the three-axis satellite, the solar array requires a mechanism to deploy the stowed array following launch and then orientate it appropriately to track the Sun. These deployment mechanisms may be of a simple telescopic construction, or of the 'Astromast' variety. In the Astromast the carbon-fibre members of the mast are deformed by coiling them into a stowage container. They may then be driven out by a screw mechanism to the undeformed (extended) configuration.

Tensioning wires are then required to achieve an acceptable minimum fundamental frequency of the array largely because of AOCS requirements. On three-axis stabilized

Table 11.2   Solar array performance figures

Panel	Type	Power (kW)	kg/kW	W.m^2
Hughes: FRUSA	Flexible fold-out	1.1	29.2	67
AEG: DORA	Flexible rollout	6.6	24.9	82.5
SEPS	Flexible fold-out	25	15.2	130
TC1	GSR fold-out	1.1	47.3	66

vehicles power take-off from the array generally, but not always, requires a rotary degree of freedom between the satellite and the array, in order that pointing requirements of the array and the payload may be met. This requires two elements—the mechanical rotation device to allow the body to move relative to the array (with appropriate sensing systems) and an electrical power take-off device. The provision of power using a solar array clearly has many design interactions with the rest of the spacecraft system.

Table 11.2 summarizes performance characteristics for a variety of solar arrays including both flight-proven systems and also development systems.

11.4   FUEL CELLS

A fuel cell converts the chemical energy of an oxidation reaction directly into electrical energy, with minimal thermal changes. A major advantage is its flexibility from a system viewpoint. For example, it provides power during both day and night, and the fuel has a high energy density and thus provides a compact solution compared with a solar array. The evident disadvantage is the need to carry fuel.

For space applications the hydrogen/oxygen fuel cell has been used, a product of whose reaction is water. This is clearly useful for manned missions. A schematic diagram of such a cell is shown in Figure 11.12.

The voltage which appears at the terminals of an ideal cell is given by

$$E_r = \frac{-\Delta G}{nF}$$

where ΔG is the change of Gibbs free energy occurring in the reaction, n is the number of electrons transferred and F is the Faraday constant (product of Avogadro number and elementary charge) equal to $9.65 \times 10^4 \, C \, mol^{-1}$. For the reaction of the H_2/O_2 cell two electrons are transferred per mole of water formed and ΔG has the value of $-237.2 \, kJ/mole$ at 25°C. The reaction takes place spontaneously.

Thus the reversible voltage of the ideal cell is $237.2 \times 10^3/(2 \times 9.65 \times 10^4) = 1.229$ V.

In practice this is not realized because there are various irreversibilities, termed polarization losses. Figure 11.13 shows a typical current–voltage curve for a hydrogen/oxygen fuel cell. Initially, as soon as a current is drawn from the cell, a rapid drop in voltage occurs. This is associated with the energy required to activate the electrode reactions.

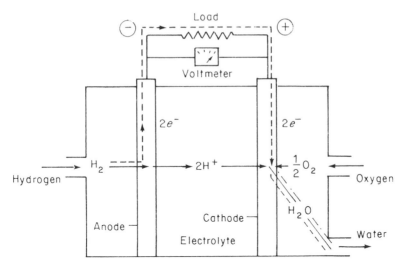

Figure 11.12 Schematic of a hydrogen/oxygen fuel cell. At the enode/electrolyte interface, hydrogen dissociates into hydrogen ions and electrons. The hydrogen ions migrate through the electrolyte to the cathode interface where they combine with the electrons that have traversed the load [2] (Reproduced with permission from Angrist, S.W. (1982) *Direct Energy Conversion*, 4th edn, Allyn and Bacon, New York.)

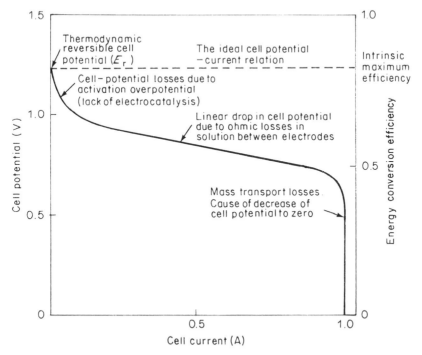

Figure 11.13 Typical cell potential and efficiency–current relation of an electrochemical electricity producer showing regions of major influence of various types of overpotential losses [14]

For the H_2/O_2 fuel cell these are three-phase: gas (fuel), solid (electrode), and liquid (electrolyte—this can also be a solid). It is necessary for the reactants to be chemisorbed onto the electrode, a process which requires breaking and forming new chemical bonds, and it requires energy; hence the voltage drops. The process is called activation polarization. The magnitude of voltage drop is given by the Tafel equation:

$$\Delta V_{ACT} = a + b \ln J$$

where J is the current density at the electrodes, and a and b are temperature-dependent constants for the reaction/surface description.

As the current drawn increases, a linear voltage–current relation is noted. This is simply the resistive nature of the electrolyte.

At high current values problems arise due to the transport of reactants to the reaction sites, a feature which is particularly important at porous electrodes since pressure gradients are set up which limit flow rates. Additionally, species concentrations are not uniform and thus ionic species can create a back e.m.f. This concentration polarization provides the ultimate limit on current density that may be achieved by a fuel cell. All of these electrochemical polarization processes are common to both batteries and fuel cells, and their voltage–current characteristics are very similar.

Early fuel cell systems were primarily based upon the technology of solid polymer electrolyte (SPE). For the Gemini series 1 kW was produced at a specific power of 33 W/kg, within a volume of 0.05 m³, and the objective to extend missions to greater than four days was achieved. However, the water produced was not of drinking quality because of degradation of the fuel-cell membrane.

The Apollo system, also used for Skylab, was based upon matrix aqueous alkaline technology and achieved a power level of 1.5 kW at a specific power of 25 W/kg. It had to operate whilst the vehicle was on the lunar surface, at a temperature greater than 394 K. The selected system, a Bacon fuel cell, operated at 505 K and 60 lb/in².

Shuttle developments, also based upon the alkaline technology, have improved the specific power by an order of magnitude, ~ 12 kW, 275 W/kg. Further, the start-up time for this cell is 15 minutes with shut-down being instantaneous, whereas for Apollo, 24-hour start-up periods were required with 17 hours shutdown. Table 11.3 summarizes past and present fuel-cell status. Regenerative fuel cells wherein water is also electrolyzed, are not yet space-proven.

Table 11.3 Performance summary of fuel cells for space use

System	Specific power (W/kg)	Operation
Gemini	33	
Apollo	25	
Shuttle	275	2500 hr at P$_{ave}$
SPE technology	110–146	> 40 000 hr
Alkaline technology	367	> 3000 hr
Alkaline technology	110	> 40 000 hr
Goal (lightweight cell)	550	

11.5 RADIOISOTOPE GENERATORS (RTG)

The operation of a radioisotope generator is based on the thermoelectric effect noted by Seebeck, that it is possible to generate a voltage between two materials, A and B (either conductors or semiconductors) if a temperature difference is maintained (see Figure 11.14). This is analogous to a thermocouple. Practical RTG space systems utilize two semiconductor materials—one p-type, the other n-type—in order to exploit the effect.

The power output from such a device is a function of the absolute temperature of the hot junction, the temperature difference that may be maintained between the junctions, and also the properties of the materials. Because such devices are relatively inefficient (less than 10%) one major problem in their design is removing waste heat.

The heat source used in space systems is derived from the spontaneous decay of a radioactive material. As this decays it emits high-energy particles which can lose part of their energy in heating absorbing materials. Suitable fuels are listed in Table 11.4, which shows the half life ($\tau_{1/2}$) for each of the fuels, namely the time required for the amount of a given radioactive isotope in a sample to halve. Thus over a period to time t, the power available from such a fuel decreases by an amount given by

$$P_t = P_0 \exp\left(\frac{-0.693}{\tau_{1/2}} t\right)$$

where P_t is the power at time t after some initial time t_0. The table indicates that high specific power levels are available from sources with shorter half lives (and hence shorter duration missions).

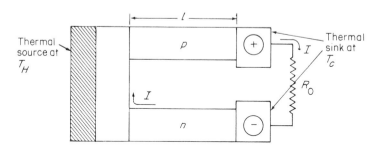

Figure 11.14 Schematic diagram of a semiconductor radioisotope generator (Reproduced with permission from Angrist, S.W. (1982) *Direct Energy Conversion*, 4th edn, Allyn and Bacon, New York.)

Table 11.4 Possible fuels and their performance for radioisotope generators [2]

Isotope	Fuel form	Decay	Power density (W/g)	$\tau_{1/2}$(yr)
Polonium 210	GdPo	α	82	0.38
Plutonium 238	PuO_2	α	0.41	86.4
Curium 242	Cm_2O_3	α	98	0.4
Strontium 90	SrO	β	0.24	28.0

Table 11.5   RTG system performance

Name	Power (W)	kg/kW
Galileo Probe/Ulysses:	285	195
(GPHS RTG Late 1980s)		
Nimbus/Viking/Pioneer	35	457
(SNAP 19 mid 1970s)		
Apollo Lunar Surface	25	490
Experiment:		
(SNAP-27 Early 1970s)		
SNAP 9A 1960s	73	261

The advantages of RTGs over other systems include the following points:

1. They provide independence of power production from spacecraft orientation and shadowing.

2. They provide independence of distance from the Sun (deep space missions are possible).

3. They can provide low power levels for long periods of time.

4. They are not susceptible to radiation damage in the Van Allen belts.

5. They are suitable for missions with long eclipse periods, e.g. lunar landers.

SNAP-19 (System for Nuclear Auxiliary Power), which powered the Viking lander vehicle to Mars, had a specific power of 2.2 W/kg, with a thermal/electric efficiency of ~5%. The output electrical power was 35 W. Table 11.5 summarizes data on the present generation of RTG systems.

11.6 BATTERIES

Batteries have been used exclusively for the secondary power system, providing power during periods when the primary one is not available. As a back-up for a solar array this means that the batteries must provide power during eclipses, and that the array must recharge the batteries in sunlight.

In GEO operation eclipses only occur during the two equinoctial periods, producing eclipse seasons for the spacecraft. These last some 45 days at each equinox. Initially they are short (~minutes), but lengthen to a maximum of 1.2 hours before decreasing again. The total of ~90 eclipses thus occur irregularly with significant periods of time when no battery operation is required. With more than 22 hours of sunlight available in each day, a trickle charge solution is possible.

In LEO, on the other hand, the spacecraft may be in eclipse and thus require battery power for 40% of each orbit. Although the precise duration will depend on orbit inclination, it is fairly regular, and the eclipse cycle results in typically 5000–6000 charge/

discharge cycles of the battery per year. This results in the array power sizing needing to be nearly twice the nominal load requirement.

In summary LEO operation requires a large number of low-depth discharges, whereas in GEO a few deep discharges suffice. This inevitably influences battery type, resulting in the present trend of using nickel–cadmium (Ni–Cd) or silver–cadmium (Ag–Cd) cells for LEO operation and nickel–hydrogen (Ni–H) cells for GEO operation. Cell cycle life, specific weight (kW h/kg) and volume (kW h/m^3) all influence the acceptability of a particular battery technology. However, work on more exotic materials, eg Li–SO$_2$, is continuing and thus it is to be expected that alternative technologies will soon be implemented on spacecraft (see for example the annual GSFC battery workshop proceedings [4]). Table 11.6 summarizes the performance characteristics of a number of cell technologies.

The detailed electrochemistry of batteries is covered in [11] and [12]. The main function of battery operation which is of importance to spacecraft design is the way in which the reliability and charge efficiency are related to charge control. Parameters of critical importance are the charge/discharge rate, the depth of discharge, the extent of overcharging and the thermal sensitivity to each of these parameters. Figure 11.15 summarizes some available data on Ni–Cd batteries. A further feature not indicated in these graphs is the changing performance of a battery after cycling, specifically the change in voltage–current characteristics. The predominant effect is that the charge control system must be flexible if long missions are to be successfully executed. One other notable

Table 11.6 Performance and goals of battery technologies for space use

Type	Goal	Performance
Ni–Cd	LEO $\left\{\begin{array}{l} 28.6 \text{ W h/kg} \\ 30\,000 \text{ cycles} \\ 5 \text{ years} \end{array}\right.$	LEO $\left\{\begin{array}{l} 4\text{–}9 \text{ W h/kg} \\ <30\,000 \text{ cycles} \end{array}\right.$
	OR	OR
	GEO $\left\{\begin{array}{l} 900 \text{ cycles} \\ 10 \text{ years} \end{array}\right.$	GEO $\left\{\begin{array}{l} 11 \text{ W h/kg} \\ \sim300 \text{ cycles} \\ \sim3.5 \text{ years} \end{array}\right.$
Ag–Zn	53 W h/kg	53 W h/kg
	450 cycles	60% DOD
	5 years (GEO)	2.5–4.5 years
Ni–H$_2$	48–55 W h/kg	
	900 cycle/10 yr	~33 W h/kg
	6000 cycle/1 year	>650 cycles
Ag–H$_2$	66 W h/kg	
	900 cycle/10 yr	~66 W h/kg
	6000 cycle/1yr	>900 cycles @ 75% DOD
Li–FeS	187 W h/kg	
	>1000 cycles	<77 W h/kg
	~10 years	>500 cycles
Na–S	220–264 W h/kg	~120 W h/kg
	~10 years	170 cycles

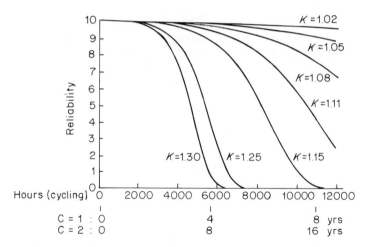

Figure 11.15 Ni–Cd cell reliability as a function of overcharge factor. Hours cycling is related to operation for two cases: $c = 1$, charge rate is battery capacity $\frac{1}{20}$ A/A h; $c = 2$, charge rate is battery capacity $\frac{1}{10}$ A/A h [16] (reproduced by permission of European Space Agency)

feature of battery ageing is the effect of hysteresis on the battery capacity. Figure 11.16 demonstrates the loss of this charge capacity over several cycles. It has been noted that if a battery is completely discharged, then capacity may be regained. Whilst this process may result in reverse polarization problems, battery reconditioning before an eclipse season is regularly used for GEO spacecraft.

Whilst the charge/discharge rate may be controlled in a fairly simple manner through current regulation, the monitoring of the state of charge in a battery is more complex and deserves attention here. The principal methods which may be used to sense charge-state are cell voltage, cell temperature, or cell pressure. It should be noted that the level of full

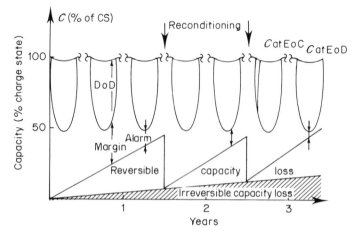

Figure 11.16 Battery reconditioning via complete discharge to improve battery capacity. Both reversible and irreversible capacity loss occurs [16] (reproduced by permission of European Space Agency)

charge noted by each of these methods results in a different level of overcharging. Pressure and temperature sensing results in overcharging by 20–30%, whereas voltage sensing may indicate 10–20% overcharge [13]. A principal problem with voltage sensing arises due to the voltage–charge–temperature characteristics which may cause significant errors in determining the state of charge of the battery.

REFERENCES

[1]   *SPS Concept Development and Evaluation Programme Reference System Report*, US, DOE and NASA DOE/ER 0023 (October 1978).
[2]   Angrist, S.W. (1982) *Direct Energy Conservation*, Allyn and Bacon, New York.
[3]   Patterson, W.C. (1976) *Nuclear Power*, Penguin Books, London.
[4]   Sheibley, D.W. (1983) Regenerative H_2–O_2 fuel cell––electrolyser systems for orbital energy storage. NASA CP-2331, 23–38.
[5]   Schreger, A. (1974) Design and stability analysis of a PWM shunt regulator (S3R). ESRO-SP 103, 97–102.
[6]   O'Sullivan, D. and Weinberg, A. (1977) The sequential switching shunt regulator (S^3R). ESA SP 126, 123–31.
[7]   Eggers, G. (1985) AC buses for LEO—a viable alternative. ESA SP-230, 17.
[8]   Rauschenbach, H.S. (1980) *Solar Cell Array Design Handbook*, Van Nostrand.
[9]   Taylor, H., Simpson, A.F., and Dollery, A.A. (1984) CMX-50: a new ultra-thin solar-cell cover for lightweight arrays. ESA SP-173, 211–14.
[10] Mawira, D. (1982) Advanced rigid array. ESA SP-173, 9–14.
[11] NASA RP 1052 (1979).
[12] NASA SP 172 (1968).
[13] Lechte, H. (1977) How essential are advanced techniques for control and protection of NiCd batteries on GEO missions? ESA SP-126, 263.
[14] Bockris, J.O'M. and Srinivason (1969) *Fuel Cells and their Electrochemistry*.
[15] Loferski, J.J. (1973) *10th IEEE Photovoltaic Specialists Conference*, 1.
[16] Montalenti, P. (1977) Software/hardware interface in control and protection of space batteries. ESA SP-126, 271–8.

12 THERMAL CONTROL OF SPACECRAFT

Ross Henderson

Department of Aeronautics and Astronautics, University of Southampton

12.1 INTRODUCTION

In the design of spacecraft, thermal control is needed in order to maintain structural and equipment integrity over long periods of time. It has been recognized since the conception and design of the first space vehicles that a prime engineering requirement is a system for temperature control that permits optimum performance of many components. In fact, if it were possible to operate equipment at any temperature, there would be no need for thermal control. The spacecraft thermal balance between cold space (4 K) and solar, planetary, and equipment heat sources is the means by which the desired range of equipment and structural temperatures is obtained. With the total spacecraft balance set, subsystem and component temperatures can be analysed for their corresponding thermal requirements.

Reliable long-term performance of most spacecraft components takes place at or near room temperature. Temperature extremes may also be required in the form of cryogenic fluids stored at less than $-180°C$ or in the form of nuclear generators operating at greater than $500°C$. Temperature limits for typical spacecraft components are shown below in Table 12.1.

Heat is generated both within the spacecraft and by the environment. Components producing heat include rocket motors, electronic devices, and batteries. Heat from the environment is largely the result of solar radiation; initial ascent heating effects are largely compensated for by the use of suitable shrouds in the case of expendable vehicles to achieve orbit conditions. Thus, any form of spacecraft thermal control makes use of suitable materials or devices to achieve a balance between the heat absorbed and the heat emitted, resulting in some required temperature.

12.2 THE THERMAL ENVIRONMENT

An important characteristic of the space environment is its high vacuum, or the virtual absence of atmospheric pressure [1]. The following discussion assumes an Earth-orbiting spacecraft but applies to any planetary body with an atmosphere.

Spacecraft Systems Engineering. Edited by P.W. Fortescue and J.P.W. Stark
© 1991 John Wiley & Sons Ltd

Table 12.1 Temperature tolerances of typical spacecraft components

Electronic equipment (operating)	−10 to +40°C
Batteries	−5 to +15°C
Fuel (e.g. hydrazine)	+9 to +40°C
Microprocessors	−5 to +40°C
Bearing mechanisms	−45 to +65°C
Solar cells	−60 to +55°C
Solid-state diodes	−60 to +95°C

The thermal conductivity of the Earth's atmosphere is a function of atmospheric temperature gradients and is independent of variations in pressure or density at altitudes below 90 km. However, above 90 km the molecular mean free path becomes comparable to the distance in which the temperature gradient varies appreciably, and thermal conductivity ceases to be pressure independent. By 300 km altitude, convective heat transfer is negligible. Another way of expressing this is to say that aerodynamic heating can be considered to be negligible above 300 km altitude. Table 12.2 indicates the aerodynamic heat flux associated with example Earth satellites; examination of the figures shows that at 150 km it is approximately equal to the solar constant (at Earth orbit). From 150 km to 300 km it reduces by two orders of magnitude, thus the spacecraft aerodynamic shroud should not be separated below approximately 150 km altitude.

Hence, heat transfer is by radiation only and the actual physical temperature of a spacecraft is determined by the exchange of energy by means of the following:

- direct solar radiation;
- solar radiation reflected from nearby planets (albedo radiation);
- thermal energy radiated from nearby planets;
- radiation to deep space from the spacecraft.

Table 12.2 Aerodynamic heating of Earth satellites

Altitude	Aerodynamic heat flux ($W\,m^{-2}\,h^{-1}$)	
	Circular orbit	Eccentric orbit ($e = 0.13$)
100	170 000	17 000
150	1300	130
200	160	16
250	39	3.9
300	15	1.5
350	7	0.7
400	3.7	0.4
450	2.2	0.2
500	1.2	0.1

The spacecraft will experience thermal equilibrium when the radiant energy received from the first three sources listed above is equalled by the emitted energy [2]. The temperature at equilibrium is called the radiation equilibrium temperature and is used in the initial analysis of spacecraft design to establish system feasibility.

It should be noted that in the case of spacecraft remote from planetary objects, only the direct solar radiation and spacecraft radiation to space become significant.

12.2.1 Solar radiation

The solar radiation parameters of interest to the thermal designer are (1) spectral distribution, (2) intensity, and (3) degree of collimation. The spectral distribution may be considered constant throughout the solar system and the solar irradiance, or spectral energy distribution, resembles a Planck curve with an effective temperature of 5800 K which means that the bulk of the solar energy (99%) lies between 150 nm and 10 μm with a maximum near 450 nm.

The solar radiation falling at right angles on an area of 1 m^2 at a solar distance of 1 AU (149 598 200 $\pm$ 500 km), is approximately 1371 $\pm$ 5 W/m^2, and is called the solar constant. A solar radiation intensity J_s for any distance D from the Sun can be expressed as

$$J_s = \frac{P}{4\pi D^2} \tag{12.1}$$

where P is the total power output from the Sun, 3.8×10^{25} W.

The intensity variation given equation (12.1) is indicated in Table 12.3. The angle subtended by the Sun is 0.53° at the average Earth–Sun distance.

Planetary (or normal) albedo is the fraction of the incident solar radiation returned from a planet. Earth's albedo varies with its surface conditions. For clouds it is 0.8; for green areas (forests, fields) it ranges from 0.03 to 0.3 [4]. Average values are given in Table 12.4.

Albedo irradiation of a spacecraft depends upon the bearing angle β between the local vertical and the rays of the sun. Maximum illumination occurs when $\beta = 0$, but since the

Table 12.3 Planetary solar constants [3]

Planet	Solar radiation intensity (percentage of solar constant* at 1 AU)
Mercury	667
Venus	191
Earth	100
Mars	43.1
Jupiter	3.69
Saturn	1.10
Uranus	0.27
Neptune	0.11
Pluto	0.064

*1371 $\pm$ 5 W/m^2 with a 3.4% seasonal variation caused by the varying Earth-Sun distance.

Table 12.4   Average planetary albedo, a [3]

Mercury	0.06	Saturn	0.42
Venus	0.61	Uranus	0.45
Earth	0.34	Neptune	0.52
Mars	0.15	Pluto	0.16
Jupiter	0.41	Moon	0.07

irradiation varies as the square of the distance, the intercepted illumination area and the isotropic nature assumed for scattered radiation from a planetary atmosphere (lambertian surface), it varies in a complex manner.

The albedo contribution J_a to the total radiation input to a spacecraft can be expressed in terms of a visibility factor F, which is the fraction of the total albedo (a) which is actually intercepted by the craft. Thus

$$J_a = J_s a F \qquad (12.2)$$

The dependence of F on the altitude h and the bearing angle β is shown in Figure 12.1.

For most heat-transfer calculations the spectral distribution of the Earth's albedo may be assumed to be equivalent to that of the Sun, and for engineering calculations this applies to the other planets too.

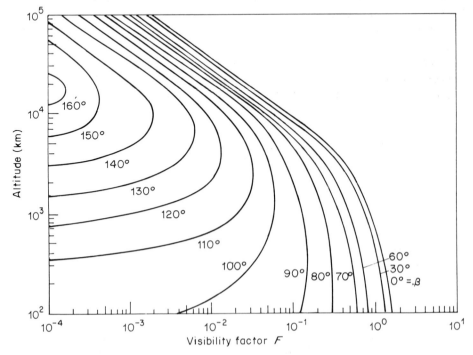

Figure 12.1   Spacecraft albedo irradiation

12.2.2   Planetary radiation

The planets of the solar system behave as black body radiators, each with an emission of energy that satisfies their planetary heat balance equations.

The thermal radiation of the Earth and atmosphere is predominantly infra-red with a wavelength greater than $1.5\,\mu$m. Its values vary seasonally and with latitude; however, the quantities of interest for satellite thermal control are the long-term averages. Thermal time constants are great enough to mask the effects of short-term, localized variations in Earth thermal emission. Only small errors ensue when the annual average emission is used for any season. Also, because of the effect of heat storage, the change in absorbed solar radiation causes only small seasonal variations in the average emission of the whole Earth. At satellite altitudes it can be assumed that all incident radiation emanates from the total cross-sectional area of the Earth. For time increments greater than 3 hours, Earth's thermal emission is taken to be $237 \pm 21\,\text{W/m}^2$. Curves for the Earth and atmospheric emission spectra are reproduced in Figure 12.2 where the solid curve is the approximate radiation from the Earth and atmosphere. The two Planck curves (dotted) presented relate to the radiation from the Earth's surface (288 K black body) and the radiation from the atmosphere in spectral regions where the atmosphere is opaque (218 K black body).

The temperature of a satellite depends upon absorbed energy or the spectral distribution of incident energy and the spectral radiation characteristics of the satellite surface. However, the total Earth emissive power and total absorptance for long-wavelength energy can be used to calculate satellite heating. Absorptance for most materials shows

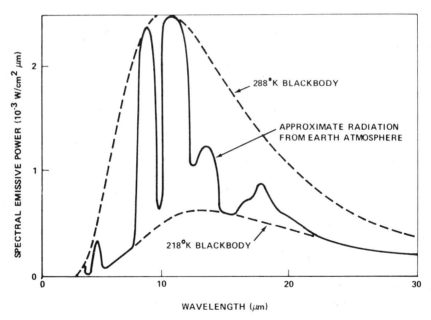

Figure 12.2   Typical spectral emissive power curve for the thermal radiation from Earth. (*Note*: The 288 K black-body curve approximates the radiation from the Earth's surface, and the 218 K black-body curve approximates the radiation from the atmosphere in those spectral regions where the atmosphere is opaque.)

only small variations with wavelength beyond 8 μm, and, as illustrated in Figure 12.2, nearly all energy emitted by the Earth-atmosphere system is beyond 8 μm.

12.2.3 Spacecraft heat emission

The spacecraft itself radiates heat into space as a blackbody having a certain value of emissivity. For practical purposes space can be considered as a black body at 0 K. It should be noted that this heat transfer takes place from the total surface area of the vehicle, $A_{s/c}$.

12.3 SPACECRAFT THERMAL BALANCE

The previous section has outlined the thermal environment in which a spacecraft operates. In addition to these external inputs there are internal heat sources represented by the equipment operated inside the spacecraft body and dissipating energy as heat. The only effective means of achieving temperature control is to adjust the spacecraft radiation balance so that the absorbed energy is balanced by the radiated energy at the required isothermal mass temperature. An ideal isothermal mass temperature is taken to be 25°C ± 10°C, with a maximum temperature envelope of 10°C. These requirements establish the possibility of temperatures within a spacecraft ranging from 5 to 45°C. The fundamental parameters establishing the required degree of radiation balance are internal power dissipation and heat generation, and the ratio of absorptance to emittance (α/ε) of the external surface.

Absorptance (α) is the ratio of radiant energy absorbed by a body to that incident upon it. For a specific material it depends on the nature of the radiation source. Emittance (ε) is the ratio of energy emitted by a surface to the energy emitted by a black-body radiator at the same radiation equilibrium temperature. A black body is a theoretical body (an ideal emitter) that radiates energy at the maximum possible rate per unit area at each wavelength for any given temperature. It absorbs all radiant energy incident upon it (i.e. $\alpha = 1$).

The α/ε ratio for various surface materials under solar radiation is listed in Table 12.5. Casual application of the data should be avoided as these values are averages. Precise calculations must be based upon spectral energy distributions at the required equilibrium temperature. The emittance values (ε) are temperature dependent. Values of α/ε for spacecraft materials under solar radiation are widely available in spacecraft thermal design handbooks. The thermal designer should use these for the particular design problem in hand.

Theoretically, it is easy to adjust the temperatures of the spacecraft to the required values by selecting the emittance and absorptance of the external surfaces. The major difficulty arises, however, from the variability of the heat inputs and outputs. The solar input and albedo disappear during eclipse periods. Some spacecraft missions involve trajectories with large changes of the distance to the Sun and corresponding changes in the insolation factor or the intensity of radiation received.

The attitude of the spacecraft can change radically, according to each particular mission and mission phase, and so it presents different surfaces and areas to the same external

Table 12.5 α/ε for a selection of surfaces and finishes [5]

Surface	Absorptance α	Emittance ε	α/ε
Gold plate on Al 7075	0.3	0.03	10.0
Polished aluminium	0.35	0.04	8.75
Polished beryllium	0.4	0.05	8.0
Gold on aluminium	0.26	0.03	6.5
Polished stainless steel	0.5	0.13	3.85
Polished copper	0.28	0.13	2.2
Grafoil	0.66	0.34	1.9
Vapour-blasted stainless steel	0.6	0.33	1.8
Gold/Kapton/aluminium	0.53	0.42	1.26
Epoxy black paint	0.95	0.85	1.12
Acrylic black paint	0.97	0.91	1.07
Epoxy white paint	0.2	0.85	0.24
Acrylic white paint	0.22	0.88	0.25
Acrylic white paint after u.v. irradiation	0.5	0.88	0.57
Silicon solar cell, bare	0.82	0.64	1.3
Silicon solar cell, silica cover	0.82	0.81	1.0
Silicon solar cell, silica cover, blue filter	0.78	0.81	0.96
Silicon solar cell, silica cover, red filter	0.7	0.81	0.86
Kapton (5 mil)/aluminium	0.48	0.81	0.6
In_2O_3/Kapton/aluminium	0.4	0.71	0.56
Quartz fabric/tape	0.19	0.6	0.3
OSR (quartz mirror) silvered Teflon	0.08	0.81	0.1
FEP (5 mil)/silver	0.11	0.8	0.14
FEP (2 mil)/silver	0.05	0.62	0.08

heat flux; additionally this variation is dependent on the overall configurational design, leading to shadowing of variable surface areas of the spacecraft. The heat dissipation inside the craft can also be highly variable for certain missions and thus complicate the thermal design task.

It becomes clear that, although the basic theory of radiation heat transfer is essentially simple in its application to the spacecraft in its environment, the application of that theory is highly complex in the variability of heat inputs and outputs with time.

Internal to the spacecraft, heat transfer is both by radiation and conduction. It is the basic aim of the thermal design to limit the radiant and conductive heat transfer from the exterior surface into the internal structure and subsystems. For certain high power dissipating components, however, it may be necessary to install controlled heat leaks for eventual emission into cold space.

12.3.1 View factors

The higher the satellite altitude the less significant are the effects of planetary radiation and albedo as the radiation intercepted by the satellite area is proportional to the solid angle of the emitting surface [6]. The geometric relationship for the calculation of spacecraft irradiation is shown in Figure 12.3.

The radiation per unit area incident on the spacecraft is the planetary emission multiplied by the solid angle defined by the spacecraft altitude and the tangent to the Earth's surface. In evaluating this we can define the view factor F_{12} between two surfaces of area A_1 and A_2 (Figure 12.4) as

$$F_{12} = \frac{1}{A_2} \int_{A_1} \int_{A_2} \frac{\cos \phi_1 \cos \phi_2}{\pi s^2} dA_1 \, dA_2 \qquad (12.3)$$

where ϕ_1 is the angle which the normal to the incremental surface dA_1 makes with the direction from dA_1 to dA_2, and ϕ_2 is similarly defined. In Equation (12.3) F_{12} accounts for the radiation intercepted by A_2 due to radiation from A_1.

It can be seen from this that F_{12} is very general and can be applied not only to the

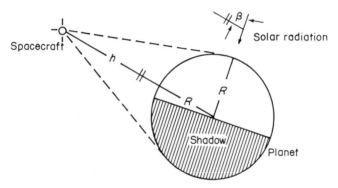

Figure 12.3   Geometric relationship for calculation of spacecraft irradiation

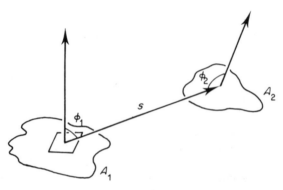

Figure 12.4   View factor geometry between surfaces A_1 and A_2

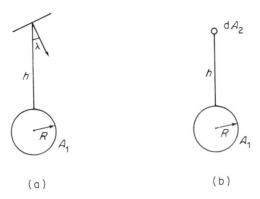

Figure 12.5 Geometry of a small vehicle modelled (a) as a flat plate, and (b) as small sphere, orbiting a planet A_1

interception of radiation by the vehicle from external sources, but can also be applied to radiative heat transfer between components on-board the vehicle which are at different temperatures.

For a small vehicle orbiting a spherical planet a particularly simple form for F_{12} is obtained. If the vehicle is modelled as a small flat plate (Figure 12.5a), the view factor is

$$F_{12} = \frac{\cos \lambda}{(1 + H)^2} \qquad \text{if } \lambda + \sin^{-1}\left(\frac{1}{1 + H}\right) < \frac{\pi}{2} \tag{12.4}$$

and for a small sphere (Figure 12.5b),

$$F_{12} = \frac{1}{2}\left[1 - \frac{\sqrt{(H^2 + 2H)}}{1 + H}\right] \tag{12.5}$$

where $H = h/R$.

The view factor of equations (12.4) and (12.5) is commonly used to assess the albedo radiation at the spacecraft. However, unlike planetary radiation which is continuously incident on the spacecraft, the albedo radiation is derived from insolation and is only incident on the day side of the orbit. Since the planet as seen from the spacecraft may be partially in shadow, the view factor can be derived as before, but by integrating only over the unshadowed area, which complicates the evaluation considerably [7]. To account for the partially shaded area, the view factor from albedo radiation must be reduced by a factor K ($0 < K < 1$) to account for the partial shade:

$$F = KF_{12}$$

where F is now the visibility factor shown in Figure 12.1.

In orbit, a spacecraft is obscured from the Sun where it is in the shadow of the planet. This shadowing and time in the sunlight varies according to the angle β between the orbital plane of the satellite and the plane of the ecliptic. The maximum and minimum average heat load over the orbit is determined respectively when β is a maximum and when $\beta = 0°$.

Bodies in space cast shadows that consist of two discernible regions, the penumbra and the umbra, which are assumed to lie in a conic projection centred on the ecliptic plane in the case of the Earth. At the times of the equinoxes, the planes of the ecliptic and the equator are aligned, and an Earth satellite with proper orbit geometry will experience maximum exposure to these regions. There are insignificant differences in the projected cross-sections up to the altitude of the stationary satellite in equatorial orbit. Beyond this altitude, the phenomenon becomes more pronounced, until at lunar distances the space-craft dwells in penumbra almost as long as in umbra. The effect is to expose it to a solar constant changing from maximum value to zero and back to maximum.

Between the solstice and equinox, solar insolation varies as the cosine of the angle of the axis of rotation of the Earth to the ecliptic (maximum $\pm 23°$). This must also be accounted for in establishing geometrical view factors. However, hot and cold cases, i.e. thermal extrema, can be determined at orbit noon and midnight for a spacecraft in a planetary orbit.

12.3.2 Equilibrium temperature

It is clear from previous discussion that the detailed analysis of the spacecraft thermal design is extremely complex. However, the start point of any design process is to assess the basic boundary conditions of the problem, in fact the extremes of temperature or the hot and cold cases. An estimate of a spacecraft's temperature T (K) may be based upon the solar absorptance α and emittance ε, by expressing the condition for a balance between the heat input and its output:

$$A_{s/c}\varepsilon\sigma T^4 = A_s\alpha J_s + A_p\alpha J_a + A_p\varepsilon F_{12}J_p + Q \tag{12.6}$$

where A_s, A_p are the projected areas of the spacecraft in the directions of the Sun and planet respectively.

The expression assumes the following:

(a) the Stefan–Boltzmann fourth power law, namely for a black body the heat radiated is σT^4 per unit area, and so for a body whose emittance is ε it is

$$J = \varepsilon\sigma T^4 \ \text{W/m}^2 \tag{12.7}$$

(b) For the absorption of power by the satellite from the nearby planetary body we observe that two radiations exist; one of short wavelength, the albedo radiation, and a long-wavelength (infra-red) radiation—the thermal planetary radiation. The absorption coefficients are different such that at the short wavelength the absorptance is α as for solar absorptance but for the long wavelength the surface absorptance α equals the surface emittance ε. This assumption is valid for average temperatures of planetary orbiting spacecraft as its skin temperature is similar to the effective black-body planetary temperature.

(c) The spacecraft surface is perfectly insulating.
(d) The internal power Q is distributed evenly over the entire surface.

From equation (12.6) it follows that

$$\sigma T^4 = (\alpha/\varepsilon)((A_s/A_{s/c})J_s + (A_p/A_{s/c})J_a) + (A_p/A_{s/c})F_{12}J_p + Q/\varepsilon A_{s/c} \tag{12.8}$$

Table 12.6 Martian and spacecraft data

Martian parameters		Spacecraft parameters	
Radius, R	3.4×10^6 m	Absorptance, α	0.2
Solar radiation, J_s	590 W/m^2 (Table 12.3)	Emittance, ε	0.1
Albedo, a	0.15 (Table 12.4)	Radius	2 m
Radiation intensity, J_p	145 W/m^2 (Table 12.5)	Altitude, h	0.3×10^6 m
		Dissipated power, Q	500 W

When averaged over a complete orbit, the Earth emission and the internal power dissipation terms are generally small compared with insolation. The dominant terms of equation (12.8) indicate that the fourth power of the satellite skin temperature is proportional to the product of the insolation and the ratio of the solar absorptance to infra-red emittance.

Example The following is an example in which a spinning satellite is in orbit in the equatorial plane of Mars. For simplicity it is assumed that this is the ecliptic plane. The data for the Martian environment and for the spacecraft are shown in Table 12.6.

For the hot case assume that the whole of the visible surface of Mars is exposed to the Sun; so $K = 1$, and $F = F_{12}$. From the data, $H = \frac{3}{34} = 0.088$. Then equation (12.5) leads to $F_{12} = 0.3$. From equation (12.2) the albedo radiation intensity is $J_a = 26.6$ W/m^2. The spacecraft's projected areas to Sun and Mars are equal; $A_s = A_p = 4\pi$ m^2, and the spacecraft's total area is $A_{s/c} = 16\pi$ m^2. Using the Stefan–Boltzmann constant $= 5.67 \times 10^{-8}$ W/m^2, it follows from equation (12.6) that the temperature is 325 K.

The equilibrium temperature (cold case) can also be evaluated as above when the solar irradiation factors are zero, leaving only planetary radiation and internal energy dissipation as the inputs. Thus in the shadow of Mars (at orbital midnight) the temperature would be 210 K. For an orbit in the ecliptic plane the spacecraft will spend 63% of its period in the sunlight, and by forming a weighted average of the sunlit and shaded cases a mean temperature of approximately 282 K results. The different weighting factors which albedo and thermal radiation actually present are ignored for simplicity.

The above case deals with a spherical spacecraft, where integration of the thermal flux is simple. But only piecewise integration is possible for other shapes, where the cosine of the angle between the satellite skin and the area element must be included in the integration.

The above example has assumed an isothermal surface; however, spacecraft attitude can lead to one surface being more exposed to solar and planetary flux than another. Spinning a spacecraft helps to achieve an isothermal condition leading to simplified thermal design criteria.

12.3.3 Skin temperature

The satellite equilibrium temperature is a matter of thermal balance; the only heat sink is the satellite skin, from which the heat is removed by radiation. The skin temperature can be calculated from the heat balance. In order to determine the temperature of a satellite with some precision, complete orbital data are necessary. Exact temperature prediction is an involved process requiring a computer due to the large numbers of calculations involved. For many design purposes, it is sufficient to assess the temperature for maximum and minimum heat exposure (hot and cold cases). The satellite temperature will then take some value between these two extremes.

The temperature of the skin will not be constant but vary with time. Equation (12.6) can be written in a differential form to account for this. In terms of the masses M_E, M_s of heat-dissipating units and the spacecraft skin respectively, their specific heats C_E, C_s, and their temperatures T_E, T_s, this becomes

$$A_s \alpha J_s + A_p \alpha J_a + A_p \varepsilon F_{12} J_p + Q - A_{s/c} \varepsilon \sigma T^4 - M_E C_E \frac{\partial T_E}{\partial t} - M_s C_s \frac{\partial T_s}{\partial t} = 0 \quad (12.9)$$

In the equilibrium state

$$\frac{\partial T_E}{\partial t} = \frac{\partial T_s}{\partial t} = 0$$

and we arrive at equation (12.6) once again. It is clear that the principal method available to control spacecraft temperatures is to adjust the α/ε ratio of the surface. If this alone is insufficient, then constraints on the spacecraft operational profile, e.g. attitude constraints and/or the physical configuration, e.g. shadowing limitations on exposed areas, will be necessary.

The temperature changes of internal heat-dissipating equipments depend on the temperature difference between these equipments and the skin (conduction), the difference of the fourth powers of the temperatures (radiation), and on the internally dissipated power; the resulting equation is (for heat transfer in a unit area):

$$M_E C_E \frac{\partial T_E}{\partial t} = h_c (T_s - T_E) + h_r (T_s^4 - T_E^4) + Q \quad (12.10)$$

where h_c and h_r are the thermal conductivity and radiation coefficients.

The simultaneous solution of equations (12.9) and (12.10) describes the temperature of the skin and the internal equipment. In order to minimize the equipment temperature fluctuations, M_E, C_E should be maximized and h_c and h_r minimized. However, h_c and/or h_r must be kept to a reasonable value because heat transfer is the only method of removing the heat dissipation, Q. Low conductive heat transfer by minimization of h_c can be achieved by making the support structure between the outer shell and the equipment of a poor thermal conductor (or thermally isolating the unit). h_r can be minimized by using highly polished surfaces. It is useful to maintain T_s fairly constant by giving the spacecraft outer shell high thermal inertia (i.e. high value of $M_s C_s$). This will assist in minimizing surface temperature variations during changing orbital conditions, particularly thermal shock on entering and leaving eclipse conditions. Heat transfer by radiation from internal equipments depends to a high degree on the 'geometry' of the configuration.

It should be noted that spacecraft thermal calculations deal with only a few kilowatts, but spacecraft have small heat capacities, leading to trouble even with small heat imbalances. With such small quantities of energy, conduction along structural components of small cross-section is extremely effective in transferring heat. The formation of 'hot spots' as heat conduction ($\propto T$) and radiation ($\propto T^4$) gradients drop sharply from high heat-dissipating units along supporting structures. However, temperatures are generally sufficiently low for radiative heat transfer between spacecraft components to be small but not always negligible.

12.4 THERMAL ANALYSIS

A thermal model of the spacecraft is necessary for the complex configurations of most designs. The structure may consist of many shells, wires, struts and fasteners of different sizes, shapes and materials, and it may be spinning, a dual-spinner or three axis stabilized. Thermal forecasts make use of simple laws for the conductive and radiative heat transfer between equipments, but simulation is needed to support the analysis. [8]

The conductive heat transfer Q_c (W) may be expressed via Fourier's law, in terms of the thermal conductivity h_c (W m^{-2} K^{-1}), the conductor's cross-sectional area A (m^2), and the temperature difference $T_1 - T_2$ (K), as

$$Q_c = h_c A(T_1 - T_2) \tag{12.11}$$

This indicates that conductive heat transfer can be controlled by means of the thermal conductivity. Conduction coefficients for a wide range of materials are listed in standard reference texts. The effective coefficient varies with the type of material, its surface finish, and the contact pressure. Quite often the designer must achieve good electrical but poor thermal conductivity, and this is commonly achieved by using copper mesh cloth shims. Increases in thermal resistance across joints due to the effects of vacuum must be accounted for in the design process.

Radiant heat transfer Q_r (W) between a source at T_1 K and a sink at T_2 K may be expressed as

$$Q_r = \sigma A_2 F_{12} E(T_1^4 - T_2^4) \tag{12.12}$$

where A_2 is the area of the sink
          E is the effective emittance of the radiant system
and σ is the Stefan–Boltzmann constant, 5.67×10^{-8} W m^{-2} K^{-4}.

The view factors F_{12} for radiant heat transfer can be found for many cases in standard reference texts [e.g. 5].

The overall effective emittance, E, relates the source and sink emissivities to the effective viewing areas of both elements. Except for the case of a small sink completely surrounded by a large source, where E is taken as ε_i, E can be determined from

$$E = \frac{1}{(1/\varepsilon_i) + (1/\varepsilon_r) - 1} \tag{12.13}$$

where ε_i = sink emittance
        ε_r = source emittance.

Analytical thermal models have now grown to become large modelling devices; in the European Space Programme typical spacecraft thermal models have grown from approximately 100 nodes with 150 couplings (1969) to over 500 nodes with over 13 000 couplings (1984). Such complexities, driven by larger spacecraft with complex subsystems, have

Table 12.7   ESABASE/THERMAL software

ORHPL	allows mathematical modelling representing all internal and external exchanges within the spacecraft and between it and space. An orbit plotting facility is included within the application.

Viewfactor and heat-rate calculation programs accessible by ESABASE/THERMAL are:

VWHEAT* is a modular radiative heat exchange program consisting of three main modules:

- VUFACT for view factor calculation;
- RADCON for calculating inter-node radiation constants;
- ROHEAT for calculation of solar and planetary heat input.

MATRAD* from Matra-Espace, France, is a radiative heat exchange program based on a Monte Carlo ray-tracing technique. It consists of two modules:

- MATVIF for calculation of view factors and internode radiation constants
- MATFLUX for calculation of solar and planetary heat input

RAVFAC* for calculation of diffuse radiation view factors.

Reformatting programs accessed by ESABASE/THERMAL are:

- PRTPCH* for conversion of VWHEAT or MATRAD output to SINDA format
- PCHESA* for conversion of VWHEAT or MATRAD output to ESATAN format.

Thermal network analysis programs accessed by ESABASE/THERMAL are:

- ESATAN* (European Space Agency Network Analyser)
- ESACAP* (European Space Agency Common Analysis Package)
- SINDA

All three programs handle both steady-state and transient analysis, and any mode of heat transfer (conductive, radiative or convective) can be accommodated.

*These programs are not supplied with ESABASE. They must be obtained from their owners.

made the thermal analysis of spacecraft expensive operations but have not necessarily led to improved thermal design. This is primarily due to the total system interface inherent in the thermal control subsystem, the limited tools available to the engineer to control the temperature environment and the inevitable dynamic aspect of the design conditions as the typical spacecraft development programme progresses. The analytical process begins with small scale lumped parameter/analogue models coupled with maximum and minimum equilibrium temperature assessments to determine the design problem areas.

There are many software packages available to the designer. ESABASE [9] supports a wide variety of space-specific analyses, of which those in Table 12.7 are for thermal applications.

12.5 THERMAL DESIGN

There are two basic approaches to the design of a spacecraft's thermal control system — passive and active. The former operates by using appropriate materials and surface finishes so that the temperature remains within acceptable limits over the range of geometries and irradiation levels experienced. Early spacecraft and many modern ones use this technique.

Active systems use mechanical or thermoelectric devices which power and/or have moving parts. They are as a consequence inherently less reliable than passive systems.

Frequently a system combines both control concepts, overall control being passive but augmented with active control for equipments having a close temperature tolerance. It is quite common to have a subsystem described as 'passive with thermostatic heaters'.

For very large spacecraft such as large platforms, Shuttle, Spacelab, Eureca, etc. thermal control is active, consisting essentially of a central thermal transfer bus, a fluid loop transporting the heat from individual equipments to radiators. Such a modular design concept is attractive in that it becomes less specific to a particular spacecraft. However, its mass, power requirement and complexity mean that the simpler but mission-specific designs will continue to be used for the immediate future.

12.5.1   Passive thermal control systems

Thermal conditions can be satisfied by a passive system if the spacecraft orientations and equipment power dissipations are known, so that a combination of surface α/ε can be chosen for operation within the temperature limits. A component requiring low temperature in the solar-Earth environment can use a highly reflecting, high-emitting white surface (low α, high ε); similarly, a component operating at a high temperature might use a high-absorptance α and a low-emitting metallic surface ε. Most spacecraft employ combinations of surfaces to produce the required temperatures for specific equipment areas. Early spacecraft were characterized by material mosaic patterns on the surface to produce an average temperature intermediate between those produced by the individual material coatings. The last twenty-five years have produced wide ranges of proprietary conductive and non-conductive materials, coatings and paints with α/ε values from 0.08 to about 10.0 [5] (see Table 12.5).

Ultra-violet radiation and the impact of the interplanetary plasma may degrade or even upgrade the surface emissivities and absorptivities of thermal control surfaces. Care in selection, handling, evaluation, and testing is essential for good design [10].

Passive thermal control surface materials may be conveniently divided into four general categories [11].

1. *Solar reflector.* Surface which reflects the incident solar energy whilst emitting infra-red energy, low absorptance α, high emittance ε, low α/ε. (Energy is not reflected in the sense used in geometrical optics—rather it is emitted. However, it is common to express emission in terms of reflection.)

2. *Solar absorber.* Surface which absorbs solar energy whilst emitting only a small percentage of the infra-red energy, high α, low ε, high $\alpha/\varepsilon > 1$.

3. *Flat reflector.* Surface which reflects the energy incident upon it throughout the spectral range from ultra-violet to far infra-red, low α and ε, $\alpha/\varepsilon = 1$.

4. *Flat absorber.* Surface which absorbs the energy incident upon it throughout the spectral range from ultra-violet to far infra-red, high α and ε, $\alpha/\varepsilon = 1$.

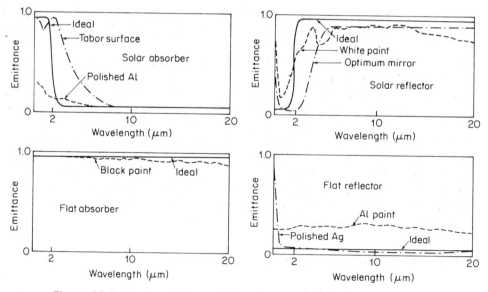

Figure 12.6   Representative spectral emittance curves for four ideal surfaces

The curves in Figure 12.6 represent both idealized surfaces and the surfaces of practical optimum available materials. In interpreting them the reader should recall the wavelength ranges of solar and planetary irradiations, particularly for the solar reflector case. The solar absorptance α_s of a spacecraft surface can be approximated by the thermal radiation properties in the range 0.3 μm to 2.5 μm as over 95% of the Sun's emitted energy falls in this range. The spectral range in which a surface emits energy is determined by the temperature of the surface. However, for the temperature ranges commonly associated with spacecraft (approximately 5–50°C) the wavelength range of most (92%) of the emitted energy is between 4.5 μm and 40 μm (less than 0.5% of the effective black body radiated energy is below 4.5 μm).

The emitted energy shows an angular dependence and for opaque surfaces (i.e. zero transmittance—typical of most, if not all, spacecraft materials) the directional reflectance varies with the angle of incidence. The ratio of the hemispherical emittance ε_H to the normal ε_N for an ideal diffuse emitter (a Lambertian surface) is equal to 1. For metallic surfaces the ratio $\varepsilon_H/\varepsilon_N$ is greater than one, and for dielectric surfaces may vary from 0.92 to slightly above 1 within the wavelength limits stated above. For initial analytic determination it can be assumed, with little error, that normal emittance (or a Lambertian surface) applies for the spacecraft materials. However, this assumption must not be made for certain detailed analyses where tight tolerances, or a thermally critical design applies.

Solar reflector

The solar reflectors, characterized by very low α/ε ratios, are generally white paints. Practical operational ratios are in the vicinity of 0.2 although 0.1 or less has been claimed for some paints. Values of less than 0.1 are obtainable using second surface mirrors (SSMs)

or optical solar reflectors (OSRs). Generally solar reflectors cause cold surfaces under solar radiation. Inorganic materials (zinc based) have given excellent performance in near-earth orbits but degrade severely in the interplanetary environment. Organic-based paints, usually with silicone binders and ZnO or TiO_2 pigments, have been shown to be stable even under the ultra-violet radiation which can lead to a doubling of the solar absorptance of paints with operational lifetimes of a few years. The use of acrylic and epoxy materials has dramatically improved the stability of solar reflector surfaces to ultra-violet radiation. They are commonly used on large or complex configurations areas of spacecraft, such as antenna dishes.

SSMs (or OSRs) are of the form of silver- or aluminium-coated quartz. The surface reflects incident solar energy, whilst the front surface controls the emittance. These mirrors must be attached to the spacecraft surface by suitable space-stable adhesives. They posses α/ε values in the range of 0.06–0.14 and exhibit complete stability in the space environment. Configurationally, application to large or complex surfaces is not practical; it is also costly. Their main application is in small area radiator surfaces occasionally mounted via heat pipes (see below) connected to high-power dissipating electronic units.

Solar absorber

The solar absorbers generally used are polished, electroplated metals or specially prepared metal foils suitably attached to a surface. Solar absorptance values are generally less than 0.5; emittance values are very low, from about 0.3 to 0.1, resulting in high values of α/ε, typically greater than 1. The high temperatures resulting find their application where equipment temperatures must be kept high, such as skirt surrounds near sensors, or thruster jets.

12.5.2 Additional techniques

Passive thermal control will generally be effected through proper selection of geometries and materials with fixed thermal radiation characteristics. However, other approaches are used and a few more commonly employed techniques are outlined below.

Superinsulation

Superinsulation is the general name given to insulation blankets that are built up of multiple separate sheets arranged to alternate low-emittance surfaces with low-conductance barriers. Each layer serves either as a radiation shield or as an insulator, or it may combine the two functions. Multi-layer blankets operate most effectively in vacuum and have found wide application in spacecraft thermal control design. Several varieties are commercially available. Multilayer blankets consisting of 0.25 mil Mylar or Kapton sheets (for higher temperature applications), with an aluminized surface 0.001 mil thick vapour-deposited on one side, have gained general acceptance [12].

In the application of these blankets, no spacers between sheets are used; spacing is maintained by crinkling. The aluminized sheets thence touch at discrete points rather

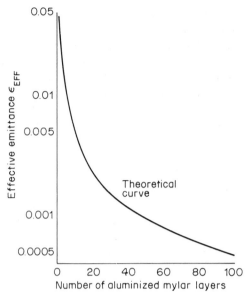

Figure 12.7 Effective emittance versus number of aluminized Mylar layers (theoretical)

than across the whole surface, reducing the amount of conducted heat through the blanket.

The theoretical emittance of the superinsulation blankets shown in Figure 12.7 assumes no significant contact between the blanket sheets, and an effective vacuum between them, and no significant heat leakage through their edges. In practice the sheets touch each other to a greater or lesser extent, particularly at the edges or perforations; pressure may exist between layers, caused by trapped air or sublimation in vacuum, and all may conduct or radiate heat, thus reducing the effectiveness of the blanket. Actual values of emittance are usually much higher than the theoretical ones.

In the implementation of superinsulation optimum spacing of sheets is typically 20–25 sheets per centimetre of blanket thickness; no significant reduction in heat leakage is practical by increasing the number of sheets.

Multi-layer insulation is frequently used in spacecraft configurations to provide iso-thermal compartments for sensitive electronics or instruments such as scientific telescopes requiring low thermal gradients or employing cryogenically cooled detectors [13] to protect internally mounted electronics from high heat inputs, e.g. solid rocket apogee boost motors centrally mounted in the spacecraft body, and in general whenever there is a requirement to minimize heat losses, or gains to all, or part, of the spacecraft configuration.

Heat pipes

The use of a thermal wick to cool spacecraft components has found common application, particularly for maintaining temperature control of high-power dissipating components within communication subsystems. A heat-transfer fluid is contained by a housing that contains capillary channels. The fluid is evaporated at the equipment to be cooled and is condensed at the spacecraft skin where the heat transfer is completed by radiation into cold space via a solar absorber surface. The condensate in the condenser section returns by capillary pressure along the wick to the evaporator, where the continuous cycle is completed by boiling off the fluid. Such devices have an extremely high effective thermal conductance and near isothermal operation.

Heat pipes are ideal for isothermalization and temperature control of equipment with a varying duty cycle in a varying environment. They are frequently employed in an active thermal control mode coupled with thermostatic heaters to reduce the heat-transfer capability of the fluid by evaporating fluid and thus allowing the temperature of the equipment to rise.

Fluid exhaustion at the evaporator limits the energy-transfer capability and the lifetime of heat pipes. The capability of available passive devices is in the range of tens of watts. To transfer larger amounts active control loops are used (see below).

Many types of heat pipe devices are available to the designer with widely different technical characteristics. Their design and application is a dynamic development area [14, 15].

Solar cell temperatures

The efficiency of solar cells decreases rapidly with increasing temperatures and means must be found to keep them as cool as possible. When they are mounted on deployed panels the back of the panels can be designed to radiate heat to cold space. Consequently it is not too difficult to keep the maximum temperatures below about 50°C. When the cells are mounted on the spacecraft body, however, radiation from the back cannot be so effective. In the case of small, spin-stabilized spacecraft the solar cells are not always sunlit and hence do not heat up as much. The spin further eases the problem of thermal design by equalizing the lateral heat inputs, thus contributing to the achievement of an isothermal interior.

The spectral response of solar cells is such that an appreciable amount of the solar radiation is not of wavelengths that are converted into electrical energy by the photo-voltaic action of the cell. In a silicon cell, for example, wavelengths less than 0.4 or greater than 1.1 contribute very little to the power output. If a filter that will not transmit these wavelengths is used on the bare silicon, less thermal energy will be absorbed by the cell. Filters of vacuum-deposited multi-layer coatings are used to reduce the absorptance of the solar cell material by as much as 50% whilst reducing the power output by less than 10%.

Thin glass or fused silica covers are also used to increase the emittance, allowing more effective re-radiation of solar energy by up to a factor of three with a subsequent reduction in the energy available for power conversion, usually of the order of 2% [5].

12.5.3 Active thermal control systems

Active thermal control systems are of various designs and are not amenable to a general presentation. Various examples are described below.

Their inherent disadvantage is their use of valuable system resources such as mass, electric power and telemetry. Additionally most of them involve an increased operational risk, and hence they decrease overall system reliability. Often systems have not performed as well as expected. This can usually be traced to improper choice of locations and types of sensor, controller and regulator devices, and the neglect of dynamic response analysis for prediction of system performance.

Although heat transfer and thermodynamic analysis employ well-proven techniques with a high success rate, control and flow-rate parameters are usually based on hot and cold case definition and heat load considerations. This does not lead to an optimum or even a well-controlled system.

Thermostatically controlled heaters

Heaters constitute probably the simplest and most obvious thermal control device. A passive thermal design may lead to the lower level of temperature excursions due to cycles and orbit variations lying below the permissible temperature limit of a particular sub-system or equipment. The heater may then be used to maintain the temperature within tolerance. Such devices are frequently used to maintain battery temperatures during eclipse periods. Another application is to maintain the temperature of components of the on-board reaction control system.

Variable external radiation devices

Mechanical means such as pinwheels, louvres, or shutters may be used to overcome problems caused by sudden or large changes in irradiation. The devices use moving parts to change the apparent irradiation, i.e. physically change the effective α/ε of the external surface.

The simplest form consists of fins made of bimetallic materials, which are mounted on external surfaces as shown in Figure 12.8. The fins and the spacecraft have relatively high α/ε properties, exposed when the fins are closed. As the fins' temperature increases they 'open' and expose a relatively low α/ε surface. Thus the temperature increase is arrested [16].

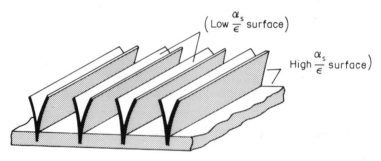

Figure 12.8 Active thermal control system with bimetallic fins. (*After* J. A. Wiebelt)

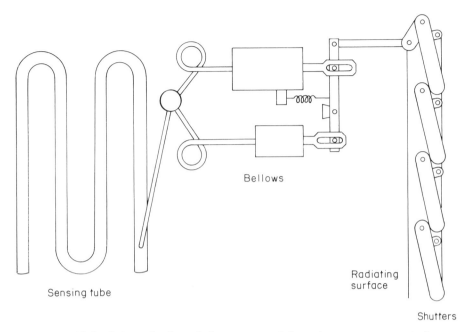

Figure 12.9   Schematic of gas-bellows-actuated dynamic temperature controller

The use of fins has a further property in that the external surface of the spacecraft is increased, creating a larger radiating surface which may increase the loss or gain of energy. Static radiator fin surfaces have been used in conjunction with heat pipes to achieve a high heat loss. Pinwheels were commonly used in early spacecraft; one form consisted of an aluminium mask in the shape of a Maltese cross, which was rotated relative to a

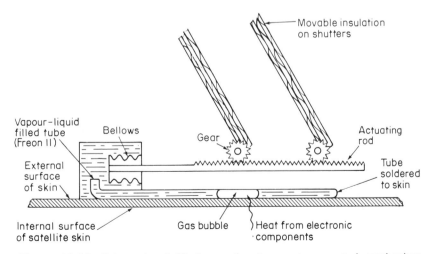

Figure 12.10   Schematic of Nimbus active temperature control mechanism (General Electric)

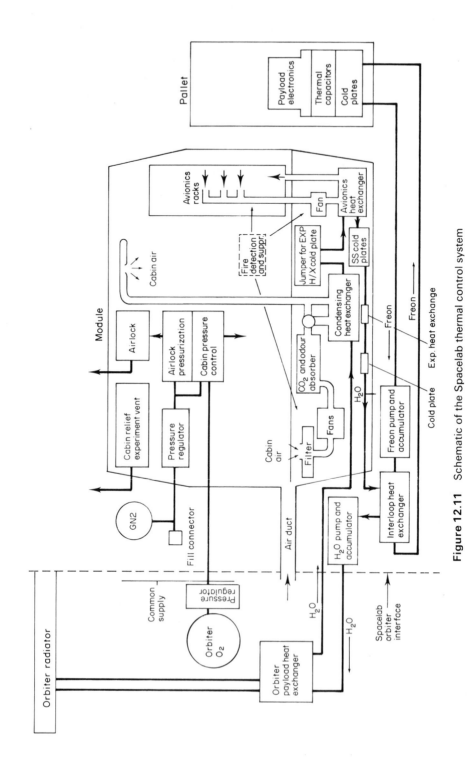

Figure 12.11 Schematic of the Spacelab thermal control system

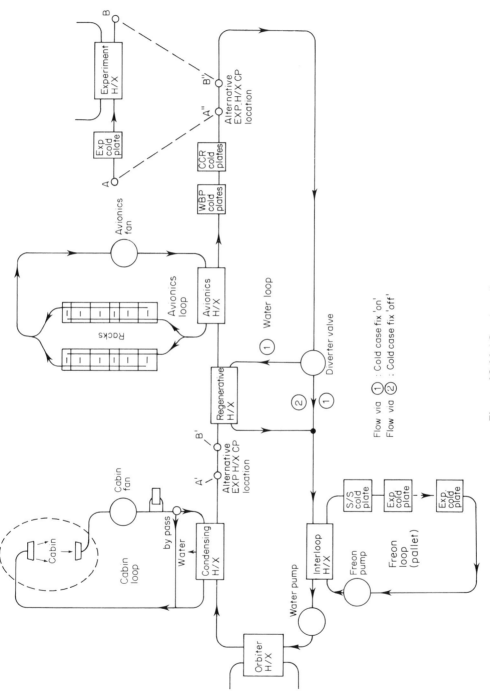

Figure 12.11 (*Continued*)

similarly shaped white-painted cross on the equipment body. The mask was actuated by a bimetallic strip, thereby exposing the white surface and cooling the equipment bay.

Louvre-type systems operate on similar principles and have been employed on low-altitude satellites such as the NIMBUS series [17]. Their actuators have been bimetallic springs, expanding gas bubbles as depicted in Figure 12.9, a gas bellows type as shown in Figure 12.10, or simply thermostatic. Such a device was flown on the Swedish Viking spacecraft.

A similar device was on ESA's Giotto spacecraft. A 'roller blind' was actuated by a stepper motor which rolled a goldized kapton foil across the skin of a portion of the external surface (0.6 m^2 in three locations) thereby achieving variable α/ε properties. The device was not automatic, but was initiated by ground control [18].

Thermoelectric cooling

Thermoelectric devices use the principle that if current flows in a circuit of two dissimilar metals, then heat is developed at one end and absorbed at the other.

Peltier heat pumps remove heat from the equipment to be cooled. It is pumped from the cold to the hot junction along with the heat to which the electric power degenerates. Commercial Peltier devices are usually of a plate form and have typical heat-flow densities of 130–650 W/kg. Their use is limited to relatively low cooling loads and small temperature differentials. Peltier cooling is best applied where accurate control of temperature and/or spot cooling is required.

Fluid transport systems

The final class of thermal control system is that employing active fluid loops to transport heat from heat-producing equipment to radiator panels. These 'refrigeration' systems are employed in large spacecraft, where high heat loads are encountered. Such systems have been primarily associated with manned vehicles where fluids such as air and water need to be present.

The Shuttle Orbiter, Spacelab and the evolving free-flying platforms such as EURECA are examples of systems using active fluid loops based primarily on variable-conductance heat pipes (VCHP). The Spacelab thermal control concept is depicted in Figure 12.11 [19]. Future space stations with their potentially high heat loads and manned capability will use active fluid loops in both centralized and decentralized forms of heat transfer. These systems require resources, such as mass, power, and volume, which are at least an order of magnitude higher than those available for normal single-mission spacecraft. However, their potential is such as to promise many advantages over those systems previously discussed.

12.6 THERMAL DESIGN VERIFICATION

Since there is little chance of repairing a spacecraft once it is launched, its thermal design must be completely tested via a sequence ranging from materials, components, and equipments, and culminating in a full system test under solar simulation and vacuum conditions.

The procedure for the selection of the thermal design, and specifically the thermal control materials, requires consideration of many factors at all levels of the spacecraft system through a series of steps:

- A spacecraft thermal load analysis is made for the entire system.

- Detailed analyses are made for component/equipment areas where precise temperature control is required.

- Based on these analyses and the mission requirements, the thermal design is formulated in terms of materials, geometries and devices. Where major difficulties arise, the thermal requirements may dictate system re-evaluation in terms of orbital geometry, operational constraints, selection of components/equipments, verification testing, configurational design and system budgets.

- Application and development test work is performed as required [20].

- The complete spacecraft engineering model or a thermal model is subjected to development tests. This phase is increasingly being omitted as Man gains confidence in analytical techniques and experience in certain classes of spacecraft design.

- Qualification tests are conducted for the spacecraft and components/equipments [9].

- Acceptance tests are conducted for the spacecraft and components.

12.6.1 Space simulators

The culmination of all testing occurs when a complete spacecraft is subjected to a thermal vacuum test. Facilities for this are extremely expensive to build, operate and utilize and there are only a few of them available in Europe for larger spacecraft [21]. Simulation of solar, albedo and Earth radiation has reached adequate levels in modern facilities although the angle of collimation cannot be accurately simulated for radiation via planetary atmospheres. Vacuum levels of 10^{-9} mm Hg (empty) are usually achieved, leading to potentially good space simulation. The main problem is becoming the sheer size of the present generation of spacecraft types [22].

12.7 A SPACECRAFT THERMAL EXAMPLE: GIOTTO

The thermal control designs employed on most spacecraft are by their nature unique to each specific application. This chapter has outlined the techniques of analysis to determine the requirements and those techniques and devices available to control temperatures as required.

Constraints imposed on the thermal control subsystem tend to be quite severe, mass allocations of about 2% (spin-stabilized) to a maximum of 5% (three-axis-stabilized) of dry system mass are typical [23]. Electrical and telemetry resources tend to be even more limited. There are no standard overall subsystem designs, but an example can assist in highlighting how design techniques are employed on present-day craft.

RADIATORS

1. TOP RADIATORS
   o silvered Teflon SSM covered aluminium honeycomb panel in three sections 1.05 m² area primarily at Transponders & EPCs
   o circumferential distribution trimmed by top blankets
   o panels supported by aluminium honeycomb cylinders mounted on top platform inner ring

2. SKIRT RADIATOR
   o 0.25 m² of spacecraft skirt honeycomb panels covered by optical solar reflectors (OSRs)
   o circumferential distribution selected for static internal gradient compensation
   o 0.18 m² additional shutter controlled OSR area
   o the remainder of the skirt panels and experiment protrusions is covered with blankets and suitable coatings

3. TWT RADIATOR
   o separate panel, 0.12 m² total area silvered teflon SSM supported by a bracket from the top platform
   o insulated from the spacecraft by MLI and low conductance stand-offs
   o doubler under collector side of TWTs which are mounted on the radiator lower face

4. ON-OFF SHUTTER SYSTEM
   o three 0.06 m² size shutter mechanisms located at the spacecraft skirts
   o goldised kapton foil, allowing free view to space of skirt mounted OSRs, when rolled-up and absorbing solar heat input when rolled-out via goldised side facing space
   o internally mounted TCU box provides electric circuitry for operation of the mechanisms

BLANKETS & FOILS

5. TOP BLANKETS
   o white PCB-Z painted outer layers for top radiator efficiency
   o conical blanket between support cylinder and despin motor unsupported and provided with "quick release" fasteners for despin motor access during launch preparation, 0.05 mm aluminium RF shielding foil included
   o 20 layers crinkled 0.25 mil mylar package with kapton cover layer

6. SOLAR ARRAY BLANKETS
   o cut-outs provided for TC diode boards
   o 20 layers 0.25 mil crinkled mylar package

7. SKIRT, MAIN CONE & DUST SHIELD BLANKETS
   o 20 layers 0.25 mil crinkled mylar package, 30 layers at dust shield
   o black conductive painted kapton outer layers

8. TOP BLANKETS
   o kapton package, 20 layers crinkled 0.3 mil thickness

9. THRUSTER CAN BLANKETS
   o 9 layers dimpled titanium foil
   o gold foil cover layer
   o inconel locking wire attachment to titanium canisters for blanket support

10. NOZZLE CLOSURE MECHANISM FOIL
    o aluminium foil closing gap between inner bumpershield and TPS nozzle in GTO

11. TPS NOZZLE EXIT COVER
    o closing TPS nozzle aperture during GTO phase, coating selected solely for TPS temperature control purposes
    o blown off during firing

HEATERS

12. TWT HEATERS
    o redundant heater circuits controlled via PDU switches and TCU-IPD switch

13. PLATFORM HEATERS
    o on top, main and experiment platforms
    o dissipations selected for axial and circumferential gradients compensation for various phases

14. TANK HEATERS
    o long strip heaters used to obtain good thermal contact between tank surface and heater foil
    o positions selected to avoid internal tank gradients, in 3 parallel redundant circuits

15. RCE PIPE LINE HEATERS
    o Dale resistors on stand-offs in two parallel redundant circuits
    o Tayco spiral heaters on longer line sections, other line sections are protected by multilayer blankets and aluminium foil for axial conductance increases

16. TPS HEATERS
    o positions selected to obtain flattest possible substrate for good thermal contact to TPS nozzle throat
    o redundant heater circuit provided

LOW CONDUCTANCE INTERFACES
   o solar array/structure
   o TPS/structure
   o dust protection system/structure
   o despin motor/antenna dish
   o batteries/main platform
   o RCE tanks & pipelines/structure
   o thruster cans/structure
   o camera and starmapper/structure

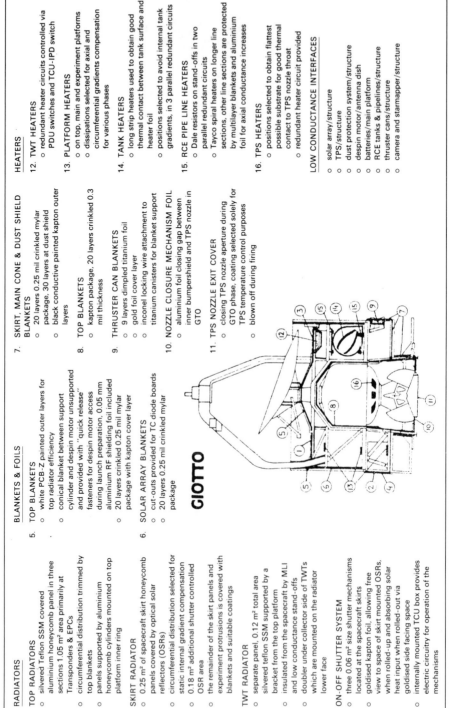

GIOTTO

Figure 12.12   The Giotto thermal control subsystem (ESA)

Giotto, the interplanetary scientific spacecraft, spin-stabilized with de-spun platform, employed a typical, mostly passive system. The following are the major requirements for its thermal control subsystem:

- Make maximum use of passive thermal control methods and of internal heaters where required.

- Maintain operating temperature limits as follows:

$-10/+40°C$	typical (subsystem units);
$-5/+30°C$	batteries ($+50°C$ at end of comet encounter);
$-10/+50°C$	battery regulator units,
$-0/+70°C$	travelling wave tubes (TWTs) operational;
$+5/+45°C$	RCE equipment (hydrazine);
$-20/+40°C$	de-spin mechanism.

- Maintain a margin of $\pm 10°C$ between computer-model predictions and design limits ($\pm 6°C$ for reaction control equipment subsystem).

- Leave a 10 W margin between available subsystem power and that used by heaters during the mission.

- Limit the use of battery power for thermal control to extreme cases, limit depth of discharge to 50%.

- Minimize design sensitivity to solar aspect angle and solar constant.

- Guarantee surface electrical conductivity and grounding to avoid surface charge build-up, particularly at experiment sensors.

The overall Giotto thermal control subsystem is summarized in Figure 12.12 and results in an overall mass of 19.5 kg.

The major spacecraft system characteristics were as follows:

Mass	960 kg (launch)
	514 kg (cometary encounter)
Size	1.9 m (diameter)
	3 m (height)
Spin rate	15 r.p.m.
Solar array power	160 W (at 1 solar constant).
RCE system	Four tanks containing 69 kg of hydrazine fuel.
Overall configuration	Spin-stabilized with a de-spun communication antenna platform.

REFERENCES

[1] Garrett, H.B. and Pike, C.P. (eds.) (1980) Space systems and their interactions with Earth's environment. *Progress in Astronautics and Aeronautics*, **71**. AIAA, New York.

[2]   Weiner, F.R. and Van Valkenburgh, C.C. (eds.) (1969) *SAE Aerospace Applied Thermo-dynamics Manual* (2nd edn). SAE Committee AC-9.
[3]   NASA (1983) *Space and Planetary Environment Criteria Guidelines for Use in Space Vehicle Development* (1982 Revision). TM 82501, Volume 2.
[4]   NASA (1983) *Space and Planetary Environment Criteria Guidelines for Use in Space Vehicle Development* (1982 Revision). TM 82478, Volume 1.
[5]   ETSI (1979) *Spacecraft Thermal Control Design Data, Handbook*, Volume 1, ESA (TST-02) Iss.1.
[6]   Juul, N.H. (1979) Diffuse radiation view factors from differential plane sources to spheres, *ASME J. Heat Transfer*, **101** (3), 558–60.
[7]   Doenecke, J. (1969) Thermal radiations absorbed by a partially obscured spacecraft, *Astronautica Acta*, **15** (2).
[8]   ESA (1983) *Environmental and Thermal Control Systems for Space Vehicles*, Session V, Toulouse. ESA SP 200.
[9]   ESA (1988) *ESABASE, A most versatile and flexible system engineering tool.* ESA BR-54.
[10] ESA Specification for the measurement of the thermo-optical properties of thermal control materials at ESTeC. PSS-16/QRM-OST.
[11] NASA (1969) *Space Materials Handbook* (3rd edn). NASA SP-3051.
[12] NASA (1974) *Thermal Performance of Multilayer Insulation.* CR 134477.
[13] Hopkins, R.A. (1979) Design of a one-year lifetime, spaceborne superfluid helium dewar, *9th Intersociety Conference on Environmental Systems*, San Francisco.
[14] B. & K. Engineering, Inc. (1979) *Heat Pipe Design Handbook.* NASA Contract NASS 23406.
[15] Marcus, B.D. (1979) Theory and Design of Variable Conductance Heat Pipes, NASA CR 2018.
[16] Wiebelt, J.A. (1966) *Design Considerations for Thermostatic Fin Spacecraft Temperature Control.* NASA CR-500.
[17] London, A. (1965) *Thermal Control of the NIMBUS Satellite System.* General Electric MSD, AIAA Unmanned Space Meeting.
[18] ESA (1983) *Environmental and Thermal Control Systems for Space Vehicles*, Session III. Toulouse, ESA SP 200.
[19] ESA-NASA (1988) *Spacelab Payload Accommodation Handbook*, SLP 2104, Iss. 2. Rev. 14/Appendix C, Rev. 5.
[20] Schlitt, R. *et al.* (1981) Problems associated with thermal testing of large heat pipe systems for space application. *Proc. IVth International Heat Pipe Conference*, London.
[21] Brinkman, P.W. (1982) *Environmental Test Facilities used in Europe for ESA Satellite Programmes.* NASA CP 2229.
[22] Tan, G.B.T. and Walker, J.B. (1982) Spacecraft thermal balance testing using infra-red sources, *12th Space Simulation Conference.* NASA CP 2229.
[23] Collette, R.C. and Herdan, B.L. (1977) Design Problems of Spacecraft for Communications Missions, *Proc. IEEE*, **65**, 3.

13 TELE-COMMUNICATIONS

Howard Smith

Department of Electrical and Electronic Engineering, Portsmouth Polytechnic

13.1 INTRODUCTION

13.1.1 The development of telecommunications satellites

Long before artificial Earth satellites became a reality, their potential in the field of telecommunications had been appreciated by visionaries such as Arthur C. Clarke [1]. And in the years following the launch of Sputnik 1, the demand for global communications systems was one of the main driving forces—along with military and political considerations—in the rapid development of space technology.

The birth of telecommunications by satellite can be seen, perhaps, in the launch of Telstar I in 1962. Telstar I permitted, for the first time, transoceanic communications by satellite. By today's standards it was, of course, a very modest affair. Just under a metre in diameter and weighing 77 kg at launch, the satellite had one channel with a 50 MHz bandwidth, providing about 12 telephony circuits. But it was the start of a revolution in international communications.

The low elliptical orbit of Telstar I limited usage for transatlantic communications to three or four half-hour periods in each day. The first successful geosynchronous communications satellite was Syncom II, used for experimental transmissions between America and Japan. This again had a rather limited capacity but the design, after further improvements, became the basis for the first commercial communications satellite, Early Bird or, as it was later called, Intelsat I. This satellite, launched in 1965, was similar in size and mass to Telstar I but was capable of providing 240 telephony circuits or a high-quality television channel.

The first world-wide satellite communications system was not established, however, until 1969 when Intelsat III satellites were in position over the Atlantic, Pacific and Indian Oceans, giving coverage of all parts of the globe other than the polar regions.

The subsequent development of satellite communications has been a response to a dramatic increase in the demand for international telephone and video traffic. This is illustrated in Figure 13.1. Since international communications by satellite became a

---

Spacecraft Systems Engineering. Edited by P.W. Fortescue and J.P.W. Stark

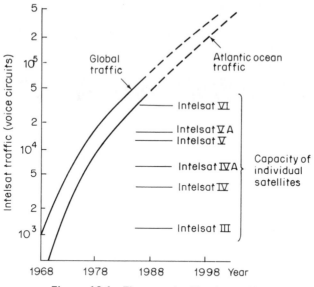

Figure 13.1   The growth of Intelsat traffic

Table 13.1   The growth of the Intelsat series of satellites

Satellite	First successful launch	Approximate mass in orbit (kg)	Primary power (W)	Approximate capacity (voice circuits)	Number in GEO
Intelsat I	1965	39	45	240 (or 1 TV)	1
Intelsat II	1967	87	75	240 (or 120 + 1 TV)	3
Intelsat III	1968	137	120	1200 (+1 TV)	5
Intelsat IV	1971	720	550	3500 (+2 TV)	7
Intelsat IV A	1975	795	700	6250 (+2 TV)	5
Intelsat V	1980	1012	1800	12 500 (+2 TV)	8
Intelsat V A	1985	1098	2100	15 000 (+2 TV)	4
Intelsat VI	1989	2200	2800	33 000 (+4 TV)	2

The channel capacities shown are typical values. These are generally less than the theoretical maximum signalling rates, particularly for the larger satellites. The number of telephone circuits depends not only on the bandwidth allocated to television channels but also on the distribution of telephone channel allocations between various countries or spot beams.

commercial reality the traffic has increased by about 20% per year and, although the rate of growth may now be reduced by increased competition from optical fibres, an upward trend is expected to continue at least for the foreseeable future. This extraordinary increase in demand has been matched by a rapid increase in the size, power capability and traffic capacity of the spacecraft as well as in the number of satellites in simultaneous use. Table 13.1 shows this development for the Intelsat series. A satellite of the principal current series (Intelsat V) has a capacity of about 12 500 telephone circuits and two colour television channels (or a much larger number of television channels with reduced telephony capacity). The Intelsat VI satellites, of which the first was launched in 1989, are even more powerful, with a capacity of about 33 000 telephone circuits and four television channels.

13.1.2   The role of the communications payload

Some early experimental systems used orbiting objects as passive reflectors or scatterers in order to achieve intercontinental communications. It quickly became apparent, however, that in order to meet the demand for high traffic capacities it would be necessary to use active satellites containing transponders which receive the signals transmitted from the ground and amplify them before re-transmitting them to Earth at different frequencies. Figure 13.2 is a very much simplified block diagram of such a link. The power amplification provided by the satellite is typically in the region of $10^{12}–10^{15}$ (120–150 dB).

The most striking difference between a satellite link and a terrestrial link is in the length of the transmission path, typically about 50 km for a terrestrial microwave link and 36 000 km from an Earth station to a satellite in geostationary orbit. The immediate result of this enormous increase in path length is that the most critical parameter in a satellite system is normally the available transmitter power—particularly on the downlink, where the transmitter power is in turn one of the major demands on the primary satellite power. Nevertheless, the advantage of continuous visibility of the satellite at both the transmitting and the receiving ground stations has ensured that the great majority of communications satellites are in geostationary orbit.

There are other orbits which have been used, particularly where coverage of high latitudes is required. For instance satellites of the Russian Molniya series are in elliptical, high-inclination zero-drift orbits. During the slow apogee transit, near GEO altitude, the satellite appears almost stationary in the sky, allowing continuous operation for up to about two thirds of the orbital period (see Chapter 5).

The simple system illustrated in Figure 13.2 is but one of a large variety of telecommunications applications of artificial satellites. The main types of system, are as follows:-

- trunk telephone and television;

- broadcast signals;

- communication with mobiles;

- data communications (between ground stations);

- data relay (between other satellites and ground stations).

A single satellite may include transponders for more than one of these functions.

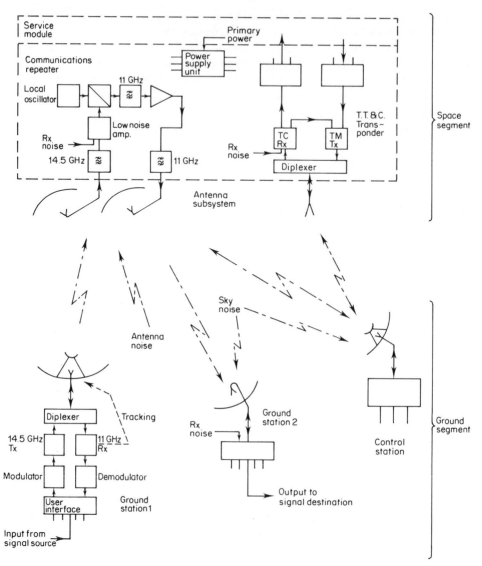

Figure 13.2　Simplified block diagram of a satellite communications link (using the 14.5 and 11 GHz bands)

13.1.3 System constraints

The design of a satellite communications system involves considerations not normally required for a terrestrial system. In general these design constraints can be divided into three broad categories: customer requirements, technical requirements, and international regulations.

Customer requirements form the starting point of the design, and are as many and as varied as the customers themselves. Among the parameters that one might expect the customer to specify would be:

- type of signals (voice, television, data etc.);

- capacity (i.e. number of channels of each type)—or bandwidths and frequencies;

- coverage area served by the satellite and the site(s) of the control station(s);

- uplink and downlink signal strength and quality—or types of ground terminals to be served;

- connectivity between different channels and traffic routes;

- availability (i.e. times of day/year, permitted outage times etc.);

- lifetime (typically 7 years for satellites currently in service but likely to increase to about 10 years or more for future designs).

There may be a further requirement that some of these parameters can be altered in orbit by command from a ground station and there may also be other limitations such as a need for a high level of security, both against illicit interception and decoding of signals and against jamming and illegal telecommands.

As with all major engineering projects, however, the most important constraint is usually that of cost.

The *technical constraints* which apply to a terrestrial link generally apply also to a satellite link, but with some additional restrictions. Of those factors which are peculiar to satellite systems the most significant are those imposed by the payload mass, the available power and the need for equipment to survive the launch phase and operate in the space environment.

It is here that cost factors are often crucial. In most terrestrial systems it is possible to guarantee adequate overall performance by allowing generous margins in the design. Such extravagance in the case of a satellite system will result in an increase in mass and power consumption and hence increased cost.

International regulations for communications systems are required to control possible interference between different systems and to ensure compatibility between the various national systems which may need to be connected end to end.

The International Telecommunications Union (ITU), which is now part of the United Nations Organization, exists for the purpose of defining and enforcing international standards. More than 160 nations are signatories to its Convention which has the status of a formal international treaty. Within the ITU there are two consultative bodies which collate information and formulate policy in relation to telephony and radio communications respectively. These are the Comité Consultatif International de Téléphonie et de Télégraphie (CCITT) and the Comité Consultatif International de Radiocommunication (CCIR). A third body, the International Frequency Resistration Board (IFRB), is concerned specifically with the coordination of frequency allocations with a view to avoiding harmful interference between systems.

Each of the consultative committees meets in full session about once every four years to consider documents submitted by various national working groups. The assembly reports provide a wealth of information for the communications satellite system designer. The *CCITT Red Book* [2] is of relevance in so far as it defines standards for telephony traffic and data transmission but of more general application to satellite communication is the *CCIR Assembly Report* [3]. In particular, Volume IV-1 is devoted entirely to documents and recommendations concerned with communications satellites, and Volumes IV/IX-2 and X/XI-2 respectively contain material on frequency sharing with terrestrial systems and on direct broadcast satellites.

The recommendations of the CCIR are in turn considered by the World Administrative Radio Conference (WARC) which may incorporate them into ITU Radio Regulations [4]. These are binding on signatories of the ITU Convention.

Among the important system parameters subject to international regulation are the available frequency bands, the orbital location and the maximum permitted power flux density at the Earth's surface. Table 13.2 shows frequencies (1–35 GHz) assigned to satellite communications as of the end of 1987. Most of the satellite frequencies are shared with other services. Full details are given in References [4] and [5].

So far, the bulk of telephony and television traffic has been in the 6 and 4 GHz bands but serious overcrowding in this part of the spectrum has led to increasing use of 14.5 and 11 GHz. Signs of overcrowding are already appearing at these higher frequencies in some areas and there is growing interest in the 30 and 20 GHz bands. So far, however, transmissions at these frequencies have been mainly of an experimental rather than a fully commercial nature.

The position of a satellite in the geostationary orbit is negotiated through the IFRB. The main requirement is that there should be sufficient separation between locations to allow a ground station antenna of reasonable size (and hence beamwidth) to discriminate between adjacent satellites. This implies a separation of about 2° (or more) of longitude between satellites using the same frequency. Some parts of the orbit are relatively empty, but in others, such as the Atlantic region, where traffic density is very high, the choice of location may be very restricted.

In addition to the regulatory bodies set up under the United Nations Organization there are several international consortia whose aim is to coordinate and rationalize the commercial exploitation of communications satellites. Of these, the most important is the International Telecommunications Satellite Consortium (INTELSAT), which owns the satellites listed in Table 13.1 and leases their capacity to a variety of users including, principally, the national post, telegraph and telecommunications organizations. Since its inception in 1964 INTELSAT has expanded rapidly from eleven to over a hundred member countries. It now serves over 160 different territories and carries about two thirds of international telephone traffic. Its dominance has not, however, gone entirely unchallenged. Russia has never joined INTELSAT and has instead set up its own organization, INTERSPUTNIK, serving mainly the countries of the Eastern Bloc. More recently, a number of regional organizations have been established—notably EUTELSAT, which provides services mainly to Western Europe and the Mediterranean region. And finally, a rather more specialized service is provided by the International Maritime Satellite Organization (INMARSAT), which already offers global communications with shipping and is currently showing an interest also in aeronautical and land mobile users.

In what follows we will be concerned primarily with the technical aspects of the system design. Later we will look at some of the units which make up the telecommunications payload but first we must examine some of the principles underlying the operation of the system as a whole.

Table 13.2 Frequency allocations for telecommunications satellites[1]

Service	Frequencies (GHz)	Direction	Scope[2]
Fixed satellite	2.500– 2.690	down	R
	3.40 – 4.20	down	W
	4.50 – 4.80	down	W
	5.00 – 5.25	up	W[3]
	5.725– 7.075	up	W/R
	7.25 – 7.75	down	W[4]
	7.90 – 8.40	up	W[4]
	10.70 –11.70	down	W
	11.70 –12.75	down	W/R
	12.50 –13.25	up	W/R
	14.00 –14.50	up	W
	15.40 –15.70	up	W[3]
	17.70 –21.20	down	W
	27.00 –31.00	up	W/R
Mobile satellite	1.530– 1.559	down	W[5]
	1.610– 1.6605	up	W[5]
	2.500– 2.535	down	R
	2.655– 2.690	up	R
	7.250– 7.375	down	W[4]
	7.900– 8.025	up	W[4]
	14.00 –14.50	up	W[5]
	19.70 –21.20	down	W
	29.50 –31.00	up	W
Broadcast service	2.500– 2.690	down	W[6]
	10.70 –11.70	up	R
	11.70 –12.75	down	W/R
	14.00 –14.80	up	W
	17.30 –18.10	up	W
	22.50 –23.00	down	R
Space operation	1.427– 1.429	up	W
	1.525– 1.535	down	W
	1.750– 1.850	up	R
	2.025– 2.120	up	W/R[7]
	2.200– 2.290	down	W[7]
	7.125– 7.155	up	R
Intersatellite links	22.55 –23.55	—	W
	32.00 –33.00	—	W
Amateur satellite	5.830– 5.850	down	W
	10.45 –10.50		W
	24.00 –24.05		W

[1] Data links for e.g. meteorological, scientific and navigational satellites are not included.
[2] W: allocated on a world-wide basis
  R: allocated on a regional basis
  W/R: part of the frequency band is world-wide and part regional.
[3] For use with aeronautical mobile services.
[4] The bands 7.25–7.75 and 7.9–8.4 GHz are used by military satellites.
[5] Maritime or land mobile: 1.530–1.544, 1.6265–1.6455, 2.655–2.690 GHz.
  Land mobile: 1.555–1.559, 1.6565–1.6605, 14.00–14.50 GHz.
  Aeronautical mobile: 1.545–1.555, 1.6465–1.6565 GHz.
  Distress and safety operations: 1.544–1.545, 1.6455–1.6465 GHz.
[6] For community reception.
[7] Also intersatellite links for space operation.

13.2 TECHNIQUES OF RADIO COMMUNICATIONS

13.2.1 Introduction

Although most of the important elements of a satellite communications link are shown in Figure 13.2, it must be emphasized that this is a very much simplified picture. In particular, only two communications ground stations are shown and the system is a simplex (i.e. one-way) link. Nevertheless, the block diagram will serve to illustrate some of the features which are common to practically all systems. In fact the extension to two-way (duplex) operation is usually trivial. The same satellite transponder will normally carry both outgoing and return traffic provided that the same uplink and downlink frequencies are available at the two ground stations.

13.2.2 Modulation

Types of modulation

The signals to be transmitted by a communications system normally consist of a band of rather low frequencies, ranging for instance from a few tens of hertz to a few kilohertz in the case of speech or from a few tens of hertz to a few megahertz in the case of television.

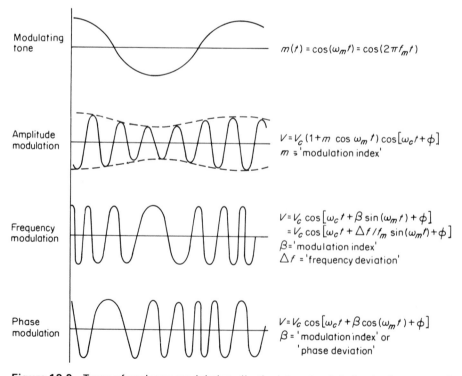

Modulating tone
$$m(t) = \cos(\omega_m t) = \cos(2\pi f_m t)$$

Amplitude modulation
$$V = V_c (1 + m \cos \omega_m t) \cos[\omega_c t + \phi]$$
$$m = \text{'modulation index'}$$

Frequency modulation
$$V = V_c \cos[\omega_c t + \beta \sin(\omega_m t) + \phi]$$
$$= V_c \cos[\omega_c t + \Delta f / f_m \sin(\omega_m t) + \phi]$$
$$\beta = \text{'modulation index'}$$
$$\Delta f = \text{'frequency deviation'}$$

Phase modulation
$$V = V_c \cos[\omega_c t + \beta \cos(\omega_m t) + \phi]$$
$$\beta = \text{'modulation index' or 'phase deviation'}$$

Figure 13.3 Types of analogue modulation. (In the interests of clarity the frequency of the carrier is shown very much reduced compared with that of the modulating signal.)

These *baseband* frequencies—those which constitute the original signal—are unsuitable for direct transmission as radio waves.

For transmission purposes the signal is imposed on a 'carrier' wave of much higher frequency—a process known as *modulation*. This is represented in Figure 13.2 by the *modulator* block in ground station 1. If we represent the high-frequency signal by a cosine wave,

$$V = V_c \cos(\omega_c t + \phi_c) = V_c \cos(2\pi f_c t + \phi_c) \tag{13.1}$$

the baseband signal may then be represented by a variation with time of either the carrier amplitude (V_c), its frequency (f_c) or its phase (ϕ_c).

Figure 13.3 illustrates these three basic types of modulation for the simple case where the baseband signal is itself a cosine wave. Some points to note are as follows:

1. In the case of amplitude modulation (AM) the receiver normally responds to the 'envelope waveform', that is to the magnitude of the signal amplitude, irrespective of sign. It follows that the modulation depth must not exceed 100% (in the case of sinusoidal modulation, the modulation index must not exceed unity).

2. There is an evident similarity between frequency modulation (FM) and phase modulation (PM). Both may be described as 'angle modulation'.

For the case of a cosinusoidal modulation signal $m(t) = \cos(\omega_m t)$, phase modulation gives

$$V = V_c \cos[\omega_c t + \beta m(t) + \phi] = V_c \cos[\omega_c t + \beta \cos(\omega_m t) + \phi] \tag{13.2}$$

while frequency modulation gives

$$V = V_c \cos\left[\int (\omega_c + \Delta\omega m(t))\, dt\right] = V_c \cos\left[\omega_c t + \frac{\Delta\omega}{\omega_m} \sin(\omega_m t) + \phi\right] \tag{13.3}$$

where $\Delta\omega$ is the peak frequency deviation (in rad/s) and $\beta = \Delta\omega/\omega_m = \Delta f/f_m$ is the peak phase deviation (in radians) or modulation index.

Analogue and digital signals

The modulated carrier waves illustrated in Figure 13.3 are examples of analogue signals. They are characterized by the fact that the instantaneous value of the baseband signal may lie anywhere within a certain range.

An increasing proportion of telecommunications traffic consists of digital signals, that is signals which can take only a finite number of discrete values—often only two values, corresponding to the binary digits 0 and 1. When the signals are in this form the three types of modulation are known as *amplitude-shift keying* (ASK), *frequency-shift keying* (FSK) and *phase-shift keying* (PSK).

Within these general categories there are many variants whose different characteristics make them suitable for differing applications. Some of the more commonly used techniques are illustrated in Figure 13.4 where, for instance, ASK is shown as 'on–off keying', with a binary 1 represented by full amplitude and a 0 by zero amplitude. FSK is illustrated by continuous-phase *fast frequency-shift keying* (FFSK) in which there are no discontinuous changes in phase at the frequency transitions and the two frequencies are separated

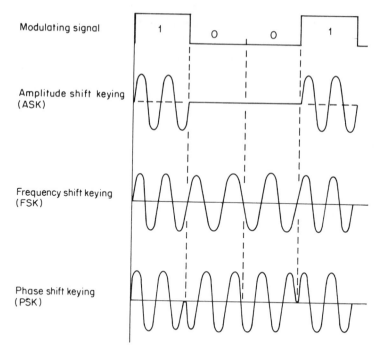

Figure 13.4   Examples of digital modulation. The examples illustrated are:

  ASK—on–off keying.
  FSK—fast frequency shift keying.
  PSK—phase reversal keying.

(In the interests of clarity the frequency of the carrier is shown very much reduced compared with that of the modulating signal.)

by half the switching frequency. This choice of frequency separation has certain advantages concerned with the ease of detection of signals in the presence of noise (see Sections 13.2.4 and 13.2.5).

PSK often uses two phases separated by π radians (as illustrated in Figure 13.4) and is then known as *phase-reversal keying* (PRK). However, another important type of PSK uses four phases separated by $\pi/2$ radians. This is *quadrature phase-shift keying* (QPSK). More complicated multi-phase or multi-phase and multi-amplitude schemes may also be used, particularly where there is a need to achieve high data rates in a limited bandwidth.

In order to demodulate directly encoded PSK signals the receiver needs a knowledge of the zero of phase. This may be achieved by periodically transmitting an identifiable phase reference signal. Alternatively, the signal may be subjected to 'differential encoding' in which the values of the digits are represented by the presence or absence of phase transitions at the start of the bit intervals. The absolute phase is then no longer relevant.

Digital signals occur naturally in, for instance, communication between computers. It is also possible to convert analogue signals into digital form. Already a great deal of telephone traffic is transmitted by digital techniques and digitized television is in limited use.

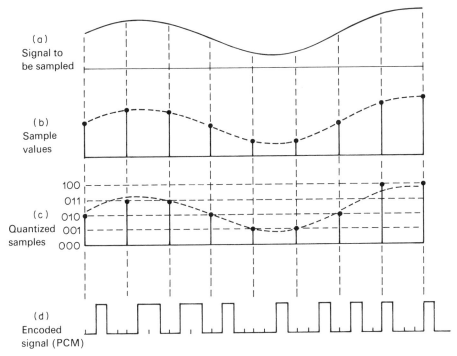

Figure 13.5 Sampling and digitization. The process illustrated here is an example of pulse code modulation (PCM). Each sample value is represented by a binary number with, in this case, three digits and an extra zero to assist in synchronization. (a) Signal to be sampled, (b) sample values, (c) quantized samples, (d) encoded signal (PCM).

The conversion of analogue signals into digital form may be described in terms of three distinct steps—sampling, quantization, and encoding.

The sampling process is illustrated in Figure 13.5(a, b). Clearly, if it is known that the signal varies with sufficient smoothness between the sample times, the entire signal can be reconstructed with considerable accuracy from the sample values. It can be shown that this is guaranteed if the signal to be sampled contains no frequencies greater than half the sampling frequency.

The process of quantization involves rounding off each sample to one of a finite number of allowed values. This introduces errors which cannot be compensated in subsequent processing of the signal. In order to reduce this 'quantization noise' to an acceptable value it may be necessary to increase the number of allowable signal levels. However, the greater the number of levels, the greater the number of digits that are necessary to identify a particular level and hence the greater the required bit rate. This tends to enhance the signal degradation caused by mechanisms other than quantization (see Section 13.2.5) and the system designer must seek a balance which results in best overall signal quality.

The final step, that of encoding, may take many forms. The simplest is to express each sample value directly as a binary number and then combine the numbers (normally with some extra digits for synchronization) to form a long sequence of binary digits. This is the system usually understood by the term *pulse code modulation* (PCM). However, a number of more subtle techniques may be introduced at this stage with a view for instance to making the signals less vulnerable to transmission impairments, providing encryption or modifying the spectrum. Thus digital modulation is in many ways much more flexible than its analogue counterpart.

Spectrum and bandwidth

In the design of a telecommunications system one of the most important signal parameters is the bandwidth or range of frequencies which must be transmitted. It is this (among other things) that determines the design of the various filters shown in Figure 13.2 The effect of modulation of a carrier wave is to produce frequencies other than that of the unmodulated carrier and the resultant spectrum depends both on the baseband signal and on the type of modulation in use. Some particular examples are considered below.

In the case of *amplitude modulation* by a cosinusoidal signal the spectrum may be inferred from standard trigonometrical identities:

$$V_c(1 + m \cos \omega_m t) \cos \omega_c t = V_c[\cos \omega_c t + (m/2) \cos(\omega_c + \omega_m)t + (m/2) \cos(\omega_c - \omega_m)t]$$
(13.4)

The components at frequencies $(\omega_c + \omega_m)$ and $(\omega_c - \omega_m)$ are known as *sidebands*, and the term $V_c \cos(\omega_c t)$ is the *carrier component*. Increasing the modulation index results in increased sideband levels but no change in the range of frequencies present in the signal. This is illustrated in Figure 13.6(a).

More generally the spectrum of any signal is given by its Fourier transform. For amplitude modulation by a baseband signal $m(t)$, the modulated carrier is $m(t) \cos \omega_c t$ and the Fourier transform is

$$\mathscr{F}[m(t) \cos \omega_c t] = \tfrac{1}{2}M(\omega - \omega_c) + \tfrac{1}{2}M(\omega + \omega_c)$$
(13.5)

where $M(\omega) = \int_{-\infty}^{\infty} m(t) \exp(-j\omega t) \, dt$ is the Fourier transform of $m(t)$.

The spectrum of the modulated signal therefore contains bands of frequencies above and below the carrier frequency ω_c, and above and below $-\omega_c$. In practical terms positive and negative frequencies are identical. The required channel bandwidth is twice the base bandwidth (irrespective of modulation depth).

The spectrum of a *frequency-modulated signal* is considerably more complicated. In the case of cosinusoidal modulation with a small modulation index, β, the presence of sidebands similar to those for AM may again be demonstrated by standard trigonometrical identities. If, however, the index is not small, the spectrum of FM is in striking contrast to that of AM. Fourier analysis of expressions such as $\cos(\omega_c t + \beta \sin \omega_m t)$ yields an infinite set of components at frequencies $(f_c \pm n f_m)$ (where n takes all integral values) and with amplitudes which can be evaluated from Bessel functions of the first kind. The number of side frequencies having significant amplitude (and hence the effective bandwidth) increases with increasing β (Figure 13.6b).

When the baseband is not sinusoidal, the side frequencies are no longer simply related to the baseband frequencies. For instance if the baseband contains two frequencies f_1, and f_2, the modulated signal contains all frequencies of the form $f_c \pm n f_1 \pm m_{f2}$ (n, m are integers). In strictly mathematical terms the bandwidth is infinite but there is a useful

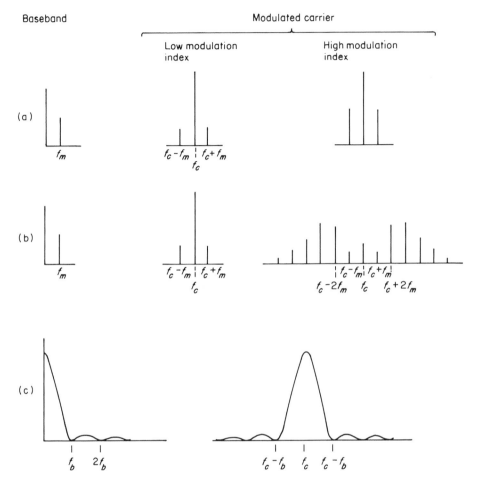

Figure 13.6 Spectra of modulated signals: (a) Amplitude modulation (amplitude spectrum); (b) frequency modulation (amplitude spectrum); (c) phrase reversal keying (power spectrum). The AM and FM spectra show modulation by a single tone of frequency f_m. The PRK spectrum is for a random binary sequence at a bit rate of f_b bits/s

semi-empirical expression (Carson's formula) for the bandwidth containing about 98% of the power:

$$\text{Carson bandwidth} = 2(\Delta f + f_m) \tag{13.6}$$

where Δf is the peak frequency deviation and f_m is normally taken to be the highest frequency present in the baseband. Frequency modulation with index sufficiently small so that only the first-order sidebands are significant (β less than about 0.5) is known as *narrow-band FM* (NBFM), while for larger β the term *wide-band FM* (WBFM) is used.

In the case of *PSK* the spectrum depends both on the type of modulation (PRK, QPSK etc.) and on the encoding procedure used in producing the bit stream. As an example we may consider phase-reversal keying by a random binary signal (that is by a random sequence of 0s and 1s occurring with equal probability). A signal of this type does not have a Fourier transform but we may define a power-density spectrum $S(f)$ which is the power per hertz of bandwidth at the frequency f. For the random binary signal:

$$S(f) = V^2 T(\sin \pi f T)^2/(\pi f T)^2 \qquad (13.7)$$

where V is the signal voltage and T is the length of one bit.

Since PRK may be regarded as a type of ASK (the amplitude is switched between $+A$ and $-A$), the spectrum of the modulated signal is similar to the baseband spectrum but centred on the carrier frequency. This is illustrated in Figure 13.6. Strictly speaking, the overall bandwidth is infinite because, in this case, the baseband has infinite bandwidth. However, the power in the 'tails' of the spectrum is not very significant, the first subsidiary maxima already being some 13 dB below the central peak. In practice, therefore, it is usual to transmit at most that part of the spectrum which lies between the first zeros. In many cases less will suffice. This has the effect of partially smoothing out the phase transitions, though without destroying the baseband information completely. Since $1/T$ is the number of bits per second, the effective bandwidth for the random binary signal is between one and two times the bit rate.

If the binary signal results from digitizing an analogue signal, the bit rate depends both on the base bandwidth and on the required level of quantizing errors. For instance, for speech (bandwidth 4 kHz) quantized to 256 levels the bit rate is 64 kbit/s since there must be at least 8000 samples per second with 8 binary bits per sample. Digital television, on the other hand, requires bit rates of the order of 100 Mbit/s. Thus PCM uses much more bandwidth than the corresponding analogue signal.

13.2.3 Multiple access

It is a requirement of most systems that several users (in some cases very many) can pass signals through the satellite simultaneously. There are two main techniques for doing this, one normally for analgoue signals and the other exclusively for digital signals. In *frequency-division multiple access* (FDMA) each ground station transmits its signals on a different carrier frequency, and at the receiving station the signals are separated by frequency-selective filters. The signals in this case are often analogue in nature.

In the case of digital signals, *time-division multiple access* (TDMA) may be used. Here each user transmits a short burst of digits in a particular time-slot within a repeating time-frame. Other users' time-slots are interleaved. Synchronization signals are included so that the receiver can select those time-slots which contain a particular 'message'. The concentration into short bursts has the result that each signal occupies the whole of the transponder bandwidth. However, no two signals occur at the same time.

A third method, *spread-spectrum multiple access* (SSMA), is of more specialized application, mainly in military systems. Each signal uses the full satellite transponder bandwidth but the signals may well all be present at the same time. In one version (*code-division multiple access*—CDMA—or direct-keying SSMA) the carrier signal is modulated twice, first by the required signal and then by a pseudo-random sequence of binary digits at a much higher bit rate. A receiver using the correct pseudo-random code can undo the effect of the second modulation and so recover the original signal. It does not, however, recover the signals of other users who have employed different codes.

13.2.4   Noise

Telecommunications would present few problems were it not for the presence of 'noise' in all electrical systems. In radio communications electrical noise is a result of the random thermal motions of atoms and electrons in matter, which reveal themselves as small randomly varying electromotive forces and currents. Each resistive element in a circuit is a source of thermally generated electrical power of kT watts per hertz of bandwidth, where $k = 1.38 \times 10^{-23}$ J K^{-1} (Boltzmann's constant), and T is the temperature (K) of the resistor. Quantum mechanics predicts a power spectral density given by the expression

$$P_0(f) = \frac{hf}{\exp(hf/kT) - 1} + hf/2 \qquad (13.8)$$

where $h = 6.625 \times 10^{-34}$ J s is Planck's constant. If $hf \ll kT$ this reduces to $P_0(f) = kT$.

The essential differences between a noise voltage and a signal are illustrated in Figure 13.7. The signal has a regular quasi-periodic structure whereas noise is essentially irregular and unpredictable. The r.m.s. value may be used as a measure of noise magnitude but a complete specification must include the probability density function (PDF), that is the probability $p_n(x)\,\delta x$ of the instantaneous voltage lying in the range from x to $x + \delta x$. For many sources of noise the PDF is approximately Gaussian, in which case $p_n(x)$ takes the form

$$p_n(x) = \frac{1}{\sigma_n \sqrt{(2\pi)}} \exp\left(\frac{-x^2}{2\sigma_n^2}\right) \qquad (13.9)$$

where σ_n is the r.m.s. value.

When two independent sources of noise are combined it is their powers (or mean squared values) which must be summed. In particular the noise powers contained in different frequency ranges are additive, so that for a constant power spectral density, $P_0(f) = kT$ W/Hz, the total power in bandwidth B is given by

$$P = kTB \qquad (13.10)$$

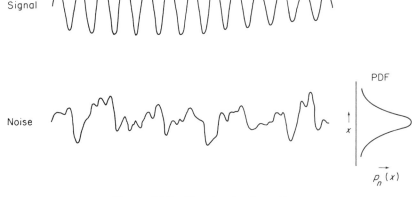

Figure 13.7   Signal and noise voltages

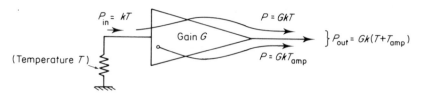

Figure 13.8   Definition of noise temperature

The importance of noise is that it sets a fundamental limit to the sensitivity of a telecommunications receiver. The crucial factor which determines the overall performance of the system is not just the signal power itself, but rather the ratio of signal power to noise power (the *signal-to-noise ratio*, SNR).

In any communications system there are many possible sources of noise. However, the signal power reaches its lowest level at the end of the transmission path and it is therefore the noise generated in the input circuit of the receiver which is most critical in determining the system performance. For this reason noise is shown in Figure 13.2 as an extra input to each of the receivers—a convenient fiction which has much the same effect as the reality.

There are two commonly used ways of expressing the 'noisiness' of an amplifier (or radio receiver). In the first, the expression for thermal noise power available from a resistor, $P_0(f) = kT$, is used to express the noise output as an equivalent temperature. The concept is illustrated in Figure 13.8.

If the amplifier were an ideal noise-free device the noise power density kT delivered by the source resistance would give an output power density GkT. In practice the output is greater than this and we can write

$$P_{out}(f) = Gk(T + T_{amp}) \qquad (13.11)$$

where T_{amp} is known as the *noise temperature* of the amplifier. The noise temperature is thus the additional noise power originating in the amplifier expressed as an equivalent increase in the source temperature. (Note that noise temperatures, like noise powers, are additive.)

In the second method of expressing the noise performance of the amplifier we assume that (in the arrangement of Figure 13.8) the source resistance is at a standard temperature, T_0. Were the amplifier noise-free, the output power spectral density would then be $kT_0 G$ but in reality it is greater than this by a factor F, the noise factor (or noise figure)

$$P_{out}(f) = FGkT_0 \qquad (13.12)$$

From (13.11) and (13.12) it can be seen that

$$F = 1 + T_{amp}/T_0 \qquad (13.13)$$

The standard choice of T_0 is 290 K—a typical 'room temperature'.

It can be shown that the overall noise temperature T_{tot} of an amplifier chain (Figure 13.9), is

$$T_{tot} = T_1 + T_2/G_1 + T_3/G_1 G_2 + \cdots \qquad (13.14)$$

and, in terms of noise factor,

$$F_{tot} = F_1 + (F_2 - 1)/G_1 + (F_3 - 1)/G_1 G_2 + \cdots \qquad (13.15)$$

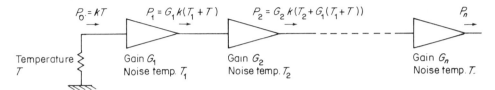

Figure 13.9 Noise temperature of cascaded units. The overall noise temperature of the amplifer chain is $T_1 + T_2/G_1 + T_3/(G_1 G_2) + \cdots$

As might be expected, the total noise temperature can never be better than that of the first stage, and moreover the total noise temperature is minimized by making the first stage gain, G_1, as large as possible.

It may happen that one of the units in the cascade is an attenuator rather than an amplifier. This usually occurs at the most critical point in the entire system—at the very input of the receiver, where the antenna feed (the transmission line or waveguide connecting the antenna to the receiver) has a small, but not negligible, attenuation. For this 'stage' the gain, G, is of course less than unity, and its noise temperature can be shown to be $T(1/G - 1)$ referred to its input (where T is the physical temperature of the cable or attenuator).

In addition to the noise originating in successive amplifier stages the overall system temperature T_{sys} will include background noise received by the antenna (usually represented by an *antenna temperature*) and a (usually small) contribution transmitted with the signal.

Antenna noise results mainly from thermal radiation from various surfaces which may be either in the 'main beam' of the antenna or in the less sensitive 'sidelobes' (see Section 13.2.8). These include for instance the Earth's surface (at about 300 K), the Moon (about 200 K) and the Sun (about 10 000 K). Also, the Earth's atmosphere emits thermal radiation when its attenuation becomes significant, as during a rainstorm (see Section 13.2.7). The situation is similar to that in a circuit attenuator except that in this case the temperature must be referred to the output. The contribution of the atmosphere to antenna temperature is

$$T_{atmos} = T(1 - 1/\alpha) \tag{13.16}$$

where α is the attenuation factor and the temperature, T, of rain is about 275 K. In the majority of geographical locations, in order to achieve the required limits on outage times, it is necessary to design the system for operation with significant atmospheric attenuation (see Section 13.2.7). Under these conditions the atmosphere may sometimes be one of the main sources of antenna noise (and indeed of system noise) for a downlink, where the receiving antenna is pointing towards the normally 'cold' sky. For an uplink, however, where the on-board antenna points towards the Earth's surface, the antenna temperature is always in the region of 290 K and is little affected by the atmospheric attenuation (though of course the signal level is affected).

In the case of a satellite downlink the transmitted signal is contaminated by noise originating on the uplink. The system is often designed so that the contribution of the uplink to the system noise is about 10 dB or more below that of the downlink. Under these conditions the transmitted noise causes a 0.4 dB degradation of downlink signal-to-noise ratio.

13.2.5   Output signal-to-noise ratio

The preceding section has outlined some of the factors influencing the system noise levels at radio frequencies (RF). The system user, however, is concerned with the signal-to-noise ratio $(S/N)_0$ on the final output of the system. Although $(S/N)_0$ must clearly depend upon the RF signal (or 'carrier')-to-noise ratio (C/N), the relationship can be complicated, depending in particular on the type of modulation in use. It is usually convenient to express C/N in terms of the signal-to-noise power density ratio, C/N_0, in which N_0 is the noise power per hertz of bandwidth. Furthermore, in comparing the performance of different types of modulation it is usual to refer to the *input signal-to-noise ratio*, $(S/N)_i$, which is the ratio of the RF signal power to the noise power in an RF bandwidth equal to the base bandwidth.

We have

$$C/N = C/(N_0 B) \quad \text{and} \quad (S/N)_i = C/(N_0 F) \tag{13.17}$$

where B is the bandwidth of the receiver and F is the maximum baseband frequency.

When referenced to the input of the receiver, N_0 is related to the system temperature by $N_0 = kT_{\text{sys}}$. Thus, for a given transmitter power, system temperature and modulating signal, $(S/N)_i$ is fixed. Differences in $(S/N)_0$ are then due to the modulation system.

For *AM systems* the relationship between input and output SNR, assuming sinusoidal modulation, is

$$(S/N)_0 = \frac{m^2}{2 + m^2}(S/N)_i = \frac{m^2}{2 + m^2}(C/N_0)(1/F) \tag{13.18}$$

Thus with the maximum allowable modulation index, $(m = 1)$, we have $(S/N)_0 = (S/N)_i/3$. However, for typical non-sinusoidal signals and for the usual modulation depths, $(S/N)_0 \approx 0.05\,(S/N)_i$. This is a rather poor performance, the main reason being that most of the transmitter power is wasted by being concentrated in the central carrier component (see Figure 13.6) which contains no message information.

At the cost of some increase in complexity of the equipment this component can be omitted (*double sideband suppressed carrier modulation*—DSBSC) and moreover the bandwidth may be reduced by also omitting one sideband, giving *single sideband* (SSB) transmission. This technique is widely used for instance in telephony, where a large number of channels must be accommodated in a limited bandwith. For both DSBSC and SSB, $(S/N)_0 = (S/N)_i$.

For *FM systems* the situation is very different. In Figure 13.10 the large vector

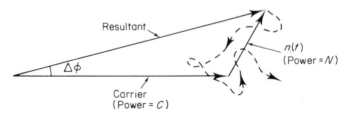

Figure 13.10   Noise in an FM system. The tip of the noise vector follows a random path such as that shown by the dotted line

represents the RF signal and the smaller, randomly varying component, $n(t)$, is the noise. The resultant signal displays random phase variations $\Delta\phi$, with an r.m.s. magnitude determined by the quantity $\sqrt{(N/C)}$—that is in effect by the RF signal-to-noise ratio.

The phase variations due to the modulation may be made much larger than the random variations (and a good $(S/N)_0$ can thus generally be achieved) by choosing a sufficiently large modulation index.

For tone modulation with index β, provided the RF signal to noise ratio is adequate, the output signal-to-noise ratio is given by

$$(S/N)_0 = \tfrac{3}{2}\beta^2(S/N)_i \qquad (13.19)$$

For more typical, nonsinusoidal signals this becomes

$$(S/N)_0 \simeq 0.3\,\beta_\alpha^2(S/N)_i \qquad (13.20)$$

where β_α is the *deviation ratio*. ($\beta_\alpha = \Delta f/F$, where Δf is the peak frequency deviation and F the maximum baseband frequency.) Since we may choose $\beta_\alpha \gg 1$, this is a very substantial improvement over AM.

Naturally, this improvement has to be paid for, and the price in the case of FM is an increase in receiver bandwidth. And as the bandwidth increases so does the in-band noise power, so that C/N decreases. But the 'FM improvement' occurs only when C/N is above a threshold value which for conventional demodulators is about 11 dB, but for the type of demodulator (phase-locked loop) used in most Earth stations may be about 8 dB. When C/N falls below this threshold the output signal-to-noise ratio declines rapidly and

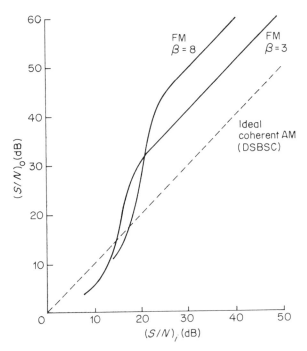

Figure 13.11 Output signal-to-noise ratio for frequency modulation by a single tone

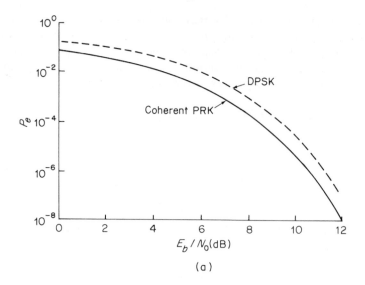

(a)

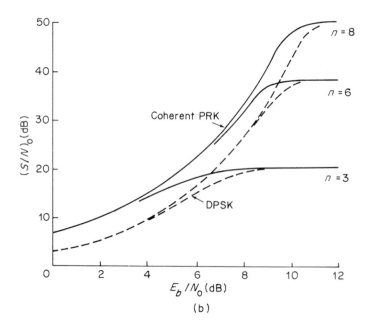

(b)

Figure 13.12   The effect of noise on digital systems. (a) Bit error rate vs E_b/N_o for two types of system. In coherent PRK the phase of each received bit is compared with that of a reference carrier derived separately from a long sample of the signal. In differential PSK (DPSK) the phase of each bit is compared with that of the preceding bit. White Gaussian noise, perfect synchronization and optimum detection are assumed. (b) Output SNR for simple PCM assuming the error rates shown in (a). n is the number of bits per sample

eventually becomes less than that available using AM. Thus, when using the above formulae to estimate the transmitter power required for a given system performance, it is important to check also that the C/N will be above the threshold level. Both the FM improvement and the threshold effect are shown in Figure 13.11, which also includes the ideal ('coherent' suppressed carrier) AM case for comparison.

In *digital systems* the mechanism by which system noise affects the output of a communications link differs markedly from the analogue case. The demodulator must now contain one or more threshold detectors which allocate to each received bit one of the permitted values. If the bit is correctly identified, the noise has no effect whatsoever. Occasionally, however, the noise voltage is large enough to cause the output to lie on the wrong side of the detector threshold and so result in an error in interpretation. The probability of error in any one bit—known as the *bit error rate* (BER)—depends on the product of the received signal power and the length of the bit interval, i.e. on the received energy per bit. For typical transmission systems the BER, P_e, may lie in the range 10^{-3}–10^{-9}. The dependence of bit error rate on the ratio of bit energy to noise power density (E_b/N_0) is shown for two types of demodulator in Figure 13.12(a). These are ideal theoretical curves. In practice the value of E_b/N_0 needed for a given error rate may be 1–2 dB higher than shown.

The error rate on the final data stream (or the SNR on the reconstituted analogue signal) depends both on P_e and on the system of encoding. For PCM (Section 13.2.2) with sinusoidal modulation the output SNR is given by

$$(S/N)_0 = \frac{3(2^{2n-1})}{1 + 4P_e(2^{2n} - 1)} \tag{13.21}$$

where n is the number of binary bits per sample and P_e is the bit error rate. This result is illustrated in Figure 13.12(b) for the error curves of Figure 13.12(a). For small values of E_b/N_0 the performance is system noise limited but as the C/N_0 improves, a stage is reached where the quantization noise (see Section 13.2.2) is dominant and there can be no further improvement in $(S/N)_0$ without an increase in the number of quantization levels.

13.2.6 Choice of modulation and access systems

The output SNR is one of the main factors in determining the choice of modulation technique. Other factors of primary importance are the equipment complexity, the payload electrical efficiency, the spectral occupancy and the general system efficiency, flexibility and adaptability.

If simplicity of the equipment and efficient use of spectrum were the only significant requirements, the natural choice would be AM. However, this is rarely the final choice. AM requires that the amplifiers, including in particular the high-power amplifier feeding the antenna, are linear in order to avoid signal distortion. But the efficiency of linear amplifiers (defined as the ratio of RF power output to D.C. power consumption) is low. AM therefore makes heavy demands on the primary satellite power.

In contrast, if FM is chosen, the transmitter may be operated in a more efficient non-linear mode and moreover, as indicated in Section 13.2.5, if wideband FM is used, a specified $(S/N)_0$ may be achieved with considerably less transmitter power than is required for AM. This technique has been widely used for intercontinental trunk communications routes. Large numbers of telephony channels are first assembled using single sideband modulation and frequency-division multiplexing and this composite baseband signal is then frequency-modulated onto the microwave carrier. Alternatively the assemblage of telephony channels may be replaced by a single television channel or other wideband signal. In a major ground station the implied complexity is acceptable. However, when many small Earth stations are involved, it is more usual to employ NBFM with a single channel per microwave carrier (SCPC). Although NBFM avoids the strict linearity requirement of AM, there is some loss of efficiency when a single wide-band transponder is used in an FDMA mode. When two or more frequencies are passed through a nonlinear system the output in general contains 'intermodulation products' at all possible sums and differences of the input frequencies and their harmonics. Particularly troublesome are the third-order products, since some of them (those of the form $f_1 + f_2 - f_3$ or $2f_1 - f_2$) may lie inside the transponder passband and so form a type of co-channel interference. It is usually necessary to work at reduced efficiency in order to keep the in-channel carrier-to-intermodulation power ratio (C/I) above a specified value—typically 15–20 dB—-with the transponder fully loaded.

The most effective way of avoiding intermodulation products is to use TDMA. In this system no two signals are amplified at the same time, so there is no intermodulation even when the transmitter is operated in a highly nonlinear mode. There is a trend towards increased use of TDMA in most multi-channel systems, a trend which is reinforced by the rapid advance in the technology of digital integrated circuits and the move towards digital transmission in the terrestrial 'tail-links'. The increased transmitter efficiency is partially offset by the need to transmit extra pulses for synchronization, for channel identification and to give the demodulators time to lock on to each signal in turn. When the signals originate in analogue form, TDMA suffers from the fact that PCM requires more bandwidth than analogue modulation. However, there is a compensation in the much greater flexibility of TDMA in switching and routeing signals and in multiplexing signals of differing bandwidths and differing types in a variety of configurations.

13.2.7 Radio propagation

Radio waves consist of a system of electric and magnetic fields which travel through free space at a velocity of about 3×10^8 m/s. The transit time for a round trip to a satellite in geostationary orbit and back is thus about 0.24 s—much greater than the propagation delays normally encountered in communication links. For many systems this delay is unimportant, but it cannot always be ignored in packet-switching systems (in which different message segments may take different routes), in systems depending on or providing accurate radio location, or in systems involving some form of closed-loop control. The effect is particularly marked if two links (Earth-satellite or inter-satellite) are connected in tandem.

Polarization

In propagation through uniform isotropic media the electric and magnetic fields are at right angles both to each other and to the direction of propagation. In a *plane-polarized* wave the direction of the electric field lies in a fixed plane. If a second wave has its plane of polarization orthogonal to the first, then the two waves will propagate independently and indeed may carry quite different signals.

If two orthogonal plane polarized waves are combined in quadrature, a different type of polarization results. The electric field vector at any point in the propagation path now rotates, the tip (in a geometrical representation) following a circular path if the two plane-polarized waves are of equal amplitude. This is known as *circular polarization*. It may be either right- or left-handed, depending on whether the vector rotates clockwise or anticlockwise, when viewed by an observer looking in the direction of propagation. More generally, if the two plane-polarized waves are not of equal amplitude, orthogonal and in quadrature, the result is *elliptical polarization*.

Propagation in the Earth's atmosphere

For most of their journey the signals from a satellite propagate through what is essentially free space, and they travel without change, apart from the steady diminution in intensity at a rate $1/r^2$ as the distance r from their source increases.

In the last few kilometres, however, as they pass through the Earth's atmosphere, they encounter phenomena of an unpredictable nature which may significantly affect the system performance. These atmospheric effects can be divided into two categories, those which occur in the ionosphere and those which occur in the troposphere. Ionospheric effects are very important at low and medium frequencies but in general their magnitude varies as $1/f^2$ and at the microwave frequencies used for satellite links they can normally be ignored. Tropospheric propagation phenomena, however, can be of considerable significance. There are three main effects—refraction, attenuation and scintillation—but of these it is usually only attenuation that can cause serious problems to the system designer.

Atmospheric refraction causes a slight shift in the apparent elevation of the satellite, whose magnitude depends on the elevation as well as on the atmospheric pressure and water vapour content. For a standard atmosphere, the shift is given approximately by

$$R = \frac{0.02}{\tan[e + 0.14 + 7.32/(e + 4)]} \text{ degrees} \qquad (13.22)$$

where e is the true elevation in degrees. Minor variations caused by differing climatic conditions can readily be accommodated by causing the ground station antenna to track the satellite.

Attenuation in the troposphere has two causes. The first is molecular absorption by gases, illustrated by Figure 13.13. This shows the attenuation coefficients as functions of frequency for oxygen and water vapour, which together account for nearly all of the gaseous absorption in the atmosphere. Both curves display strong absorption bands, and although these occur at higher frequencies than have so far found extensive use in satellite communications, the water vapour band centred on 22.2 GHz will certainly be a significant effect at 20 and 30 GHz. It is also of some interest that the strong absorption by oxygen at 60 Ghz makes this frequency very suitable for intersatellite links, with the atmosphere providing a screen from interference from terrestrial sources. Apart from the microwave absorption bands, the background attenuation shows a steady rise with

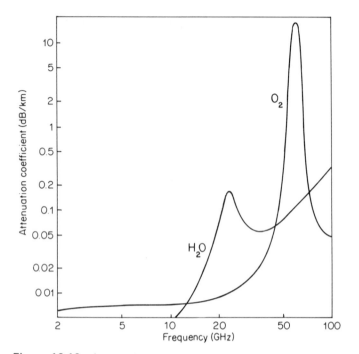

Figure 13.13 Attenuation of microwaves by oxygen and water vapour. The graphs show attenuation per kilometre for the two most significant components of the atmosphere. The water vapour density is assumed to be 7.5 g/m³ (about 40% humidity at 20°C). At frequencies other than the absorption bands, the effective vertical path length is about 5 km for oxygen and 2 km for water vapour. The attenuation at other elevations can be modelled by a cosecant law

frequency due to the tails of absorption bands in the far infra-red. This clear-air attenuation is generally negligible at 4 and 6 Ghz but is not entirely insignificant at 11 and 14.5 Ghz.

Much more dramatic attenuation effects are caused by rain. Water droplets both scatter and absorb radiation, the effect being strongest when the drop size is of the order of a wavelength. Thus in the microwave range, for which the wavelength is larger than a raindrop, the attenuation increases with frequency. Semi-empirical curves may be obtained showing attenuation as a function of frequency for various rainfall rates assuming reasonable models of drop-size distribution and of rain-cell size (see Figure 13.14). Since attenuation by rain is very variable, the system designer must seek some way of deciding what is a reasonable performance margin to allow for the occasional deep fade. It is usual for the customer to specify an allowable outage time and of this some will be allocated to loss of signal because of rain. The designer must then attempt to predict the atmospheric attenuation which will not be exceeded for more than this allocated time. If he is fortunate, experimentally determined statistics of attenuation may be available for the site of his ground station. Otherwise, he will need to use data such as those of Figure 13.14 together with the local meteorological data in order to make his predictions. For European

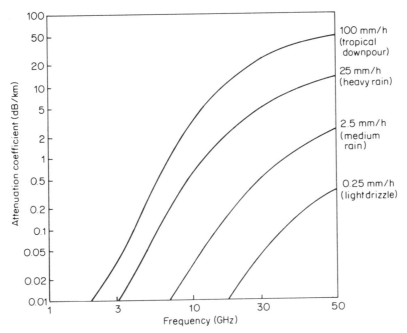

Figure 13.14 Attenuation of microwaves by rain. In order to calculate the total attenuation on an Earth–space path the attenuation coefficient must be multiplied by an effective path length which depends on location and climatic conditions. Typical values for temperate climates can be modelled by

$$l = a \operatorname{cosec}^b \theta$$

where θ is the elevation. The parameter a varies from 4 km to 2.2 km and b varies from 0.75 to 0.25 for rainfall rates between 5 and 100 mm/h

locations, typical fade margins are in the region of 2 dB at 11 Ghz and 4 dB at 14.5 Ghz, for a link reliability of about 99.9%.

Scintillation, or rapid fluctuation in signal amplitude analogous to the twinkling of stars, is normally a small effect which for most purposes can be ignored. However, at low elevations, because of the longer atmospheric path, all propagation effects are greatly enhanced and under these conditions scintillation depths of several decibels are not unusual. As a general rule satellite links are not operated at elevations less than about 10° but in some cases this cannot be avoided. The importance of scintillation is that, along with clear-air attenuation, it causes a slight reduction in capacity (and hence, in a commercial system, loss of revenue) at all times. In the long term this may be as serious as the occasional complete drop-out—especially in digital systems which may be able to adapt to short-term fades by changes in signalling rate.

Another consequence of rain is signal 'depolarization'. Because raindrops are slightly flattened, they absorb one plane of polarization rather more than the other. Since the axis of the drops is in general inclined to the plane of polarization of the signal, this differential absorption causes a slight rotation of the plane or, in the case of a circularly polarized signal, a slight ellipticity of polarization. The resulting power loss due to mismatch between the signal polarization and that of the receiving antenna is not significant. However, the presence of the orthogonal polarization, albeit at a level well below that of the wanted polarization, can have a serious impact on systems employing frequency re-use by polarization discrimination, (see Section 13.2.9).

13.2.8 Antennas

This section covers the general properties of antennas as they affect the telecommunications system. Various types of on-board antenna will be discussed later (Section 13.3.2).
   There is no need to describe transmit and receive antennas separately since the main properties of an antenna—such as gain and beam width—are the same for the two functions.

Radiation pattern and beamwidth

No antenna is strictly 'isotropic'. That is, no antenna radiates or receives signals with equal intensity in all directions. Indeed, most antennas are designed specifically so that they radiate very strongly in just one direction.
   The directional properties can be represented by a polar plot of the radiated field intensity or power as a function of direction. A two-dimensional section through this pattern is known as a polar diagram (see Figure 13.15).
   Referring to the highly directional pattern of Figure 13.15(b), it is seen that most of the radiation is emitted within a narrow range of directions known as the *main beam*. There are other directions in which radiation emerges, albeit at considerably lower lever, known as *sidelobes*. Their importance lies in the fact that they may contribute significantly to antenna noise and to interference from—and to—other systems. Figure 13.15(b) also illustrates the concept of *beam width*, which is usually measured between -3 dB points (as shown) but is sometimes specified between other limits such as -1 dB, -4 dB or between the first minima in the radiation pattern. Beamwidth is related to antenna size. Many microwave antennas consist of a large physical area, or aperture, illuminated by a single primary feed or by many small radiating elements. The most common arrangement is a paraboloidal reflector illuminated by a horn (see Section 13.3.2). The radiation pattern measured at a large distance from such an antenna is the Fourier transform of the

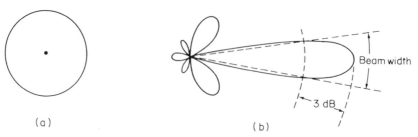

(a)   (b)

Figure 13.15 Antenna radiation patterns: (a) isotropic antenna; (b) directional antenna (large aperture)

distribution of electromagnetic field in the aperture. It follows directly from the scaling property of the Fourier transform that beamwidth is inversely proportional to antenna size. A useful rule of thumb for calculating the beamwidth is

$$\text{3 dB beamwidth} = 70\lambda/D \text{ (degrees)} \tag{13.23}$$

where λ is the wavelength and D the antenna diameter. This formula applies to a circular aperture with a typical distribution of illumination but it may be used as a rough guideline for most antennas of the 'aperture' type.

It is not usual to design for uniform illumination. By 'tapering' the illumination at the edges of the antenna it is possible to achieve lower sidelobe levels as well as reduced 'spillover' of radiation from the primary feed. The consequent slight increase in beam-width and decrease in gain (see below) is often a price worth paying.

Antenna gain

The importance of using highly directional antennas is that they provide signal *power gain* as compared with the same system using isotropic antennas.

In any given direction

$$\text{directive gain} = \frac{\text{actual power flux density}}{\text{power flux density from an isotropic radiator with the same total radiated power}}$$

$$\text{power gain} = \frac{\text{actual power flux density}}{\text{power flux density from a loss-free isotropic radiator with the same input power from the generator}}$$

It is assumed that the flux densities are measured at a fixed large distance from the antenna.

For a uniformly illuminated antenna with physical area A, the directive gain at the centre of the main beam is given by

$$G = 4\pi A/\lambda^2 \tag{13.24}$$

We may use the same relation to define an *effective aperture* A_e for any antenna. Thus:

$$A_e = \lambda^2 G/4\pi \tag{13.25}$$

Non-uniform illumination results in the effective aperture being less that the physical area. We may define an *aperture efficiency* by

$$\eta = A_e/A \tag{13.26}$$

For typical microwave antennas η is in the range 0.5–0.7. An overall antenna efficiency may be defined in the same way by using the power gain in place of the directive gain.

In a satellite system, where the coverage area is specified, the optimum antenna size depends on a trade-off between antenna gain and beamwidth. In order to optimize performance at the edge of coverage (which is usually the critical requirement in the system design) the gain in this direction should normally be about 4.2 dB less than at beam centre.

Polarization axial ratio

Most antennas are designed either for pure circular or for plane polarization but in practice the radiation is always elliptically polarized. In either case the antenna quality, from this point of view, can be expressed as an *axial ratio*. This is defined as the ratio (dB) of the powers radiated in two orthogonal planes of polarisation, where the planes are chosen to maximize the ratio.

For circular polarization the axial ratio should ideally be 0 dB. In a ground station antenna, which points directly at the satellite, 1 dB should be achievable, but for an on-board antenna the axial ratio may be as high as 2–3 dB at the edge of coverage.

Because the transmitting and receiving antennas may not have their planes of maximum gain aligned, allowance must be made for possible 'polarization loss'.

In the case of a plane-polarized antenna the axial ratio should be as large as possible in order to discriminate against signals which may be present on the other polarization. An axial ratio (or cross-polar discrimination, XPD, as it is usually called in this case) as high as 30 dB is not unreasonable.

13.2.9 Frequency re-use

It has already been pointed out that overcrowding of the spectrum has led to a gradual shift towards higher frequencies. However, since higher frequency usually means less transmitter power and increased atmospheric loss, there are considerable advantages in making maximum use of the lower-frequency bands. Many satellites now use the same frequency for two or more signals. This may be achieved either by spatial discrimination, in which two signals at the same frequency may be transmitted on separate narrow beams to different regions on the Earth's surface, or by polarization discrimination, in which signals at the same frequency may be transmitted over the same path on orthogonal polarizations. In the latter case, antenna imperfections and atmospheric phenomena may give rise to cross-polar interference. Although this can be troublesome, special receivers can use the signal in one channel to 'null out' the unwanted signal in the other, thus making the technique viable. The saving of space in the radio spectrum can be consider-able. For instance, by a combination of spatial and polarization discrimination a satellite of the Intelsat V series offers 2300 Mhz of signal bandwidth but occupies only 910 Mhz of spectrum.

13.2.10 The link budget

We are now in a position to discuss the calculation of the transmitter power requirement. We have seen that, for a given type of modulation, the overall performance depends upon the RF carrier-to-noise power-density ratio, C/N_0, at the receiver. It remains to show how C/N_0 is related to the transmitter power.

The definition of antenna gain implies that a transmitter with output power P_T asso-ciated with an antenna of gain G_T can be replaced, for the purpose of this calculation, by an isotropic radiator with output power $P_T G_T$. The quantity $P_T G_T$ is known as the *equivalent isotropic radiated power* (EIRP). This isotropic power spreads out uniformly so that the power flux density at a distance r from the source is

$$S = P_T G_T/(4\pi r^2) \qquad (13.27)$$

If the atmospheric loss is L_A, the flux density at the receiver is

$$S = P_T G_T/(4\pi r^2 L_A) \qquad (13.28)$$

The effective area of a receiving antenna with gain G_R is

$$A_R = \lambda^2 G_R/4\pi. \qquad (13.29)$$

Thus the signal power at the input to the receiver is

$$C = P_T G_T G_R(\lambda/4\pi r)^2(1/L_A) \qquad (13.30)$$

Table 13.3 Example of a link budet[1]

Transmitter output power (per carrier)	−14.4 dBW
Multiple carrier loss	0.2 dB[2]
Transmitting circuit loss	0.9 dB
Transmitted carrier power	−15.5 dBW
Transmitting antenna gain	18.0 dB
EIRP	2.5 dBW
Space loss	206.1 dB[3]
Polarization loss	0.1 dB[4]
Atmospheric and multi-path losses	4.0 dB[3,5]
Total transmission loss	210.2 dB
Ground terminal G/T	35.0 dB/K
Boltzmann's constant	−228.6 dBJ/K
Received C/N_0	55.9 dBHz
Transmitted C/N_0	69.2 dBHz[6]
Resultant C/N_0	55.7 dBHz
Carrier to intermod density ratio, C/I_0	67.0 dBHz[7]
Overall C/N_0 (including intermods)	55.4 dBHz
Required C/N_0	55.3 dBHz[8]
Margin	0.1 dB

[1] Based on the shore-to-shore link in a study of a maritime satellite system.
    The budget given here is for the 11.7 GHz downlink carrying voice traffic on a single channel per carrier basis. Global coverage is assumed.
[2] This correction allows for robbing of transmitter power by noise and intermodulation products.
[3] For the minimum elevation of 5° at the ground terminal.
[4] Antenna axial ratios: satellite 3 dB; ground terminal 1 dB.
[5] Multi-path loss is caused by destructive interference between a signal arriving at the antenna by a direct path and one transmitted by some other route—usually by ground reflection and reception in an antenna sidelobe. Multi-path is not usually very significant at this frequency even at low elevations.
[6] From the uplink budget.
[7] The ratio of carrier power to the power spectral density of intermodulation products (regarded here as a type of 'noise').
[8] From the customer's requirements.

Finally for a system temperature T_{sys} the noise power density referred to the receiver input is kT_{sys}, giving a signal-to-noise power density ratio of

$$C/N_0 = P_T G_T (\lambda/4\pi r)^2 (1/L_A)(G_R/T_{\text{sys}})(1/k) \qquad (13.31)$$

In this expression, which is known as the telecommunications equation, the factor $P_T G_T$ can be regarded as a figure of merit for the transmitter and the term G_R/T_{sys} as a figure of merit for the receiving system. Leaving aside the constant $1/k$, the remaining factors refer to the propagation path. The quantity

$$L_S = (4\pi r/\lambda)^2 \qquad (13.32)$$

is known as the *free-space loss*. We may think of it as the attenuation between two isotropic antennas separated by a distance r. It may seem curious that the space loss should depend on frequency but this is an artefact of the (slightly arbitrary) way in which we have separated out the factors related to the transmitter and to the receiver.

For a satellite in geostationary orbit (for which r is 3.6×10^7 m) the space loss from the transmitter to the subsatellite point is given in decibels by

$$L_S \text{ (dB)} = 183.6 + 20 \log_{10} f \text{ (GHz)} \qquad (13.33)$$

If the ground station is not at the subsatellite point the increased path length results in additional space loss of up to 1.3 dB, depending on the elevation of the satellite as viewed from the ground station.

In terms of the space loss C/N_0 is given by

$$C/N_0 = P_T G_T (1/L_S)(1/L_A)(G_R/T_{\text{sys}})(1/k) \qquad (13.34)$$

Since the required C/N_0 can be determined from the system specification, this expression allows us to calculate the required transmitter power, P_T. A typical link budget is shown in Table 13.3. It is usually convenient to work in decibels and for this purpose all quantities which are proportional to power are expressed in decibels relative to the appropriate unit. For instance a power of 1 W is 0 dBW (decibels relative to one watt) or 30 dBm (dB relative to one milliwatt), a bandwidth of 1 MHz is 60 dBHz (decibels relative to one Hertz) and Boltzmann's constant is -228.6 dBJ/K (decibels relative to one Joule per Kelvin).

13.3 THE COMMUNICATIONS PAYLOAD

13.3.1 The transponder system

Figure 13.16 is a simplified block diagram of a typical satellite repeater, which together with its associated antenna subsystem would make up a complete on-board transponder. Depending on the purpose of the satellite, there may be just one or perhaps many such repeaters within the same payload.

Before considering any of the details of the units we shall trace the signal's path through the transponder, listing the main subsystems that it encounters and outlining their functions. The scheme adopted is a dual-conversion payload. Some alternative arrangements will be mentioned later.

Low noise amplifier   Down converter

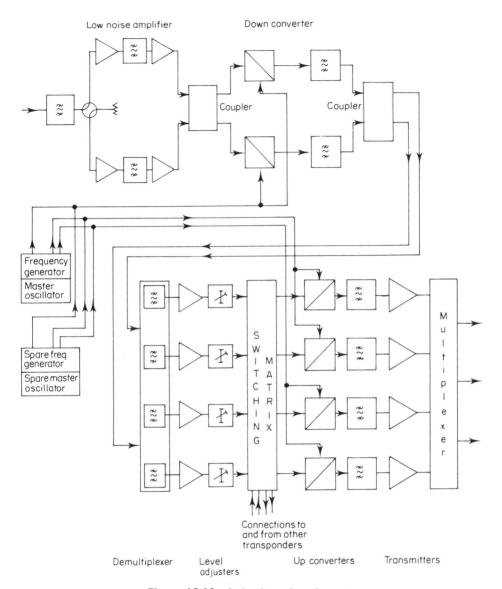

Figure 13.16   A simple on-board repeater

The signal path

1. *The antenna subsystem*'s function is to collect the incident signal power. Clearly the
    main requirements are that the aperture should be of sufficient size to receive a signal
    of adequate strength, and that it should be pointed in the right direction (an obvious
    need but the source of many nontrivial problems in implementation). In a system
    relying on frequency re-use, the antenna subsystem must distinguish between signals

on the basis of direction of arrival and polarization. Different signals on the same frequency must be presented at different output ports.

2. *The low noise amplifier (LNA)* must amplify the weak signals arriving at the antenna to a level at which they can be processed without causing degradation of signal quality by noise in the electronic circuits. In addition, the LNA usually incorporates some preliminary filtering with the main purpose of attenuating any strong signals which may be present at adjacent frequencies.

3. *The down-converter* converts the signals to a lower frequency (the *intermediate frequency*, IF) at which most of the amplification takes place. This avoids coupling between the input and the output of the repeater which would cause the amplifiers to break into oscillation. It also enables the creation of narrow-band channels which are suitable for many types of communication.

    The down-converter includes filters at both its input and its output. In Figure 13.2 the input filter is shown as part of the LNA. Its purpose is to give a more sharply defined RF passband and to reject the 'image channel' (see Section 13.3.4). The output filter rejects the many unwanted frequencies which are generated in the process of down-conversion, in favour of the wanted one at the IF.

4. *The IF processor.* The first part of the processor is normally a demultiplexer or set of filters which divide the broadband output from the down-converter into a number of separate channels. Although the proliferation of equipments in a multi-channel transponder carries a penalty in mass and cost, it also has advantages such as a reduction in intermodulation products in the high-power stages, greater flexibility in routeing signals between different antenna beams and different frequency bands, and the provision of 'graceful degradation' of the system (meaning a gradual reduction in performance as equipment failures occur rather than a sudden and total loss of capacity).

    The second part of the processor is a switching matrix to perform routeing operations and, in the case of equipment failures, to select channels which are still working. And finally there are the IF amplifiers—one for each channel—which provide most of the transponder gain.

5. *The up-converter* reverses the function of the down-converter by translating the amplified IF signals to the higher frequency at which they will be transmitted. Like the down-converter it produces many unwanted outputs as well as the one at the right frequency. There is consequently a need for further filtering before the signals pass on to the final amplifiers.

6. *The transmitters* raise the signal power to the required output level. The power amplifiers must achieve this with the maximum efficiency (minimum drain on primary satellite power) consistent with acceptable distortion. In a channelized system the signals must then pass to a multiplexer. There will then be some filtering to remove the harmonics which are always generated in the nonlinear power amplifier. If the same antenna is used both for transmitting and receiving, the transmitter harmonic filter and the receiver input filter may be combined in a single unit, the diplexer, which has the added function of ensuring that the incoming and outgoing signals are routed to their correct destinations.

Other payload units

There are several other units which form part of the communications payload but are not directly on the signal path. There is the local oscillator which provides CW (continuous sine-wave) signals used by the down- and up-converters in order to provide the required frequency shifts. Typically the unit would provide several (perhaps many) outputs, one for each frequency converter. Possibly some or all of these will need to be harmonically related.

In order to achieve stable gain and frequencies, the power supplies require regulation in addition to that provided by the spacecraft bus. This may sometimes be provided in part within the service module but almost certainly where specialized power conditioning is needed, as for instance for a travelling wave tube or Impatt amplifier, this will be provided by units within the payload.

Finally, the TT&C transponder is sometimes regarded as part of the communications payload. Indeed in some cases telecommand and telemetry signals may be transmitted within the normal communications bands and extracted or injected in the IF processor. More commonly, however, there is a separate transponder either at VHF (around 140 MHz) or at S-band (about 2200 MHz) or both.

Other types of transponder

Until fairly recently the dual-conversion repeater was in almost all cases the cheapest and most effective way of providing the necessary transponder gain. However, there is a useful reduction in complexity if the up-converter is eliminated, leaving a single conversion from the receive frequency to the transmit frequency. The bulk of the amplification and the other functions of the IF processor must in that case be performed at one of the radio frequencies (usually the down-link frequency, since this is normally lower). In the last few years, improvements in the technology of microwave circuits have made this scheme viable especially in high-capacity satellites using rather wide-band channels. At the other extreme, where for specialized applications very narrow-band channels are required, it may be necessary to use two down-conversions to reach a low enough IF for the channel filters to be realized. Other schemes are possible. For instance, an Intelsat V satellite has inputs in the 6 and 14 GHz bands and outputs at 4 and 11 GHz. All received signals are down-converted to 4 GHz and some of them are subsequently up-converted to 11 GHz. Cross-strapping is possible at 4 GHz. Thus the payload is single conversion for downlink signals at 4 GHz and dual conversion for the downlink at 11 GHz.

The scheme described above is known as a *transparent* repeater. A possible departure from this system would be on-board demodulation of the signals followed by modulation of a separate downlink carrier. This arrangement is known as a *regenerative* transponder. Its use can lead to an improvement in signal-to-noise ratio in digital systems where uplink noise is significant. This might be the case for instance if very small ground stations are in use so that uplink power is limited. In a regenerative transponder, digital signals can be 'cleaned up' at baseband so that only that part of the noise which is in the form of bit errors will be transferred to the output. Since the bit error rate is low, normally varying as a very high power of the signal-to-noise ratio there is an advantage in a system in which error rates rather than noise powers are cumulative.

The use of redundancy

As is usual in all payload systems, the communications payload includes cold spares of the most critical units. The redundancy inherent in a channelized system has already been mentioned as a source of graceful degradation. Units which are common to all or many of the signal paths, such as the low-noise amplifiers, down-converters and local oscillators, are usually duplicated. Where possible the spares are connected into the repeater using passive power splitters and combiners or hybrid couplers so that the selection of the operating unit can be effected simply by switching the power supplies on or off. The use of passive components rather than RF switches leads to greater reliability. However, since the use of any of these components involves some loss of signal (in the case of a power splitter or hybrid coupler, at least 3 dB loss) they cannot be used at the input to the receiver where the noise level is critical. In this position a low-loss switch must be used.

13.3.2   The antenna subsystem

The antenna subsystem is often a critical factor in the spacecraft design because of its impact on total mass and stability, the possible need for stowage during launch and erection in orbit, and the requirement for Earth pointing, if necessary by the provision of a de-spun platform. From the point of view of communications, the first constraint on antenna design is the required coverage area which determines the beamwidth and hence the antenna size (Section 13.2.8). Table 13.4 gives antenna diameters for three down-link frequencies and for two extremes of service area; earth coverage, corresponding to beamwidth of 17° and a (directive) gain of 18.5 dB (at the edge of coverage), and a 'spot' beam covering, say, the British Isles, with a beamwidth of about 1.5° and an edge-of-coverage gain of 40 dB. It is seen that for large coverage areas (comparable with earth coverage) and especially at the higher frequencies, antennas of quite modest size are sufficient but spot-beam antennas can be decidedly cumbersome, and they would be used only where justified by a significant improvement in the system performance. Apart from a possible requirement for a strictly limited service area, the advantage in using a spot beam usually

Table 13.4   Spacecraft antenna diameters

Frequency	Diameter (m)	
	Earth coverage	UK coverage
1.5 GHz (L-Band)	1.0	11.0
4.0 GHz (C-Band)	0.37	4.1
12.5 GHz (Ku-Band)	0.11	1.3
Beamwidth	17°	1.5°
Gain (at edge of coverage)	18.5 dB	39.8 dB

lies in the fact that in many systems the bulk of the traffic is routed between relatively few very busy nodes. If these high-density routes are served by high-gain antennas there can be a marked reduction in the transmitter power required for most of the signals and consequently a much more efficient use of the primary satellite power. The trade-off to decide on the number and size of the spot beams is clearly very complicated and involves many different aspects of the overall system design—the spacecraft structural design, the launch sequence, attitude control, power supplies, thermal control etc., as well as tele-communications capacity and market prediction.

Antenna types

1. The *horn antenna* can readily provide the small aperture needed for Earth coverage at 4 GHz or higher frequencies. In its simplest form it is a section of rectangular or circular waveguide spread outwards at the end to give the required aperture dimension (Figure 13.17a).

2. *Helical antennas* (Figure 13.17b) are 'end-fire' antennas, preferred to the horn at frequencies below 4 GHz, since even for Earth coverage a horn antenna would be excessively large. For instance Navstar uses helical antennas at 1.5 GHz to give Earth coverage from an altitude of 16 000 Km. Their use is generally limited to rather wide beams and gains less than about 14 dB.

3. *Reflectors*, such as a paraboloid illuminated by a horn, are the most satisfactory solution when a narrow beam is required. The most usual configurations are the front-fed arrangement in which a waveguide runs through or round the reflector to the focus of the paraboloid (Figure 13.17c), and the offset feed in which the reflector is a segment of a paraboloid taken from one side of the axis (Figure 13.17d). The front-fed arrangement has the disadvantage that the feed and its support structure cause blockage of the beam, reducing the gain and scattering power into the sidelobes. Moreover, the offset arrangement is often mechanically better, since the feed horn can be mounted rigidly on a face of the spacecraft. On the other hand its lack of symmetry results in poorer cross-polar discrimination when linear polarization is in use, and in the case of circular polarization causes beams of opposite polarizations to point in slightly different directions. The Cassegrain system (Figure 13.17e), which is common on ground stations, is not often used on the satellite because blockage by the subreflector usually makes it unsatisfactory for a small antenna.

    In some cases, where the service region is irregular in shape, there may be an advantage in forming a 'shaped beam' which, in a reflector antenna, is achieved by replacing the single primary feed by a cluster of horns. The reflector must be large enough to form the smallest features of the beam pattern.

4. *Phased arrays* are based upon the principle illustrated in Figure 13.18. The aperture is excited by many separate radiating elements which individually have only very weakly directive properties. Their combination may, however, have a very narrow beam because the radiation in some directions interferes constructively and in others destructively.

    The advantages of this arrangement are that one array can produce a large number of beams simultaneously and that these can be steered electronically without the need

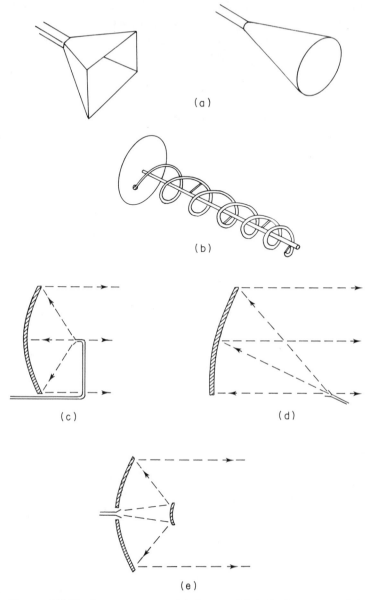

Figure 13.17 Antenna types: (a) horns; (b) helical antenna; (c),
(d), (e) paraboloidal reflector antennas: (c) front-fed; (d) offset feed;
(e) Cassegrain

for mechanical pointing systems. The distribution of transmitter power between many
output paths reduces the demand on any one power amplifier and facilitates the use
of solid-state devices. Finally, the array can be mounted rigidly and can often be made
to conform to some convenient surface.

The major disadvantage is the complexity of the associated equipment.

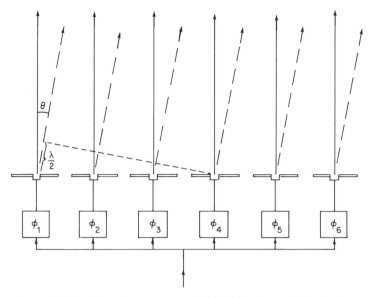

Figure 13.18 Phased array antenna. If all of the phase-shifts, $\phi_1, \phi_2, \ldots$ are equal, there is constructive interference in the broadside direction. At a small angle θ from this there is no emission since radiation from each element interferes destructively with that from an element three spacings away (for the six-element array shown here). By changing the ϕ_i the beam can be steered electronically without any movement of the antenna

13.3.3 The low-noise amplifiers (LNA)

The primary requirements for the LNA are a low-noise temperature and sufficient gain to ensure that the noise contributions from the succeeding states are negligible. A variety of amplifier types have been used in the past, including travelling-wave tubes (TWT), tunnel diodes, parametric amplifiers and transistors.

In most current designs a transistor amplifier is chosen for the 'front-end', with bipolar junction transistors generally preferred for frequencies up to about 2 GHz and GaAs FETs for higher frequencies.

The noise figures of field effect transistors increase with increasing frequency at a rate around 0.2 to 0.25 dB per GHz in the microwave range, while at the same time the gain per stage decreases. As a result, at the highest frequencies a transistor amplifier would have a noise figure in excess of what can be achieved with a mixer, using Schottky barrier diodes, followed by a low-noise IF preamplifier. Thus, for these frequencies the RF amplifier would be omitted and the mixer would become the front-end. However, transistor technology is continually improving, and new devices are already appearing which push the break-even point well beyond 30 GHz.

One further requirement for the LNA is good linearity over a wide dynamic range. Since the amplifier handles a multi-carrier signal (it is placed ahead of the demultiplexer which separates the signals into channels) any departure from linearity can give rise to intermodulation products (see Section 13.2.6).

13.3.4 Frequency converters

Any nonlinear device will serve as a frequency converter or mixer. It may be shown that in response to an input of the form $A\cos(2\pi f_1 t) + B\cos(2\pi f_2 t)$, its output will contain every frequency of the form $nf_1 \pm mf_2$, where n and m are integers. Usually the required output frequency is $f_1 + f_2$ or $f_1 - f_2$. Most conventional mixers depend on resistive diodes. Best performance is achieved when one input (the local oscillator waveform) is at a much higher level than the other, so that the signal is in effect switched on and off at the local oscillator frequency. As a general rule the local oscillator level should be at least 10 dB above the maximum signal level in order to minimize the *conversion loss* (the reduction in signal level from RF to IF), and to keep the level of intermodulation products as low as possible.

A mixer produces many output frequencies in addition to the wanted IF. It is also sensitive to frequencies other than the expected input frequency. In particular, if the input frequency is $(f_{LO} + f_{IF})$, then the mixer also responds to $(f_{LO} - f_{IF})$, which is down-converted to the same IF (and vice versa). This is known as the *image response* of the down-converter. Noise and interfering signals in the image channel must be filtered out prior to down-conversion. The levels of the various mixing products for a typical single-balanced mixer are shown in Table 13.5.

The noise temperature of a mixer is closely related to the conversion loss. To a reasonable approximation a down-converter can be regarded as an IF amplifier fed through a resistive attenuator. Semiconductor noise generated by the diodes can usually be made relatively insignificant. Thus the noise figure of the down-converter (in decibels) is given roughly by the sum of the IF amplifier noise figure and the conversion loss. Noise figures in the region of 6 dB are possible for mixers operating at 14.5 GHz but if noise performance is not critical, 8 dB would be a more comfortable value for the purpose of system design.

Table 13.5 Spurious outputs of a typical singly balanced mixer

m \ n	0	1	2	3	4	5	6	7	8
0		36	40	45	55	55	56	56	55
1	28	0	30	12	35	33	40	35	45
2	60	65	70	65	80	60	75	70	100
3	75	60	65	70	70	55	75	55	75
4	80	80	95	85	95	85	95	85	90
5	85	80	85	70	90	65	90	65	85
6	95	90	95	90	100	95	100	90	100
7	100	95	100	90	95	100	100	85	95

The entry in row m and column n gives the level of the product at frequency $mf_{in} \pm nf_{lo}$. For $m > 0$ the levels are expressed in decibels below the level of the wanted output at $f_{in} \pm f_{lo}$. For $m = 0$ the levels are in decibels below the local oscillator input level. This table is applicable for a reference signal input level $P_{ref} = -40$ dBW and for a local oscillator input -23 dBW. At a different input level P_{in}(dBW) the levels of spurious products are increased, relative to the $(f_{in} \pm f_{lo})$ level, by $(m-1)(P_{in} - P_{ref})$ dB

13.3.5   Local oscillators

Apart from the signal levels, the two most significant aspects of a local oscillator's performance are the frequency stability and the phase noise. The specification for long-term stability usually limits the frequency drift to less than about 1 p.p.m./yr, and this requires a crystal controlled source. Since crystal oscillators operate at less than about 150 MHz, frequency multiplication is necessary. A non-linear device may be used such as a varactor diode or step-recovery diode, the appropriate microwave frequency being extracted either by filtering or by phase-locking a high-frequency oscillator to the correct harmonic (see Figure 13.19). Several stages of multiplication or more complicated frequency synthesis techniques may be needed and mixers may also be used to generate outputs at higher frequencies. Since in the process many frequencies are produced in addition to the wanted one, careful filtering is required.

Short-term random frequency jitter can be regarded either as a noise-like phase modulation of the oscillator or as low-level sidebands in the spectrum of the oscillator output (see Section 13.2.2). When the oscillator is used in a frequency converter this phase noise is transferred to the signal. The sideband level is highest close to the carrier frequency. Thus phase noise has most effect where low baseband frequencies are important—particularly where low-rate digital signals are to be transmitted.

13.3.6   IF processors

The main functions of the IF processor are to provide most of the transponder gain and to define the frequency response of each channel. Thus the principal components are filters and transistor amplifiers.

At rather low intermediate frequencies (less than 200 MHz) the best choice of filter, in terms of bandwidth, frequency stability and weight is almost always some type of crystal or surface acoustic wave (SAW) device. At higher frequencies, helical resonators provide a satisfactory performance while at frequencies above 2 GHz, metal rod resonators or

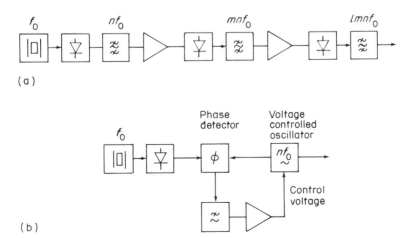

Figure 13.19 Frequency multipliers: (a) direct multiplier chain; (b) phase-locked loop multiplier

thin-film technique would probably be used. (This is a method of forming microwave circuits by etching a metal film deposited on a substrate of high dielectric constant.)

The amplifier blocks would normally use bipolar transistors except at the higher frequencies. Printed circuit board or thick-film construction (which depends on the deposition of conducting materials by a silk screen printing technique) could be used at frequencies up to 2 GHz. Above 2 GHz FETs may be used in thin-film construction.

13.3.7 Filters

The need for RF filters at various points in the transponder has already been noted. Most of these can be described by one of the labels, 'low-pass', 'high-pass', 'band-pass' or 'band-stop' (see Figure 13.20). Some of the more important RF filters in the transponder are listed below.

● The input filter: band-pass; main requirements, low in-band loss with adequate attenuation out of band.

● The channel filters: band-pass; main requirements, rapid roll-off (increase of attenuation with frequency outside the band) together with moderate loss and a sufficiently flat in-band response.

● The output (harmonic) filter: low-pass or band-pass; main requirements, low in-band loss with adequate attenuation at specified higher frequencies.

● The multiplexers and demultiplexers. In general these can be regarded as combinations of band-pass and band-stop filters. There may be a requirement for very rapid roll-off if the channels are closely spaced. The multiplexer, coming after the high-power amplifier, must have low in-band loss but this requirement is usually less stringent for the demultiplexer.

Two of the most important characteristics of a filter are the flatness in-band and the rate of roll-off. Another is phase linearity or group delay response. (Group delay is the time delay experienced by the modulation waveform.) A nonlinear phase characteristic leads to different signal components experiencing different delays, resulting in distortion

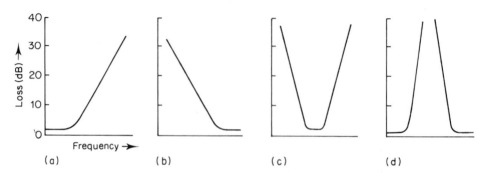

Figure 13.20   Filter types: (a) low pass; (b) high pass; (c) band pass; (d) band stop

of the modulation waveform. Thus it is usual for the system specification to impose stringent limits on the variation in group delay across the channel bandwidth.

There is much in common in the procedures for the design of the various types of filter. Any of the types shown in Figure 13.20 can be synthesized initially as a low-pass design, followed where necessary by the application of standard rules to transpose the impedances to those required for a high-pass or band-pass characteristic.

The attenuation in the pass-band depends mainly on the number of sections, n, needed to achieve the required roll-off and on the so-called quality factor, Q, of the resonators (or reactive elements for a low-pass or high-pass type). For a band-pass filter with a bandwidth B, the attenuation at the centre frequency f is given approximately by

$$L(\text{dB}) \approx 4.8(Q_L/Q_u)n \qquad (13.35)$$

where $Q_L = f/B$ and Q_u is the 'unloaded' Q of the resonators. Q_u depends on the method of implementation and may be typically of the order of 50–100 for microstrip quarter-wave lines, 2000 for quarter-wave resonators consisting of metal rods mounted in a box, or 8000 for rectangular waveguide cavities. Low in-band loss, which implies high Q, is bought at the cost of increased size and mass.

13.3.8 Transmitters

The only two acceptable means of amplifying the signals to the level required for downlink transmission are the travelling wave tube (TWT) and power transistors. The TWT is the most commonly used but as transistor technology improves, solid state amplifiers are finding an increasing number of applications, mainly because of their higher reliability and lower mass. TWTs, however, are currently the only choice of transmitter at the highest frequencies and at high power levels.

Travelling wave tube amplifiers (TWTA)

In a travelling wave tube, amplification is achieved by interaction between an electron beam and a signal travelling along an elongated electrode structure. For a significant effect, the electrons and the signal must travel at almost the same velocity, which implies that the signal must be slowed down very considerably. In low- and medium-power tubes the 'slow-wave structure' is a helix of wire held in place by ceramic rods. The signal modulates the density of the beam and the beam in turn transfers energy to the signal, which increases exponentially in amplitude as it travels along the tube to the output port. Helix tubes are capable of up to 20 or 30 W output. High-power tubes require a more robust slow-wave structure. A series of coupled resonant cavities can yield powers up to 1 kW for applications such as direct broadcast television.

A disadvantage of a TWTA is the requirement for a complicated high-voltage power supply (see Figure 13.21). The accelerating voltage is several kilovolts and must be very well regulated. The efficiency of a TWTA can be high—as much as 40–50% for the tube itself or about 35–45% including the power supplies.

Though attractive from the point of view of gain, efficiency and power output, a TWT is a rather nonlinear amplifier. When amplifying a multi-carrier signal it both generates

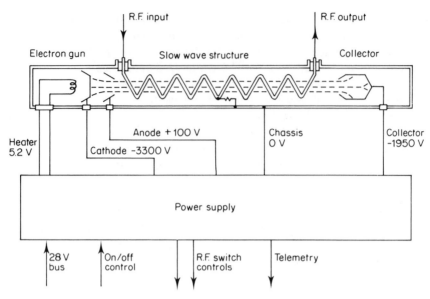

Figure 13.21   Typical TWTA configuration

intermodulation products and converts signal amplitude variations into spurious phase modulation. More linear operation can be achieved by 'backing-off' the tube to a lower power level, but this results also in a loss of efficiency.

Transistor power amplifiers

Transistors are now available which are capable of providing an output of up to a few watts at microwave frequencies, making them competitive with TWTs for many low-to-medium power applications. It is possible to combine the power from two or more transistors in order to extend their range of application.

A solid-state transmitter has the very significant advantage of eliminating the need for the TWTs complicated high-voltage power supply, giving a useful overall improvement in terms of mass and reliability. Transistors are, however, at a disadvantage with respect to gain and efficiency. The gain of a microwave transistor is so low (typically a few decibels) that several transistor stages are needed to replace one TWT. Also, the microwave power at the input to a transistor is a significant fraction of the output power, so that in computing the efficiency, the power requirements of the driver stages cannot be ignored. For a transistor amplifier operating close to saturation, the overall efficiency might be a few percent less than for a TWTA. But the efficiency of a TWTA decreases rapidly with back-off. If good linearity is required, the transistor amplifier may prove to be more efficient than a TWTA.

Solid-state amplifiers delivering up to about 70 W at around 1.5 GHz have been developed for transmissions from satellites to mobile receivers such as shipping. However, further development in this area seems likely to concentrate on 'distributed' power amplifiers, such as are required for phased arrays, in which each individual module has a fairly modest output.

Transistors are also beginning to replace TWTs in C-band transmitters for the trunk telephone and television service. The output per transponder is typically between 5 and 10 W, which is within the present capabilities of transistors at 4 GHz and will presumably become possible also at 11 GHz in the fairly near future.

Passive intermodulation and multipaction

Having achieved sufficient power from the final amplifier the designer faces two further problems associated with the output circuits and antenna.

Although these assemblies contain no active devices such as transistors or vacuum tubes, they may display a slight electrical nonlinearity caused by magnetic materials or imperfect contact between metal surfaces contaminated with dirt or a thin layer of oxide. The resulting passive intermodulation products (PIMs), though very weak compared with the transmitted signal, may be comparable in power with the received signal. If they are generated in a unit (e.g. antenna, cable or diplexer) which is common to the transmit and receive paths, they cannot be removed by filtering. They must be controlled by extreme care in manufacturing—that is by extreme cleanliness, very precise machining and high contact pressures between mating surfaces.

The second problem is a type of electrical breakdown which occurs only in high vacuum and at high frequencies (UHF and microwaves). 'Multipaction' is caused by secondary

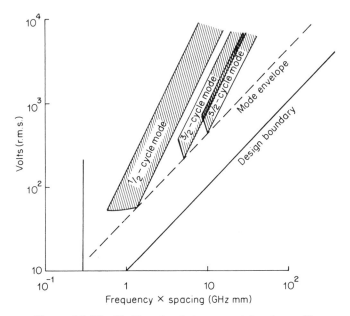

Figure 13.22 Multipaction between metal surfaces. The shaded areas show a typical theoretical arrangement of multipactor regions for parallel plate geometry. Regions are shown corresponding to modes in which there are one, three and five reversals of field during the transit time of an electron (reproduced by permission of European Space Agency)

emission of electrons from two metal surfaces between which there is an alternating electric field—an effective high-frequency short circuit. Figure 13.22 shows an example of the conditions under which simple theory predicts the possibility of multipaction between parallel metal plates. In practice the multipaction zones are found to extend somewhat beyond these theoretical limits and a generous design margin must be allowed. Where units cannot be configured to avoid operation in the multipaction region special precautions must be taken, such as filling with a dielectric foam or pressurizing with enough inert gas to prevent the formation of an electron cascade.

13.4 CONCLUSION

A modern communications satellite is a far cry from the simple transponders of Telstar and Early Bird. What seemed in the 1960s a bold and imaginative step is now a mature technology. Nevertheless, it is a technology that continues to develop both under the stimulus of advances in component technology and under the pressure of increased competition from other techniques. It would be foolhardy to predict with any confidence what new turn the developments will take but there can be little doubt that satellites will continue to have a major role in telecommunications for many years to come.

REFERENCES

[1]   Clarke A.C. (1945), Extra-terrestrial relays. *Wireless World*, **51**, 10.
[2]   CCITT Red Book (1984). VIIIth *Plenary Assembly of the* CCITT, Malaga-Torremolinos. *ITU*, Geneva.
[3]   *Recommendations and Reports of the CCIR 1986*. XVIth Plenary Assembly, Dubrovnik. *ITU*, Geneva.
[4]   *Radio Regulations* (1982). *ITU*, Geneva. (Updated 1984; further update in preparation.)
[5]   *Final Acts of the World Administrative Radio Conference for the Mobile Services* (MOB-87) (1987). *ITU*, Geneva. 1987.

14 TELEMETRY AND COMMAND

Julian Robinson

Marconi Space Systems

14.1 INTRODUCTION

The Telemetry, Command and Ranging (TC&R) subsystem provides an important two-way flow of information between a spacecraft and its ground control station(s). It therefore has transmission (downlink) and receiving (uplink) functions to perform, as well as the tasks of gathering and processing of data ready for transmission, and the processing and routeing of command data from the receiver. In addition it incorporates a transponder for ranging purposes.

The mission—orbit and type of payload—and the selected ground-control station(s) will play a considerable part in determining the nature of the subsystem's design. This is illustrated by studying Intelsat V and Eutelsat 2, both of which are in GEO and have communications payloads, ERS which is in a polar LEO and whose payload monitors the ground, and Space Station which will be in a non-synchronous LEO.

The spacecraft/ground-station link from GEO can be a continuous direct one, with the spacecraft being visible at all times from an appropriate ground station. This allows for a relaxed transmission link compared with that from a nonsynchronous orbit. ERS, for example, will be visible at its ground station for only a few minutes at each ground pass; it therefore needs on-board data storage and a rapid transfer link. Space Station requires a continuous link from its nonsynchronous orbit, and this can be provided by a two-way link with any ground station in sight, and/or by a system (TDRSS) which makes the ground link via special-purpose spacecraft in GEO.

With all spacecraft their early phases, such as during launch and intermediate orbits, will be nonsynchronous and will require special ground support before they are handed over to their dedicated ground controllers. The Intelsat spacecraft use a worldwide Comsat Launch Support Network operating on C-band, both during launch and subsequently. A dual system is proposed for Eutelsat 2, however, with the ESA network operating in *S*-band being used for initial orbits, and with control being transferred to a dedicated station operating in *Ku*-band when it is on station.

The telemetry downlink must provide the ground-control team with information about

---

Spacecraft Systems Engineering. Edited by P.W. Fortescue and J.P.W. Stark
© 1991 John Wiley & Sons Ltd

the functioning of the subsystems in the craft, so that they may detect whether it has its correct orientation, or whether any fault has developed, and if so they must be able to diagnose its cause. It may also be the channel for passing information from the payload to the ground when the mission is scientific or observational, for example, as in ERS-1. When the payload comprises communications equipment, such as on Intelsat or Eutelsat, then that equipment will serve as an alternative route for the telemetered data, once it is deployed.

The command uplink must enable the ground controller to change the role of the spacecraft, such as to reorientate it, to correct a fault, to operate a mechanism, or for other reasons. It must do so in a highly reliable way, giving confirmation that the instructions have been carried out.

The ranging transponder forms part of the system by which the ground controller tracks the spacecraft and determines its orbit.

14.2 TELEMETRY DATA FORMATTING

14.2.1 Classification of data

Telemetry data falls into three basic categories: housekeeping, attitude, and payload.

Housekeeping data, sometimes known as engineering parameter data, needs to be monitored to keep a check on the health and operating status of the on-board equipment. It can arise in many forms and some typical examples are as follows:

- *Temperatures* of equipment boxes, solar arrays, attitude control thrusters and plenum chambers, parts of the structure etc. Thermistors are customarily used to convert temperature into an analogue voltage. For high temperatures thermocouples are used and the output of a few millivolts is d.c. amplified to a level which is suitable for the telemetry encoder.

- *Pressure* in fuel tanks, plenum chambers etc. Various forms of pressure transducers are used.

- *Voltages and currents* of equipment power supplies. The rail voltages are scaled to a common full-scale range, which is often 0 to $+5.12$ V. Current monitoring may involve a variety of circuit techniques.

- *Operating status* of equipment is represented by a single bit, signifying, for example, that a particular functional mode is selected (logic 1), or deselected (logic 0). For proportional status information, such as amplifier gain settings, the single bits are grouped together into words of appropriate length.

- *Redundancy status.* Information on whether the main side A or the cold redundant side B of an equipment is in use is provided by furnishing each relevant relay or switch with a set of contacts to provide a status bit.

- *Deployment of mechanisms*, separation from the launcher, etc. A microswitch is fitted in order to provide an appropriate status bit.

On a modern large communications satellite there will be several hundred engineering parameters to be monitored and the result of every command is usually checked via the telemetry. The great majority of these will only need sampling and updating at infrequent intervals of typically 30 seconds to 2 minutes, so the bandwidth required is quite small. A bit rate of a few hundred bits per second is sufficient to transmit the total information.

Attitude data arises from a variety of sensors such as Sun and Earth sensors, star mappers, gyroscopes and accelerometers (see Section 10.5). The data can be analogue, digital, or a mixture of both.

During transfer and intermediate orbit phase the attitude and velocity will change rapidly, and frequent sampling is needed, typically from once to four times per second. For geostationary orbit operations a reduced rate may be provided, selected by command.

Although only a few channels of attitude information are required the high sampling rate needed during some mission phases may lead to a bandwidth which exceeds that which can be provided conveniently by a standard PCM data system, and a separate wide-band system may have to be provided, as on Intelsat V, for example.

Payload data is very variable and each case needs to be considered individually. Scientific and applications payloads are likely to need only a few channels of data, but their rates may be very high—possibly many megabits per second. An entirely separate high-rate system may then have to be provided, as on ERS-1. Data compression may also be required to reduce the downlink rate.

A communication payload gives rise to many channels of engineering parameter data in addition to that generated in the service module. Because of its complexity and the large amount of redundancy employed it is likely to demand considerably more channels than the service module; for example Intelsat V has a total of 520 mixed analogue and digital channels from the combined service and payload modules while the planned capacity for Eutelsat II is even greater at 840 channels, allowing for some growth margin. In both cases roughly two-thirds of these arise in the payload.

Typical of the monitoring requirements which are peculiar to a communications payload are the following:

- temperatures of travelling wave tubes and other repeater equipment;

- power supply voltages and currents for each main equipment;

- operating and redundancy status monitors for the many waveguide switches in the RF 'plumbing';

- analogue power monitors for the signal levels at each main RF equipment interface;

- telemetry monitoring of the digital gain settings telecommanded to the various channel amplifiers.

14.2.2   Influence of mission phase

The phases of a mission run from ground testing through the launch phase to the orbit phases. In each of these the telemetry plays a part.

During *ground testing* there is access to inter-unit connections, test and diagnostic connectors, in addition to the normal telemetry data. On satellites with a data bus, for example the OBDH bus used on Eutelsat II, direct access to the bus may be possible. This allows higher data rates for telemetry and command, with the benefit of faster ground testing, but RF links to the payload are still used for overall performance testing.

During *the launch phase*, satellite data is usually minimal and is restricted to a few housekeeping parameters such as battery condition, some key temperatures, RCE pressures, and deployed item status. Payloads are not switched on until in orbit and outgassing has been completed, but some telemetry activities may occur at separation.

The above data is sent down via the launch vehicle.

The *in-orbit phases* include the transfer and intermediate orbit phases of geostationary missions. In these cases telemetry contact is not continuous unless the TDRSS relay satellites are used. Spacecraft operation must be autonomous as far as possible in order to avoid the need for intervention by ground control. For example, majority voting techniques may be used instead of cold redundancy.

The payload items are not switched on until the spacecraft is in its final orbit.

14.2.3 Telemetry data encoding

All the data considered so far arises in three basic forms: analogue, digital bi-level, and digital serial. The block diagram of a typical encoder (Figure 14.1) illustrates how each type of data is time-division multiplexed into a pulse code modulated (PCM) bit stream which modulates the downlink RF carrier (see Chapter 13).

The first step in conditioning *analogue data* is to scale it to a common full-scale range, usually 0 to $+5.12$ volts. This is done at the source of the data.

Frequency components greater than half the sampling frequency need to be removed by a low-pass filter to prevent aliasing errors. It is good practice to include a simple *RC* filter in each analogue line.

After filtering, the channels are sampled in turn by analogue switches which are usually MOSFET or junction FET devices, and each sample is converted to a digital word which is mixed in with the digital data as described later. For most data an overall accuracy of about 1% is sufficient, and an 8-bit analogue-to-digital converter is used to achieve this.

Analogue commutation is invariably carried out in two stages by a mainframe multiplexer sampling at a relatively fast rate, and a slower submultiplexer sampling at a binary submultiple of the mainframe rate. The larger number of housekeeping channels which only need sampling once or twice per minute use the submultiplexer and the few channels requiring fast sampling go straight into the main multiplexer.

This arrangement is very convenient because some of the submultiplexers can be remotely located in subsystems or payloads, an arrangement which saves complexity and mass in the harness.

Digital bi-level data arising from relay contact closures, etc., is first conditioned to appropriate logic levels in which the 'off' state is represented by nominally zero voltage, and the 'on' state by a positive voltage suitable for the IC logic family used, i.e. $+5$ V for CMOS, or $+2.4$ V for TTL logic. Individual bits are then grouped together into 8-bit words and sampled by logic gates whose outputs are serialized in the parallel-to-serial converter and mixed in with the main PCM data stream.

Digital data is usually acquired in serial form, thereby simplifying the cable harness. It

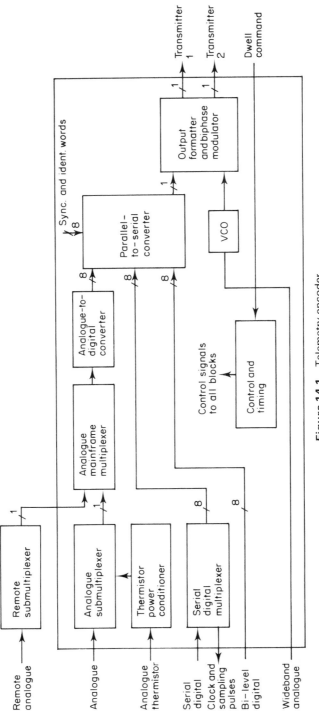

Figure 14.1 Telemetry encoder

is initially stored as an 8- or 16-bit word in a shift register located in the equipment generating the data.

In normal PCM mode the 8-bit parallel words from the analogue-to-digital converter and the serial digital multiplexer are loaded, together with bi-level status data, into a shift register in the parallel-to-serial converter, and clocked out at the telemetry bit rate as a continuous PCM bit stream. Unique sync. and ident. codes are inserted in the bit stream so that all data can be identified when it reaches the ground. Finally the bit stream is bi-phase modulated on to a coherent subcarrier at an integral multiple of the bit rate before routeing it to the two TM transmitters.

In many encoders an alternative to the above allows sampling to be concentrated on a particular word, to the exclusion of the other data. In Figure 14.1 this 'dwell' mode is achieved by loading the address of the desired channel into the control and timing block, which then sets up the input multiplexers permanently to that channel instead of cycling round all the channels.

Wideband analogue signals bypass the PCM section of the encoder altogether and frequency modulate a voltage-controlled oscillator instead, using one of the IRIG standard centre frequencies. Since the PSK subcarrier and VCO frequencies are different the PCM and FM channels can operate simultaneously.

14.2.4   Telemetry list and data format

The first step in designing the telemetry system is to draw up a telemetry list for the spacecraft. For each item this will contain the signal identification, the type of data (analogue, digital bi-level, or digital serial), the required accuracy, and the sampling rate required. This list is first established during the initial Phase A feasibility studies, and evolves with the project. It is important to allow enough spare channels at the outset to cater for natural growth as the project matures.

The next stage is to lay out the format of the PCM message, and Figure 14.2 illustrates a simple format consisting of eight frames F0–F7, each containing 64 8-bit words W0–W63. The first two words of each frame contain a fixed sync. code, 16 bits long, which the ground decommutation equipment will recognize as the start of a frame. The next word is a frame ident. channel. The decommutation station is then able to identify every channel in the format and display the parameter contained in it on the 'quick look' facility at the station. Further processing by computer provides outputs in engineering/scientific form for users.

The rest of the format consists of data channels. Note that the analogue channel A1 is sampled once in every frame, in word W3. Such a channel is known as a mainframe or prime channel, and could be used for example for a nutation sensor or other data source needing fast sampling. When a still higher rate is required the same data can be put into more than one mainframe word; such a channel is said to be supercommutated.

Housekeeping parameters requiring infrequent sampling can be subcommutated by sampling only once per format as illustrated by channels A2–A9.

By extending the principles of sub- and supercommutation over a longer sequence of frames, a wide variety of sampling rate needs can be accommodated, and once this is done the bit rate needed to provide every parameter with at least its minimum sampling rate can be worked out. The subcommutation depth of eight frames used in the above example

8-bit WORDS ———→

	W0	W1	W2	W3	W4	W5	W6		W61	W62	W63
F0	FRAME SYNC	CODE	FRAME ID CHANNEL	A1	A2	A10	DB1	≈	DS1	A14	A22
Frame F1 →	″	″	″	A1	A3	A11	DB2	≈	DS1	A15	A23
F2	″	″	″	A1	A4	A12	DB3	≈	DS1	A16	A24
F3	″	″	″	A1	A5	A13	DB4	≈	DS1	A17	A25
F4	″	″	″	A1	A6	A10	DB5	≈	DS1	A18	A26
F5	″	″	″	A1	A7	A11	DB6	≈	DS1	A19	A27
F6	″	″	″	A1	A8	A12	DB7	≈	DS1	A20	A28
F7	″	″	″	A1	A9	A13	DB8	≈	DS1	A21	A29

A = analogue channel   DS = digital serial   DB = digital bi-level

Figure 14.2 Simple data format

would not be sufficient for a typical communications satellite. For example, Intelsat V uses a depth of 32 frames.

A PCM message structure with the following specification is compatible with both NASA and ESA standards, and ensures a spacecraft's compatibility with ground stations operated by both these agencies. The requirements of other agencies are generally similar but may differ substantially in detail.

For further details and options consult the published standards:

- Binary representation: NRZ-L, NRZ-M, split phase-level, split phase-mark.
- Bit-rate range: 10–200 000 bits/s (NRZ)
  5–200 000 bits/s (split-phase)
- Bit-rate changes: Permissible during transmission only by command from a ground station.
- Telemetry video signal: Can be either the PCM waveform direct, or a PCM modulated subcarrier as below.
- PSK subcarrier: Square wave, frequency an integral multiple of the bit rate in the range 100 Hz to MHz.
- Word length: 8 bits at all times.
- Frame length: Maximum 1024 words, frames longer than 128 words shall be subject to justification and prior approval.
- Format length: 256 frames and 32 768 words maximum.
- Format changes: Changes in structure during transmission are permissible only by command from a ground station and must occur at the start of the format.

14.2.5 Packet telemetry

The Consultative Committee for Space Data Systems (CCSDS) has produced a series of recommendations for packetized data systems which are likely to be adopted eventually by most space agencies. These represent the next evolutionary step from the traditional time-division multiplex methods in use today, and will come into their own on multi-agency projects such as the Space Station, where increased inter-operability and standardization will be essential. The data flow in a packetized system is illustrated in Figure 14.3.

Application data is first encapsulated by the source into a 'source packet' by prefacing the data with a standard label known as the 'packet header'. This is used to route the data through the system and must therefore contain identification of the source and its particular applications process, the number of the packet in the sequence of packets produced by the source (so that the packets can be delivered in the right order at the data sink end of the link), and the length of the data field attached to the header. Source packets exceeding a prescribed length are segmented into several shorter 'telemetry packets' which can then be interleaved with packets from other sources as shown in Figure 14.3

CCSDS recommendation for packet telemetry

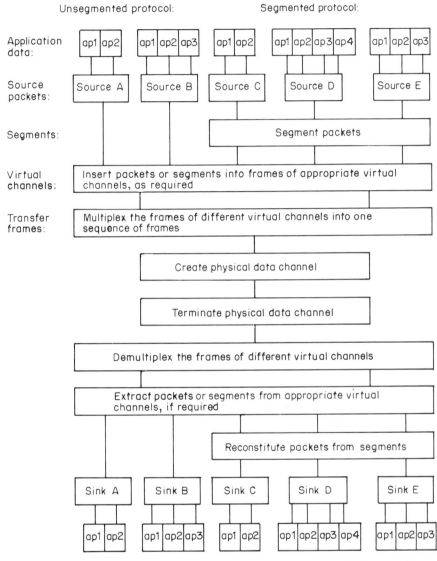

Figure 14.3 Packet telemetry data flow (reproduced by permission of European Space Agency)

The next stage is to assign each group of sources to a so-called 'virtual channel'; the packets are inserted into fixed-length frames known as virtual channel data units (VCDU) which are then multiplexed into a single transfer frame for transmission on the downlink. If necessary each virtual channel can be sampled at a different rate, appropriate to the bandwidth of that channel.

After transmission the transfer frame can be demultiplexed in the normal way, and the telemetry and source packets can be reconstructed and routed to their destinations using the headers. Labelling the source packet with all the necessary information about its routeing and interpretation is the key to the flexibility of the packet system.

14.2.6   Error control

On a communications satellite there is enough transmitter power for a bit error rate of better than 1 in 10^5 on the downlink. The loss of even a complete frame is not catastrophic, so there is no great need for correction of such errors; nevertheless an error-checking code is sometimes included in the frame.

In a typical scheme this is the 16-bit remainder which results from dividing the data portion of the telemetry frame by a specified polynomial such as $x^{16} + x^{12} + x^5 + 1$. The remainder is placed in the last two words of the telemetry frame and transmitted to the ground with the rest of the data. On the ground the complete frame including the check bits is again divided by the same polynomial, and if a nonzero result is given, then that frame is flagged as being in error. This scheme is not able to correct errors and offers no saving in transmitter power since the actual error rate on the link is unaffected.

When forward error correction is needed a scheme based on convolutional coding provides good correction capability in a Gaussian noise channel and is simple to implement in hardware. Significant further improvement (particularly with bursts of errors) can be obtained by concatenating a Reed–Solomon block code with the convolutional code, the Reed–Solomon being used as the outer code while the convolutional code is the inner one.

One application for forward error correction is in packet telemetry. This needs nearly error-free channels for the successful routeing and control of the data packets, and the above codings used either singly or together are an integral part of the standards.

Missions involving very long transmission path lengths, such as Giotto, represent another application for coding. The telemetry transmitter may then dominate the power budget, and the coding gain of several dB can be exploited to reduce the r.f. power needed for a given bit error rate.

Joint NASA/ESA telemetry coding guidelines have been issued to standardize on particular convolutional and block codes to facilitate network cross-support on joint missions. The convolutional code recommended is the constraint length $k = 7$, rate $r = \frac{1}{2}$ code that is in common use by ESA, and within NASA by the Goddard Space Flight Center and the Jet Propulsion Laboratory. Detailed specifications appear in Reference [1].

The selected Reed–Solomon code is a (255, 223) code, which can be used either non-interleaved ($I = 1$), or interleaved to a depth of $I = 5$. This means that $223I$ bytes of uncoded information are fed into the RS encoder and emerge unaltered with $32I$ check symbol bytes appended. The whole code block of $255I$ bytes is then transmitted followed by a further $223I$ information bytes, and the cycle is repeated indefinitely. The RS code is capable of correcting bursts of errors in up to $16I$ consecutive bytes, and it can be seen that the interleaved form is considerably more effective in combating long burst errors.

The Reed–Solomon code generator is a more complex proposition than the convolutional coder as it involves binary multiplication and the intermediate storage and processing of up to 1275 bits in a shift register or equivalent RAM. However, custom chips have been developed by both NASA and ESA to cut down the hardware to manageable proportions.

14.2.7 Downlink frequencies and modulation

A spacecraft uses one or more of the worldwide ground station networks for command, telemetry and ranging, at least in the early mission phases. The frequency bands are constrained to be those which are supported by the chosen ground stations.

The first step in choosing the band(s) to use is to contact the Frequency Management Office of the networks concerned, with a request for the favoured frequency band.

The second step is the selection of discrete frequencies within the allocated bands. This choice is a very complex and lengthy process due to the need for coordination with other space agencies and national P&T authorities.

Many networks provide telemetry support in the UHF band, the exception being the COMSAT Earth stations covering Intelsat missions.

Phase modulation of the RF downlink is universally used, and there is a degree of compatibility between the ESA and NASA standards which allows cross-support provided certain limitations are observed.

At low bit rates a PSK modulated subcarrier is used to prevent modulation sidebands occurring too near to the RF carrier and upsetting the phase-lock loops in the ground station receivers. At higher bit rates a preferred alternative, achieving the same result, is to modulate the RF. carrier directly with PCM-SPL data as defined in Figure 14.4

Pre-modulation filtering of the square-wave PCM bit stream is usually provided to attenuate high-order modulation sidebands.

14.3 TELECOMMAND SUBSYSTEM

14.3.1 Telecommand user interface

The three basic types of command are as follows:

- *Low-level on–off commands.* These are logic-level pulses which are used to set or reset logic flip-flops.

- *High-level on–off commands.* These are higher-powered pulses, capable of operating a latching relay or RF waveguide switch directly. Typically these may be 12–28 V pulses lasting several tens of milliseconds and capable of supplying 90 mA for relay drive, or up to 600 mA to drive an electromechanical RF switch. Separate 'on' and 'off' pulses are supplied on two different lines to drive the two coils of the latching device.

- *Proportional commands.* These are complete digital words which may be used for such purposes as the reprogramming of memory locations in an on-board computer, or for setting up registers in the attitude control subsystem.

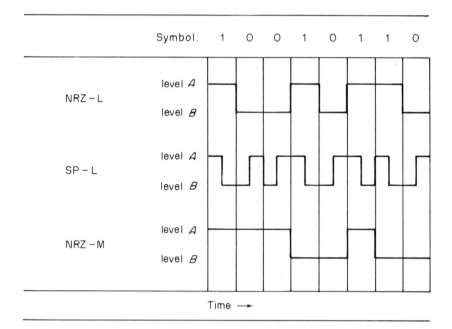

Symbol: 1   0   0   1   0   1   1   0

NRZ – L   level A

            level B

SP – L   level A

            level B

NRZ – M   level A

            level B

Time ⟶

NRZ – L   ● level A signifies symbol '1'

           ● level B signifies symbol '0'

SP – L   ● level A during the first half – symbol followed by level B
             during the second signifies symbol '1'

           ● level B during the first half – symbol followed by level A
             during the second signifies symbol '0'

NRZ–M   ● level change from A to B or B to A signifies symbol '1'

           ● no change in level signifies symbol '0'

Figure 14.4 PCM waveforms

Enough spare channels of each type need to be provided to allow for 'natural growth', otherwise a redesign may be necessary at a later stage.

14.3.2 PCM telecommand standards—Intelsat and SCF

The agencies responsible for the ground station networks publish standards covering space–ground interface requirements and procedures to be adhered to for the trasmission of commands. Their purpose is to ensure compatibility between the spacecraft and the equipment at the ground station. Conformance to these standards is mandatory and any deviations for a particular mission have to be justified and approved by the Agency before being included in the design.

In geostationary orbit, with continuous visibility from several Earth stations, it is very seldom that a sequencs of commands has to be sent within a short period of time. The Intelsat command standards are therefore based upon a command–verify–execute strategy in which each command is held in the satellite decoder and verified through telemetry before it is executed.

Figure 14.5 shows a simplified block diagram of a typical decoder for an Intelsat spacecraft. The explanation relates primarily to the Intelsat V system, but other spacecraft in the series have used fundamentally similar principles.

The command uplink signal consists of a sequence of tones which represent data 0, data 1 and execute information 'bits'. These are frequency modulated on the 6 GHz carrier and demodulated by the two command receivers. The ground operator is able to choose which receiver is to be used by selecting either of two alternative data 0 tone frequencies.

Each digital command message includes an introductory series of zeros to select the receiver and to synchronize the bit detector clock, and this is followed by a decoder address word, a command vector, and finally an on–off command word or a proportional

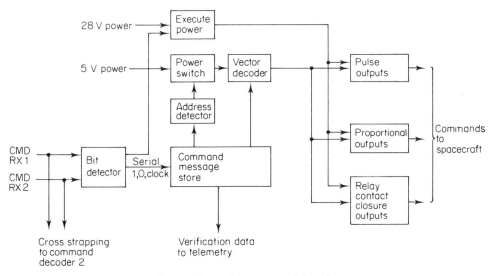

Figure 14.5 Telecommand decoder

data command word. The 0 and 1 bits comprising the message are stored in a shift register.

Each command unit has a unique address which is hardwired, and only messages which contain this address will be accepted. The data held in the command message store is telemetered to the ground for verification, and assuming this is satisfactory an execute tone burst is sent. This turns on the execute power switch, which then outputs the command to the specified user channel. The combination of power switching and the use of diode isolation and redundant components in the output circuits allows the outputs of the two command decoders to be combined in a fail-safe manner without loss of reliability due to single-point failure modes. The US Air Force SCF tracking network uses a basically similar ternary (three-tone) modulation scheme using frequency-shift keyed tones.

14.3.3   PCM telecommand standards—ESA and NASA

Agencies supporting scientific and LEO missions often need to send rapid sequences of commands in a short time period, such as during a brief ground-station pass. Although the ESA and NASA systems do make use of ground verification the structure of the telecommand message allows the checking and correction to take place inside the satellite's decoder itself.

The ESA telecommand message starts with a fixed number of acquisition bits to initialize the message—usually fifteen 'zeros' followed by a single 'one'. The message sequence proper then consists of an unbroken sequence of one or more 96-bit frames similar to that shown in Figure 14.6. A single 16-bit spacecraft address and synchronization word (ASW) is used to terminate the message. The initial 16-bit ASW is used to identify individual spacecraft and to synchronize the decoder.

A 4-bit mode selection word using 2-out-of-4 redundant coding is included in each frame after the ASW. This word is repeated once and provides means for selecting on–off pulse commands, proportional commands, or time-tagged commands for delayed execution.

Since a single-bit error will give an invalid mode word the acceptance criterion in Figure 14.6 ensures the acceptance of a command even if a single-bit error has occurred, and its rejection if more than one error has taken place. In addition a command receiver 'squelch' cuts off the input to the bit detector if the signal-to-noise ratio is too low, thus ensuring a low bit error probability (normally better than 10^{-5}) and a correspondingly high probability of frame acceptance.

The three data words following the mode word are each of eight bits and represent respectively the address of the 16-bit memory to which the data is to be sent, the first eight bits of data, and the final eight bits of data. Each data word is repeated once to increase the probability of acceptance, and four Hamming-code check bits are appended to each word to permit error detection and correction in the spacecraft. By comparing the received check bits with a set generated locally from the data word it is possible to detect all single and double bit errors and, in the case of single-bit errors only, to determine which data bit is in error and correct it.

The end-to-end probability of command rejection can be reduced to less than 1 to 10^6, and the probability of an erroneous command to less than 1 in 10^8.

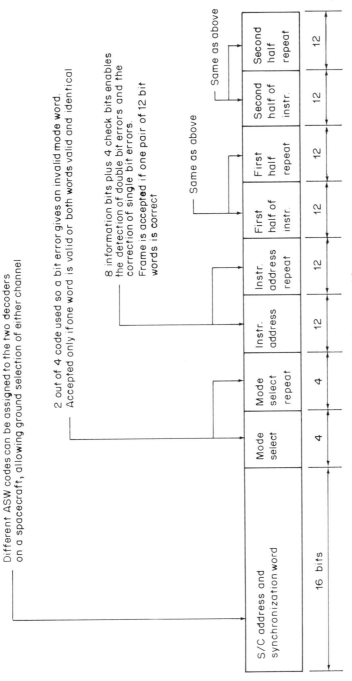

Figure 14.6 Memory load command frame structure

Spacecraft using the ESA standard are also compatible with the NASA ground network and procedures, whose modulation and coding techinques are to a large extent compatible.

14.4 RANGING

Ranging is achieved by means of a transponder which is integrated into the telemetry–command subsystem. This retransmits ranging tones through the telemetry channels in response to tones received via the command route.

On Intelsat V the output of a command receiver on a 6 MHz carrier can be connected to a beacon transmitter which phase modulates the ranging tones on to the 4 MHz downlink carrier. On Eutelsat II and ERS-1 these operations are carried out using the S-band links. The signals are tones as specified in the ESA ranging standards. These may be phase-modulated on to the uplink carrier in place of command signals, and on to the downlink simultaneously with telemetry signals. Interference on the downlink is avoided by the choice of subcarrier frequencies.

REFERENCE

[1]   *Recommendation for Space Data System Standards: Telemetry Channel Coding* (1987) CCSDS 101.0-B-2 Blue Book.

15 GROUND CONTROL

R. Holdaway

British National Space Centre, Rutherford Appleton Laboratory

15.1 INTRODUCTION

This chapter describes the main functions performed at a ground station in support of
an operational spacecraft. Examples and data are given for a number of ground station
configurations, in particular the operations environment built at the Science and Engineering Research Council's Rutherford Appleton Laboratory near Chilton, Oxfordshire.
This particular station was initially designed for operation of the NASA–UK–Dutch
Infra-Red Astronomical Satellite (IRAS) and is a good example of a low-cost operational
facility [1]

The main functions performed are highly complex, usually involving the following tasks:

1. Tracking to determine the position of the satellite in orbit.

2. Telemetry operations to acquire and record satellite data and status.

3. Commanding operations to interrogate and control the various functions of the satellite.

4. Controlling operations to determine orbital parameters, to schedule all satellite
   passes, and to monitor and load the on-board computer.

5. Data-processing operations to present all the engineering and scientific data in the
   formats required for the successful progress of the mission.

6. Voice and data links to other worldwide ground stations and processing centres.

From the above list, it is clear that there are three main components of a ground station,
namely hardware, software, and people/operations. The remainder of this chapter deals
in turn with the first two of these.

15.2 HARDWARE

The main hardware components of a ground station are an antenna, a receive–transmit
system, tape recorder(s), computer(s) and their peripherals, and a control console. This
is a basic list which does not change very much with the type of spacecraft being
controlled.

Spacecraft Systems Engineering. Edited by P.W. Fortescue and J.P.W. Stark
© 1991 John Wiley & Sons Ltd

Figure 15.1   Chilton 12 m antenna

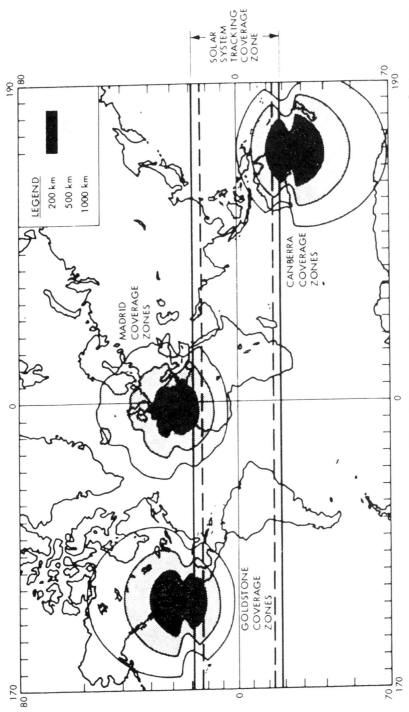

Figure 15.2 DSN 26-Metre Subnet Coverage, 200 km, 500 km, and 1000 km altitude. (*Note:* Effects of land-mask on Canberra coverage are not shown.)

15.2.1 Antennas

The main component is the antenna, whose support functions may include tracking, telemetry, command, air–ground voice and television capabilities. Figure 15.1 shows a 12 m version at Chilton, UK. Within the Spaceflight Tracking and Data Network (STDN) of NASA there are antenna systems operating at a number of different frequencies. The most used is the S-band system, of which STDN has many ground stations.

The S-band systems employ monopulse autotrack principles to maintain the antenna pointing towards the spacecraft's transmitted signal. To aid initial acquisition, a computer-controlled program mode is also available. This uses orbital prediction data to generate the required pointing angles. Initial acquisition of the spacecraft RF signal may be facilitated by means of a small, wider beamwidth acquisition parabolic antenna, mounted at the edge of the large one. This is visible in Figure 15.1. Other operating modes include manual position and velocity, slave, and manual program. The $X-Y$ mounts are capable of tracking through zenith but have a gimbal restriction keyhole near the horizon. It is usual to orientate this north to south on 9-m antennas and east to west on 26-m ones. Coverage patterns are further restricted at most stations by the surrounding terrain, as shown in Figure 15.2.

STDN ranging equipment (SRE), operating in conjunction with the multifunction receivers (MFR) and S-band exciters (SBE), provides precision range and Doppler measurements for a variety of spacecraft. For spacecraft carrying an S-band phase-locked transponder, this will provide unambiguous range data to distances greater than 500 000 km, and nondestructive Doppler data for carrier Doppler frequencies up to 230 kHz. The system is designed for spacecraft dynamics of over 15 000 m/s and 150 m/s^2. It employs sinusoidal modulation and extremely narrow-band processing techniques to provide high-accuracy data with low received signal strength. Figure 15.3 shows a typical antenna configuration, and Table 15.1 gives the location of the STDN S-band antenna stations.

Of increasing significance is the worldwide network of laser tracking stations (Table 15.2). These provide highly accurate range measurements to satellites equipped with optical retroreflectors, based upon the propagation time of a laser pulse from the tracker to the spacecraft and back. Corrections for internal system delays and refraction are made on-station. Ranging can be performed during night and day up to distances of several thousand kilometres, provided atmospheric conditions are favourable. Nominal ranging accuracy is in the order of 2 cm. Angle data is also provided. The laser telescopes have no autotrack, pointing being computer driven according to *a priori* orbit information; thus the angle measurement is only as accurate as the laser beamwidth. The laser transmitter and the sighting telescope are mounted on each side of the receiving telescope. Data from the laser tracking systems is available in two forms: raw data mailed from the remote sites on magnetic tape, and quick-look data (computed from raw data), which eliminates system delays and is teletyped to users.

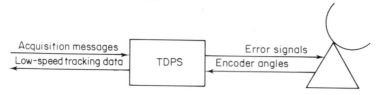

Figure 15.3 S-band antenna configuration

Table 15.1 Location of main STDN stations

Station	System	Latitude*	Longitude (E)	Height above Ellipsoid (m)
Ascension Island (*ACN*)	9 m USB	−7°57'17.37"	345°4022.57"	528
Santiago, Chile (*AGO*)	9 m USB	−33°09'03.58"	289°20'01.08"	706
	VHF GRARR	−33°09'06.06"	289°20'01.07"	706
	Interferometer	−33°08'58.10"	289°19'54.20"	694
Bermuda (*BDA*)	9 m USB	32°21'05.00"	295°20'31.94"	−33
	FPQ-6 radar	32°20'53.05"	295°20'47.90"	−35
Grand Canary Island (*CYI*)	9 m USB	27°45'51.61"	344°21'57.88"	167
Engineering Training	9 m USB	38°59'54.84"	283°59'54.84"	−1
Center, Maryland (*ETC*)	9 m USB (ERTS)	38°59'54.08"	283°09'29.21"	4
	Interferometer	38°59'57.25"	283°09'38.71	−5
Goldstone California (*GCS*)	26 m USB	35°20'29.66"	243°07'35.06"	919
	9 m USB (ERTS)	35°20'29.64"	243°07'37.45"	913
Guam (*GWM*)	9 m USB	13°18'38.25"	144°44'12.53"	116
Hawaii (*HAW*)	9 m USB	22°07'34.46"	200°20'05.43"	1139
	FPS 16 radar	22°07'24.37"	200°20'04.02"	1143
Madrid, Spain (*MAD*)	26 m USB	40°27'19.67"	355°49'53.59"	808
Merritt Island, Florida (*MIL*)	9 m USB NO 1	28°30'29.79"	279°18'23.85	−55
	9 m USB NO 2	28°30'27.91"	279°18'23.85	−55
Orroral Valley, Australia (*ORR*)	Interferometer	−35°7'32.19"	148°57'15.15"	926
Quito, Ecuador (*QUI*)	Interferometer	−00°37'22.04"	281°25'16.10"	3546
Rosman,	4.3 US	35°11'45.99"	277°07'26.96"	810
North Carolina (*ROS*)	VHF GRARR	35°11'42.02"	277°07'26.97"	810
Tananarive, Malagasy	4.3 USB	−19°01'13.87"	47°18'11.87"	1368
Republic (*TAN*)	VHF GRARR	−19°01'16.34"	47°18'11.83"	1368
	Interferometer	−19°00'31.66"	47°17'59.75"	1347
	FPS 16 radar	−19°00'05.52"	47°18'53.46"	1307
Fairbanks, Alaska (*ULA*)	9 m USB	64°58'19.20"	212°29'13.39"	339
	VHF GRARR	64°58'17.50"	212°29'19.12"	339
	Interferometer	64°58'36.91"	212°28'31.89"	282
Winkfield, England (*WINK*)	Interferometer	51°26'46.12"	359°18'09.13"	87

*A minus sign indicates South latitude.

The accuracy of laser ranging is critically dependent upon precise timing. Typically, the sequence of events begins with the 1 pulse/s output of the timing subsystem, which is generally referenced to US Naval Observatory (USNO) time. When the laser is fired, the 1 pulse/s is gated to the laser power supply, producing the high-voltage pulse required to energize the laser. A time delay occurs between the timing system 1 pulse/s output and the actual firing of the laser. This time delay must be measured and recorded for accurate time-tagging of the tracking data. A sampling of the laser output pulse is used to start the time interval unit (TIU), which counts the time in tenths of nanoseconds until the received pulse from the target terminates the count, providing accurate ranging information. Additional (system) delays are removed by calibrating the laser with a surveyed target of known distance.

Table 15.2 Location of main SLR laser tracking stations

Station			E. longitude	Latitude	Height (m)	Single-shot precision (cm)	Agency
1181	Potsdam	E. Germany	13.0652	52.3803	148	20	ZIPE
7086	Fort Davis	USA (Tx)	255.9841	30.6770	1964	7	NASA
7090	Yaragadee	Australia	115.3467	−29.0465	245	2	NASA
7105	Greenbelt	USA (Md)	283.1723	39.0206	22	4	NASA
7109	Quincy	USA (Ca)	239.0553	39.9750	1110	3	NASA
7110	Monument Peak	USA (Ca)	243.5773	32.8917	1842	4	NASA
7112	Platteville	USA (Co)	255.2740	40.1828	1505	11	NASA
7121	Huahine	Tahiti	208.9588	−16.7335	47	10	NASA
7122	Mazatlan	Mexico	253.5409	23.3429	34	12	NASA
7210	Haleakela	Hawaii	203.7440	20.7072	3069	4	NASA
7833	Kootwijk	Holland	5.8098	52.1784	94	17	THD
7834	Wettzell	W. Germany	12.8780	49.1449	661	7	IFAG
7835	Grasse	France	6.9	43.7	1320	7	CERGA
7838	Simosato	Japan	135.9370	33.5777	102	11	SHO
7839	Graz	Austria	15.4933	47.0671	540	4	Obs. Lust buehel
7840	Herstmonceux	UK	0.3361	50.8674	76	4	SERC
7843	Orroral	Australia	148.9	−35.6	950		DNM/NASA
7907	Arequipa	Peru	288.5068	−16.4657	2492	15	NASA/SAO
7935	Dodair	Japan	139.2	36.0	850		SHO
7939	Matera	Italy	16.7046	40.6488	536	15	SAO

15.2.2   Tracking and data relay satellite system (TDRSS)

The National Aeronautics and Space Administration (NASA) is in the process of making a transition from tracking and communications support of low Earth satellites with a ground-based station network to a geosynchronous relay network via the tracking and data relay satellite system (TDRSS). TDRSS comprises two operational satellites and one in-orbit spare. The two operational ones are separated by 130° of longitude, at 41°W and 171°W, and are centred about a ground terminal (WSGT) at White Sands, New Mexico (Figure 15.4). While the ground network only provides about 15% visibility coverage, TDRSS can provide 85–100% for most spacecraft.

Each tracking and data relay satellite (TDRS) is equipped with two user service antenna systems. The high-gain system comprises two steerable, 5-metre dual S/K-band antennas, known as the single-access system (SSA and KSA). The low-gain system consists of a 30-element S-band phased array, which can provide one forward link and multiple, simultaneous pseudo-random noise (PN) code division multiplexed return links; this is known as the multiple access (MA) system.

Space-to-ground communications (Figure 15.5) for command, telemetry, and user signals, are routed through a 2-m K-band space-to-ground link (SGL) antenna. During

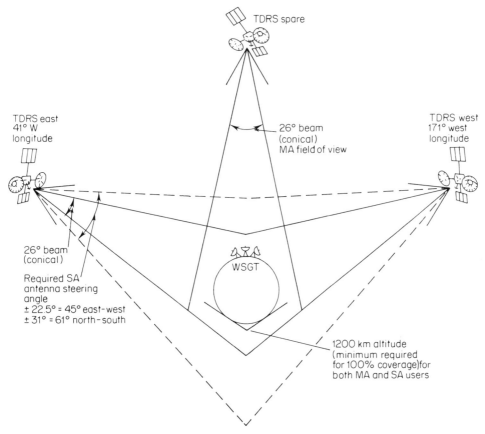

Figure 15.4   TDRSS configuration and coverage limits

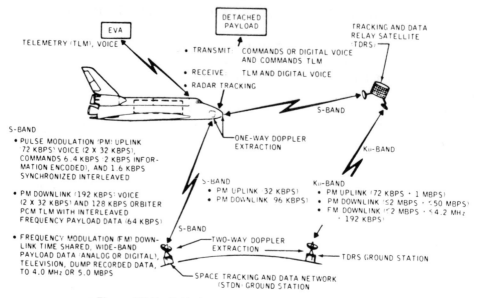

Figure 15.5   Orbital communications and tracking links

periods of maintenance and K-band outages, tracking, telemetry, and command (TT&C) can be supported via an S-band omni TT&C antenna. Future NASA communications will be provided by the previously designated Advanced Westar C-band antenna. In addition to the rotating solar array appendages the spacecraft is equipped with a solar sail for attitude disturbance compensation torques.

The WSGT is configured with three 18-metre K-band elevation-over-azimuth (az-el) antennas, a 6-m S-band az-el TT&C antenna, and roof-mounted S- and K-band simulation/verification antenna. The communications equipment in the WSGT can simultaneously support two-way communications for 6 SSA, 6 KSA, and 3 MA services, as well as a total of 20 MA return links.

User-tracking equipment can provide 9 ranging and 19 Doppler services simultaneously. The Doppler observation is a continuous count of a bias plus the Doppler frequency, resolved to 0.001 Hz at S-band and 0.01 Hz at K-band. The range observation is the four-leg round-trip light time resolved to 1 nanosecond (ns). The range measurement is ambiguous in multiples of the ranging code period, about 0.086 s (13 000 km one way). The observations are strobed on the whole second, formatted, and transmitted to NASA at selectable sample rates of 1, 5, 10, 60, and 300 seconds.

TDRS-1 was launched in June 1983. Among the users involved in testing were the bilateration ranging transponder system (BRTS), Landsat-4, Landsat-5, and the space transportation system (STS) orbiters. It is a system of four ground-based unmanned facilities that contain transponders similar to those flown on user spacecraft. It provides a set of transponders whose positions are accurately known so that ranging information can be used to determine the trajectories of the TDRSs. The facilities are located at WSGT (two transponders, three antennae), Ascension Island (two transponders, two antennae), Alice Springs, Australia (one transponder, one antenna), and American Samoa (one transponder, two antennae). TDRS-2 (launched in 1988 after the original TDRS-2 was lost in the Challenger accident) is supported by BRTS at White Sands and Ascension Island, while TDRS-3 (launched in 1989) is supported by BRTS at Alice Springs, American Samoa, and White Sands.

This geometric configuration provides the ability to determine the ephemerides of all three TDRSs to an accuracy of at least 100–200 m. Orbiting unmanned user spacecraft, currently supported operationally by the ground S-band network, provide an excellent opportunity to evaluate the end-to-end navigation accuracy of user spacecraft. The Shuttle orbiters are supported by a combination of C-band radars and S-band trackers, in addition to the TDRS-1.

15.2.3   Typical ground station configuration (Chilton, UK)

Although the antenna is the main hardware component, the other items identified above are an equally integral part of a ground station's facilities. This section gives some design background to a typical station, shown in block form in Figure 15.6. The one chosen is at the UK Science and Engineering Research Council's Rutherford Appleton Laboratory at Chilton in England. It was initially designed for operating the IRAS satellite shown in Figure 15.7, but has been used to track and receive data from many other near-Earth and high-altitude spacecraft.

The main antenna, shown in Figure 15.1, is a transportable S-band 12 m diameter, cassegrain instrument. The reflector is a hyperboloid section, made with 20 petals constructed from 2-in thick aluminium honeycomb and faced with aluminium sheet. The reflector, the radio frequency feed, cassegrain subreflector and equipment cabinets are supported on elevation over azimuth bearings at the top of a cylindrical steel pedestal. Three tubular steel legs provide support for the pedestal and, with screw jacks, allow accurate levelling of the antenna structure. The whole antenna weighs approximately 32 tons.

The RF feed mounted at the vertex of the main reflector is a complicated waveguide structure, able to transmit and receive simultaneously on 2074 and 2253 MHz respectively, in either right-hand or left-hand circular polarization. In the receive mode, three output ports are available: one is the channel containing the received signal, the other two provide error signals (one each for azimuth and elevation axes) so that, with a servo loop, the antenna can lock on to an incoming transmission, allowing very accurate tracking of selected satellites. In addition to this autotrack mode, the antenna can be driven along a predicted path. This is enhanced by the addition of a joystick control, allowing a manual search to be made around the predicted position of a satellite if it differs appreciably from the real position. A summary of technical data is shown in Table 15.3.

15.2.4   Receive/transmit system

The Chilton receivers and exciters are based on the NASA unified S-band system, and comprise:

1.   two identical receivers with a common phase reference generator;

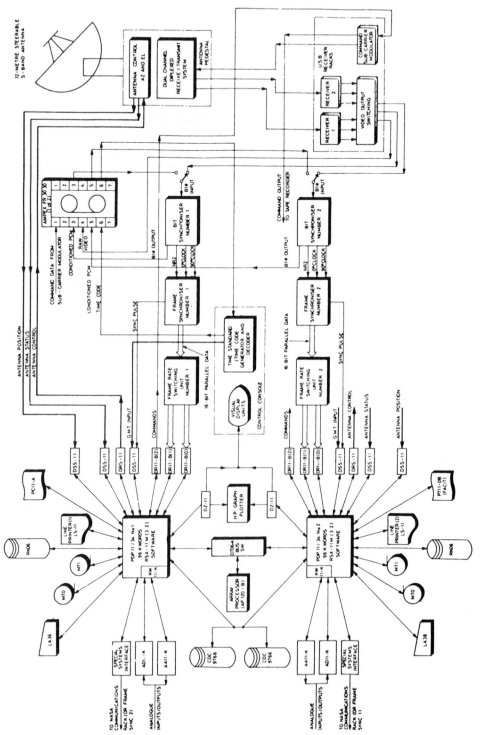

Figure 15.6 Basic elements of a typical ground station

Figure 15.7   IRAS satellite

Table 15.3   Mechanical and drive details of Chilton S-band antenna

Cassegrain configuration, elevation-over-azimuth mount	
Feed, Rantec S-band four-horn monopulse	
12 m diameter primary reflector	$f/D = 0.325$
1.06 m diameter secondary reflector	$e = 1.413$
Main reflector surface accuracy	1 mm r.m.s.
Az rotation (hardware limits)	± 265 deg
El rotation (hardware limits)	-5 to $+95$ deg
Intertia	3370 kg m^2
Wind torque (60 m.p.h.)	0.11×10^6 N m
Static friction (at antenna)	6340 N m
Gear ratio	$1700 : 1$
Encoders, one per axis, 20-bit resolution	
Electrical drive, two motors per axis (Inland type TT4501-A)	
Motor inertia	0.46×10^{-3} kg m^2
Torque constant	0.75 N m/amp
Back e.m.f. constant	78 V/1000 r.p.m.
Peak torque	110 N m
Continuous torque	24 N m
Tachometer sensitivitiy	18.9 V/1000 r.p.m.
Load resistance	> 2000 ohms
Maximum slew rate	7 deg/s
Maximum accn (software limited)	2.35 deg/s^2

PERFORMANCE FIGURES

Antenna gain (2.253 GHz)	46 dB
Beamwidth	0.7 deg
First sidelobe level	-18 dB
System noise temperature (zenith)	115 K
Estimated pointing error (for worst case of 30 m.p.h. wind)	0.05° max
Measured autotracking accuracy (22 m.p.h. wind, 35 m.p.h. gusts)	0.02° peak

2.   two identical transmitters with a common phase-modulation drive;

3.   an RF path-switching subsystem;

4.   the control and monitor subsystem;

5.   the calibration and test subsystem.

   These are physically distributed between an inner cabin on the antenna pedestal (S-band components, adjacent to the antenna feed), an outer RF cabin, also moving with the antenna, and the remainder within the Operations Control Centre about 100 m from the antenna pedestal base. Almost all of the OCC subsystems operate at 50 MHz and below (receive) and 65 MHz and below (transmit). However, a low-loss coaxial feeder is used between OCC and antenna.

   Each receiver comprises three channels. The reference/telemetry channel establishes carrier phase-lock, supports wide-band (dump) telemetry and outputs video TLM. Two angle-error channels detect the angular deviation from antenna boresight in the *X*- and

Y-planes relative to the antenna feed, and output error signals for feedback to the antenna servo drive to establish autotrack.

The common phase reference is a crystal oscillator from which both the final intermediate frequency (10 MHz) and third local oscillator (60 MHz) are derived.

The IRAS downlink signal is an RF carrier at 2253.0 MHz, with phase-modulation sidebands containing, on acquisition, the 4096 bit/s low-speed telemetry (LST) and, in the Dump mode, both 1049 Mbit/s high-speed (HST) and 4096 bit/s low-speed telemetry. The significant components of this combined phase-modulation spectrum extend to 3.5 MHz on either side of the carrier.

Each receive channel performs triple-conversion from a selected frequency between 2200 and 2300 MHz to intermediate frequency (IF) at 340, 50 and finally 10 MHz. The first local oscillator (LO), derived from a voltage controlled crystal oscillator (VCXO) and a synthesizer, both tunes and phase-locks all channels to the downlink signal.

The front-end of each reference/telemetry channel is a parametric amplifier with gain 40 dB, noise temperature 60 K, and bandwidth 120 MHz. Carrier alone is detected to provide receiver phase-lock and AGC for the three channels of each receiver. This achieves a signal-to-noise ratio of 15 dB for the IRAS HST in 6 MHz bandwidth at expected antenna signal level.

The modulation sidebands are separately phase-detected to give a video signal which is further processed to provide three basic outputs:

1. filtered and amplitude-limited LST;

2. combined HST with LST, both for pulse code modulation (PCM) conditioning;

3. raw video for direct wide-band tape recording.

Each angle-error channel converts carrier alone to 10 MHz, which is phase-detected to give a balanced voltage output as a linear function of pointing error in each axis.

Each transmitter comprises a multiplier chain to raise the phase-modulated RF drive to the uplink frequency and a stage of power amplification to produce the final RF level of 10 W into the diplexer.

An associated translator unit samples the outgoing uplink and converts this to the downlink frequency, as a test input to the down-conversion stages of the receiver. The common phase-modulation drive is derived from a VCO, tripled and modulated with the command subcarrier, which is itself modulated with command messages generated in the computer. Thus, the IRAS uplink signal is an RF carrier at 2074.6375 MHz, phase-modulated with a 16 kHz subcarrier which is phase-reversal keyed with NRZ-L PCM.

For *RF path switching* the parametric amplifier in each receiver may be connected to either main or acquisition aid antenna, and the outputs to either or both down-converters. All RF paths may be reconfigured by control from the OCC. Test and calibration signals may be connected either to the parametric amplifier, or down-converter inputs.

Receiver and transmitter VCOs may be offset for Doppler shift and fine tuning (acquisition). The receiver phase-lock loops may be zeroed, and manual or automatic gain control applied. Oscilloscope, counter and status displays monitor essential parameters. Analogue meters show autotrack errors, and RF spectrum displays enable monitoring of modulation index.

For *calibration and test*, provision is made for injecting RF signals at S-band or IF, either as a carrier phase-modulated by an IRAS simulated PCM. Broad-band noise or a precision termination may be connected to either receiver to enable system noise measurements.

A tower-borne transmitter in the far field of the main antenna is used to check antenna and receiver performance and to phase-balance the angle-error channels.

15.2.5   The IRAS satellite RF link characteristics

The RF link characteristics which have to be met by a ground station depend primarily on the specification of the satellite to be tracked. For IRAS the data is shown in Table 15.4

15.2.6   Control centre equipment

The equipment available and its configuration for operations are described here. For IRAS it consisted of two computer systems, each of which were used for controlling the 12 m S-band antenna or for digitizing and processing the satellite telemetry and command data, two wide-band instrumentation tape recorders which were used for buffer storage of high-speed PCM data prior to off-line digitizing (which can be at reduced speed), PCM bit and frame synchronizers, a time standard and a communications system. The computers are controlled from a central satellite controller's console at which point the functioning of all ground station and control centre equipment is monitored.

The two computers are PDP 11/34s manufactured by the Digital Equipment Corporation and have a duplicated configuration so that they may be used on either of the main tasks or act as a back-up for each other in the event of equipment failure. They both have 96 K words of memory in the central processing unit and are operated under RSX-11M (3.2) software.

The *wide-band instrumentation tape-recorders* used for real-time recording of satellite PCM data are FR3030s manufactured by Ampex. The tape transport speeds range from $\frac{15}{16}$ to 120 in/s. The signal record and reproduce hardware includes FM and direct electronics with electrically switched filters and equalizers for all tape speeds. Each machine has the tape speed servo-control option together with remote control, footage counter, and shuttle capability. Bandwidths range, on FM, from d.c. to 500 kHz and on direct recording up to 2.0 MHz (sinusoidal). There are seven record/reproduce channels per machine. Each recorder also has a phase optimization circuit incorporated on certain direct reproduce electronics for use when the machine is operated in reverse and optimisation of high-speed PCM (1 Mbit) data is required. The record and reproduce signals are monitored at the control console enabling the satellite controller to look at the data being recorded during a satellite pass.

There is a *central control console* which houses visual display units, a terminal for manually accessing the computers, remote control and signal monitoring of the tape recorders, standard time code generator and decoder read-out, PCM equipment status, antenna mode and receiving equipment status.

The *PCM conditioning devices* include EMR Type 720-02 bit synchronizer; EMR Type 2246 frame synchronizer (PCM conditioner); Frame-rate switching unit—project designed unit to allow alternate frames of telemetry data to be switched to alternate DMA

Table 15.4   IRAS RF link characteristics

Downlink—spacecraft parameters	
Design frequency	2253.0000 MHz
Settability	$\pm 2 \times 10^{-6}$ (± 4.506 kHz)
Temperature stability	$+10$ to $+40°$C, $\pm 5 \times 10^{-6}$
Ageing	4 years $+ 12 \times 10^{-6}$
	$\pm 3 \times 10^{-6}$/year
Doppler range	± 45 kHz Max
Doppler rate	± 400 Hz/s Max
Low-speed telemetry (LST)	4096 bits/s
Directly PSK modulated on the carrier	
Modulation index	1.0 radian
Simultaneous low speed telemetry	4096 bit/s
and high speed telemetry (HST)	1 048 576 bits/s
The LST is directly PSK modulated	
on the carrier	
Modulation index	0.15 radian
The HST is directly PSK modulated	
on the carrier	
Modulation index	1.1 radian
Transmitter output	1 W ($+2$ dB, -1 dB) at
	diplexer antenna port
Satellite antenna gain	-2 dBi-level A
	-7 dBi-level B
Polarisation	RH circular

Uplink—spacecraft parameters	
Design frequency	2074.6375 MHz
Settability	$\pm 3 \times 10^{-6}$ (± 6.244 kHz)
Temperature stability	$+10$ to $+40°$, $\pm 7 \times 10^{-6}$
Ageing	4 years $\pm 12 \times 10^{-6}$
	$\pm 3 \times 10^{-6}$/year
Doppler range	± 45 kHz Max
Doppler rate	± 400 Hz/s Max
Acquisition sweep rate	1 kHz/s $< f <$ 20 kHz/s
Acquisition sweep range	± 100 kHz
Command code	1000 bits/s NRZ-L
Subcarrier	16 kHz ± 0.1 Hz sinewave
Modulation subcarrier	PRK (phase reversal key)
Carrier modulation	PM, synchronous with
	1000 bits/s data
Modulation index	1.0 radian
Satellite receiver antenna gain	-10 dBi
Polarization	RH circular
Ellipticity	<6 dB anywhere in the pattern
Receiver threshold	-170 dBW

interfaces (DR-11Bs) of the PDP11 and so allow the PDP-11 to accept continuous telemetry at the high-speed rate (1 Mbit). Redundant bit and frame synchronizers are connected on line so that failure of the prime system does not result in loss of data. Switching and patching facilities enable a flexible configuration to be achieved.

The *time coder generator and decoder* (EES Ltd Type TCGR-103B) is driven by a remote rubidium frequency standard at 1 MHz. The 36-bit NASA time coder and parallel BCDT outputs are synchronized to standard time transmission (MSF, 60 kHz) with an accuracy of ± 1 msec. A standard frequency of 5 MHz is also fed to the frequency synthesizer in the USB receiver. Time read-out is distributed to salient points in the OCC.

The data analysis computer is an ICL 2960, used for the bulk of the pre-pass preparation and post-pass data analysis. Multiple on-line peripherals (MOP) are used for on-line communications. Two disk transports are devoted to on-line file-store and operating system files, with a further two disks being used for large data files. The fifth disk transport is used for back-up. There is considerable data traffic between the 2960 and PDP11 computers and this is provided by the two clusters of magnetic tape decks. Online terminal facilities, particularly for preliminary post-pass analysis, are provided by the 7903 communications processor. Certain key peripherals have been duplicated for resilience, notably the disk file controller and communications processor.

15.3   SOFTWARE

There are four major areas of software. The first is that on-board the spacecraft; the second is the software required for determining and predicting the orbit of the satellite; the third produces and processes uplink and downlink data, and the fourth deals with integration, testing, and configuration control. Except for the most basic spacecraft systems some aspects of each of these will be required.

On-board software resides in the spacecraft's own on-board computer. There are two basic types. Read-only memory (ROM) contains its basic instructions and safeguard modes during its operation, and is built into its computer before launch. Once launched, it cannot be changed. Random-access memory (RAM) software contains more subtle instructions for the spacecraft, giving it a greater deal of sophistication and flexibility. It can be programmed before launch, but has the added advantage that is can be modified or built up after launch by uplinking from the ground station. Data handling is covered in greater depth in Chapter 14.

15.3.1   Orbit determination and prediction

In determining and predicting the orbit of a near-Earth satellite there are usually three factors [2] which determine the accuracy required. These are the ability (1) to track the satellite correctly during a station pass, (2) to reconstruct the orbit and hence the position of the satellite at any time during the mission, and (3) to predict ahead many weeks for the purposes of mission planning. It is usually (2) which has the least tolerance on allowable errors. In order to track from a ground station, the maximum allowable error in azimuth and/or elevation can typically be $\pm 30'$ for a large antenna. This is interpreted in the worst case as a time error of about one second in the position of a satellite at 1000 km altitude. Clearly the allowable error decreases slightly with decreasing altitude.

For orbit reconstruction, the requirement is based on observations taken at times before and after a reconstructed epoch. For instance in a project which requires mapping features of the Earth, position reconstruction has to be exceptionally accurate (within metres), and this is definitely not possible using a single ground station, even with the aid of advanced techniques such as laser ranging. However, for many satellites an accuracy of ± 1 km in altitude and ± 5 km in along-track position is sufficient. For orbit predication, the required accuracy depends not only on the accuracy of orbit determination but also on the accuracy of the orbit propagator. Long-term predictions are usually required for advance planning of pass times, eclipses, attitude manoeuvres or experiment observations. Typical accuracy requirements are a few seconds per week accumulative.

For highly accurate orbit determinations to less than a 1 metre error in position, it is necessary to use laser tracking data together with highly sophisticated numerical and analytic algorithms such as those incorporated in the NASA GSFC GEODYN [3] program. For most purposes, however, a simpler orbit determination and prediction process is sufficient.

15.3.2   Other support software

This section gives an overview of the ground support organization and software needed. The system developed for the IRAS [4] satellite is used for the purpose of illustrating the case of a low-altitude satellite.

A ground station OCC provides ground operations software for three main tasks: constructing a satellite's attitude program, tracking and communicating with the satellite during the passes, and processing its engineering data. These tasks require detailed knowledge of the orbit. For instance in planning the observation schedule for an astronomy satellite it will be necessary to know where the satellite will be in the future, the antenna control software needs to know where to find it during passes, and the data processing programs need to know where it was when the data were accumulated. The data exchange between these software packages is summarized in Figure 15.8.

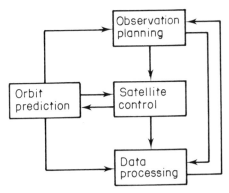

Figure 15.8 IRAS ground operations software packages

The main functions of the satellite control and data acquisition software are to control the antenna which tracks the satellite during its passes over the ground station, to command the satellite, and to receive and process data from it in real time.

A *pass schedule* is constructed as far ahead of the pass as possible. It contains a set of instructions that define the actions to be taken during the pass, the tracking data, the new satellite plan and any other data to be loaded into the computer memory. Normally, the pass schedule will be created automatically using a set of default options, but if necessary it can be constructed interactively, overriding some or all of these options.

In ideal conditions the pass schedule will control the entire pass with minimal intervention from the satellite controller. Each table entry will be interpreted by the commanding system software which will invoke other modules as necessary.

A *typical pass schedule* might include instructions such as request dumps of stored data, confirm that the old satellite plan is unchanged, or load or enable a new satellite plan. The data to be dumped during the pass may have been stored on an on-board tape recorder and in part of the computer memory.

For IRAS the data from the tape recorder is transmitted to the ground over a high-speed telemetry (HST) link and stored on analogue tape for digitization after the pass (Figure 15.9), or direct on to digital tape if the bit rate is low enough. The contents of the computer memory form part of the data which is transmitted continuously over a low-speed telemetry (LST) channel at a rate of 4096 bits/s. The rest of the frame contains command verification data and satellite housekeeping data. All the LST data transmitted during the pass is received and processed in real time.

As each low-speed frame is received, its satellite clock time is extracted and saved, together with the current value of the universal time. This set of timing points will be used during post-pass processing to calibrate the satellite clock. The command verification data from each frame are used to check that the commands sent to the satellite were received correctly. The housekeeping data contain a set of satellite status indicators and sensor read-outs, and the values of a selected group of these parameters can be extracted from each frame and displayed on a VDU, or listed or plotted. For many analogue parameters a set of engineering and working limits have been defined, and any limit violations which are detected are also displayed on a VDU.

During the pass a copy of the contents of the whole of the on-board computer memory will be built up, and new data will be loaded into some parts of it. Normally the whole memory is dumped cyclically, but during the pass commands may be sent to the satellite to request that particular areas be dumped immediately. Parts of the memory, such as the program area, should remain unchanged and will be dumped and compared with the expected contents. Parts of the memory which have been reloaded will be dumped and verified. Error-free copies of the memory areas used for the storage of engineering data will be built up.

If necessary, the satellite controller can interrupt the pass schedule and enter commands from the console. Control may be returned later to the pass schedule, either at the interrupted entry or at another. If an error is detected during processing, the satellite controller will always be informed so that appropriate remedial action can be taken.

After a pass, the ground operations software [4] sorts the scientific data for distribution to the various processing centres. It extracts and processes the engineering data in order

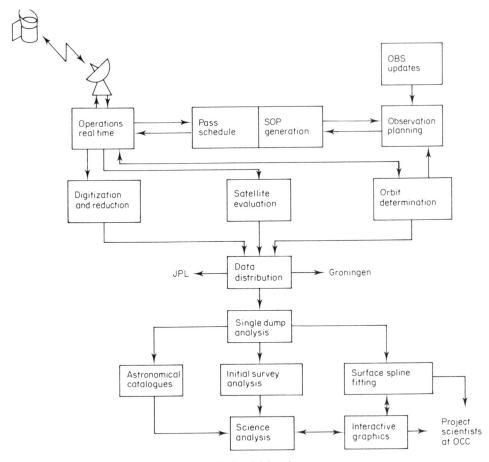

Figure 15.9   OCC software

to assess the success of the scheduled observations, and supplies the telescope and spacecraft support teams with the information they need to monitor the satellite's health and performance.

The data derived from the low-speed telemetry link include a set of timing points and a copy of the satellite event data, together with the stored housekeeping data from the computer memory. The timing data are used to derive a single timing point which correlates the satellite clock with the universal time (UT). All the data from the satellite are labelled with the satellite clock value, so an accurate correlation with UT is essential.

The event data include a record of all the attitude calibrations and changes, and this is combined with some of the stored housekeeping data to give a compendium of engineering data which is needed during analysis of the science data.

For the high-speed telemetry link one data frame per second is stored by the on-board tape recorder. During the pass, the data are dumped and recorded on analogue tape and subsequently digitized. For operational reasons, the dump may be taken in two parts with some overlap, and if the digitization program detects any duplicate frames a new index is constructed which contains pointers to just one copy of each frame.

Each frame contains science and housekeeping data. Copies of different subsets of it are written to magnetic tape for distribution to the science analysis centres, and the housekeeping data are written to tape for processing.

About 6 Mbytes of housekeeping data are stored on the tape recorder during each SOP. This is too much to check and store routinely, so samples are nominally taken each 64 seconds. For some critical subsystems parameter values are extracted from every frame. These include the cryogen flow-rate and the atttiude control subsystem data.

15.3.4   Configuration control

Controlling the design, development, integration, testing and modification of a large software system is a complicated task in any business. For a large space project the problems are compounded for a number of reasons:

1. There are interfaces not just between local system boundaries, but between systems designed and developed in different locations.

2. The requirements of the software system are regularly changing to match increasing capabilities of the hardware designers.

3. There is a need, post-launch, to be able to modify any of the software extremely quickly. A satellite cannot be turned off and put on ice until the problems is solved some days or weeks later.

Throughout the design, build, integration and test period it is necessary to keep a strict control on changes to the design of software. A number of procedures are involved and are now described (see also Figure 15.10).

Project reviews. Regular reviews of all areas of the project, including the software, are required. In the early stages there are reviews of requirements and design philosophies. Project reviews continue throughout the development stage, culminating in flight readiness and launch readiness reviews just prior to launch.

Configuration control board (CCB). A CCB needs to be formed, consisting of the software manager and subsystem managers. Their function is to approve/disapprove of all software change requests, to approve the completon of all test reports, and to approve the final delivery of all software products.

Local reviews. In-house there are regular internal reviews of software design and build. These are accompanied by walk-throughs of each program.

Test plans. The plans are written for all software items, and test reports produced on the results. All errors, or omissions from the requirements specification, are documented on non-conformance report forms, and these are also reviewed regularly to ensure progress in countering the errors.

Simulation data. Data simulators are always required, often one to simulate the real-time system, one to simulate the housekeeping data and the spacecraft orbit and environment, and one to simulate the science data if present. The three sets of simulated data are merged and a series of tests on the complete software system are run by passing the data right through the system. Initially the data is error-free; then the simulation data is variously and deliberately corrupted to be more like the expected operational situation, and to ensure the software system can cope adequately.

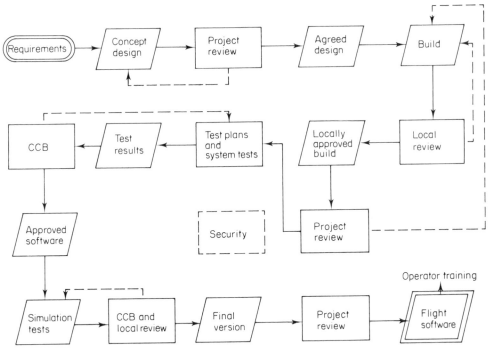

Figure 15.10 Software control mechanisms

Operator control. A number of training exercises are held, to train the OCC operators in running the different software packages. Similar exercises are held to train the data analysts in the art of identifying software and spacecraft problems. In this respect, several of the training exercises consist of 12-, 24- and 48-hour mission simulations.

Security. A sophisticated system of security is required to ensure (1) that no programs and/or files are inadvertently erased, (2) that copies of programs and files are available in case of system crashes, and (3) that no individual user can inadvertently corrupt part of his, or another individual's software. At the same time it has to be possible to make changes quickly when required. Consequently, a number of *flight directories* are set up containing the mission software. No software can be modified without prior approval of the software manager, and all such directories are *password* protected. All software modifications are made and tested in external directories before approval is given for the modified software to be incorporated in the flight directories. Security tape dumps are made at regular intervals, both before and after launch, for all software programs and files that are updated, so protecting the system against computer failure.

Post-launch control. In a perfect world there would be no need for post-launch control of the software system, but life is never perfect and space projects are certainly no exception. It is recognized that errors will appear after launch that have not shown up during pre-launch testing. It is also envisaged (correctly) that satellites do not behave in a totally predictable way, and so design changes are needed in the software. Therefore it is desirable from the outset to retain the CCB post-launch and adhere to all procedures adopted pre-launch.

People/operations. As with any large operational set-up, there is a requirement for people to run the facility, ably abetted by well-defined operational procedures. These include the following:

- Site management—for the supervision of day-to-day operations both on the ground and in orbit.

- Operations shift staff—to man the facility, often on a 24-hour/day, 7-day/week basis.

- Specialist data and engineering support staff—to plan operations, discuss results, and be on hand at times of crisis.

- Procedures documentation and user manuals—to support the operations staff.

- Schedules—to coordinate (often worldwide) activities on the spacecraft.

- Hardware and software maintenance—both preventative and curative.

- General administration—including archiving, data distribution, typing, personnel support, secretarial.

REFERENCES

[1] Bevan, H.C. *et al.* (1983) The IRAS ground station operations control centre at Chilton. *JBIS* **36**, 10–16.
[2] Holdaway, R. (1985) Orbit prediction for IRAS using vector and analytic techniques. *Acta Astronautica* **12** (3), 139–48.
[3] Martin, T.V. (1987) *GEODYN Descriptive Summary.* NSA 5-11735-MOD 65.
[4] Mount, K.E. (1983) Ground operations software at the IRAS operations control centre. *JBIS* **36**, 34–7.

16 SPACECRAFT ELECTROMAGNETIC COMPATIBILITY ENGINEERING

L. J. C. Woolliscroft

Department of Control Engineering, University of Sheffield

16.1 INTRODUCTION

A spacecraft is usually a set of complex electronic systems contained within a mechanical structure. Electromagnetic compatibility (EMC) engineering is concerned with ensuring that these systems neither suffer nor cause electromagnetic interference (EMI) in orbit (or during the test and launch phases of a programme). Solutions (if possible) for EMI problems found during a test programme are frequently extremely expensive in both money and time. Few, it any, spacecraft escape without problems during the test phase and many have suffered serious problems in orbit.

Spacecraft suffer from EMI in a variety of different forms. These range from d.c. effects such as electrostatic charging, with its potential for arcing and magnetization which will create a disturbance torque to upset the vehicle's attitude, to a.c. effects which span the frequency range and include obvious interference hazards between emitters and receptors in the communication equipment. There are also less obvious effects due to the plasma of the space environment.

The spacecraft presents a particularly severe test for EMC engineers. To start with it is physically small, and thus it is not easy to separate the many electronic systems which almost always include both high-power transmitters and sensitive receivers which have to be operated at the same time. There is also usually little spare mass for shielding. In addition, spacecraft are produced as one-offs or in series of only a few similar vehicles.

16.2 INTERFERENCE MECHANISMS

Interference paths are generally of a conductive or a radiative nature. Figure 16.1 shows an idealized situation in which just one emitter interferes with one susceptor, but of course

Spacecraft Systems Engineering. Edited by P.W. Fortescue and J.P.W. Stark
© 1991 John Wiley & Sons Ltd

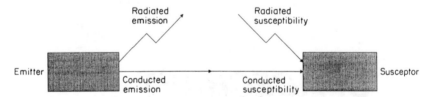

Figure 16.1   This diagram shows interference paths between one emitter and one susceptor. The 'radiated' path may involve electric or magnetic induction rather than radiation

there will usually be many more than one to consider. The level of the units which may interfere with each other varies from devices on a single circuit board to a complete spacecraft which is subject to interference from radar equipment. Systems engineering is usually concerned with the various packages and subsystems, and there are typically between 10 and 100 of these.

Conducted interference (Figure 16.1) is the unintentional conduction of signals from an emitter to a susceptor via electric connections between the two. This usually takes the form of conduction along the power distribution wiring, but may also involve the on-board data handling (OBDH) wiring or even the spacecraft structure.

'*Radiated*' *interference* (Figure 16.1) is traditionally so-called, but it is usually a misnomer. Radiation only applies to paths longer than about one wavelength—otherwise the situation is near-field, and the coupling is 'inductive'. The emitter as a source can be considered in terms of a magnetic and an electric dipole, and higher-order multipoles. The radial and azimuthal field components, and hence the near-field interference due to a dipole in free space, are

$$E_r = \frac{1}{4\pi\varepsilon_0} \frac{2P}{r^3} \cos\theta, \quad E_\theta = \frac{1}{4\pi\varepsilon_0} \frac{P}{r^3} \sin\theta \tag{16.1}$$

where θ is the azimuthal angle, P is the dipole moment, and ε_0 is the permittivity of free space. These equations show that the field strength decreases rapidly with r, the distance from the centre of the dipole, and that there is an important azimuthal effect. Multipoles of higher order than a dipole produce fields which decrease even more rapidly with distance. For radiation the power decreases according to an inverse square law. Thus one obvious technique of EMC engineering is to separate susceptors and emitters and also to consider their orientation. Generally the azimuthal effects become the antenna pattern of aerials, and the analytic expressions are much more complicated but the principles remain the same.

Plasma-coupled interference. A spacecraft is in a cavity in the space plasma, as shown in Figure 16.2. The cavity dimension is typically less than a few metres, and is of the order of the Debye length, λ_D. In terms of the electron temperature $T(\mathrm{K})$ and the plasma number density n (particles/m^3), the Debye length is

$$\lambda_D = 69\sqrt{(T/n)} \tag{16.2}$$

For space plasmas λ_D is usually of the order of a few millimetres to a few metres.

Radiated emissions from the spacecraft may

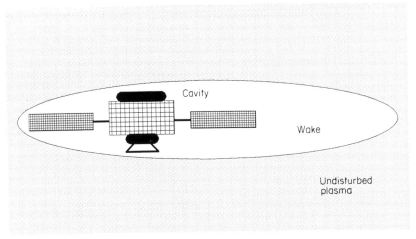

Figure 16.2 A schematic spacecraft in a cavity in the space plasma where the cavity is extended into a wake by the vehicle motion

- be reflected back to the spacecraft by the boundary of the cavity;

- be amplified by the space plasma and re-radiated back to the spacecraft;

- generate other frequencies in the plasma which are then received back at the spacecraft. These may be harmonics of the radiation or result from demodulation of the radiated emission and be harmonically unrelated.

These effects occur for frequencies up to a few times the electron plasma frequency f_{pe} (Hz) of the ambient plasma, where

$$f_{pe} = 9\sqrt{n} \tag{16.3}$$

This is generally below a few megahertz. Thus the space plasma is unlikely to be very important at VHF or higher frequencies.

D.c. magnetic fields. Establishing that no significant stray magnetic fields are produced is a complicated topic as the ambient magnetic field may only be a few nanoteslas (compared with $\sim 50\,000$ nT at the surface of the earth). The interference may be due to either the effect of permanently magnetic materials in the spacecraft or to low-frequency or d.c. currents. Techniques to overcome these include the obvious avoidance of magnetic materials (batteries and relays may be difficult) and care over the physical design of the power subsystem, particularly the solar arrays.

D.c. electric fields. The main technique to secure electrostatic cleanliness is to have around 95% of the outer surface area of the spacecraft covered with a conductive surface. This helps to avoid the practical problems of charging and sudden discharge causing damage with short-duration high-voltage pulses. Such pulses can damage semiconductor devices or can be misinterpreted as command pulses. Solar cell arrays may be made conductive by depositing a thin metallic coating (usually indium) on the cover-glasses of the cells (see Chapter 11).

16.3 PRACTICAL TECHNIQUES

There exists a range of practical techniques which are used to achieve EMC. Filters and shields are often used in equipment on the ground but their applicability to space systems is rather restricted by mass. Filters are used to lower conducted interference. Filter design is extensively treated in electronic engineering literature and in References [1] and [2]. It is important to remember when designing a practical filter that the individual devices which are used to make the filter will have stray capacitance, inductance, etc., which may cause them to have undesirable properties at other frequencies.

Shielding is used to contain or exclude radiated interference. It is worth noting that multiple layer shields usually give more effective shielding for a given shield mass. Both filtering and shielding are usually best avoided at the spacecraft system design stage. More important techniques are discussed in the next sections.

16.3.1 Grounding

Currents can flow in a ground loop and these are common mechanisms for producing an EMI problem. The simplest form of grounding for several packages which are connected together is to use a 'floating ground' concept (Figure 16.3a). The individual packages have no connection to ground, or to the spacecraft reference. This configuration may be possible in practice by the use of optical isolators or transformers, at least as far as data systems are concerned. Power systems are difficult to float.

The most common grounding philosophy for several packages within a subsystem is

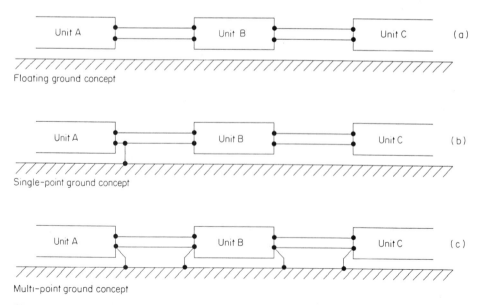

Figure 16.3 (a) Floating grounding between three units of a subsystem; (b) single-point grounding between three units of a subsystem; (c) multi-point grounding between three units of a subsystem

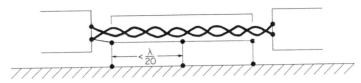

Figure 16.4 This figure shows the use of a twisted pair of wires in a harness with an overall shield

the 'single-point' concept (Figure 16.3b). It avoids the possibility of currents flowing in ground loops but it should be noted that at high frequencies the effective circuit may not be as simple as that shown. If the ground connection lead and the lengths of the ground leads between packages is not significantly shorter than $\frac{1}{4}$ × wavelength of the highest frequency which needs to be considered, the ground connection will fail to operate as anticipated.

At higher frequencies use is made of multi-point grounding (Figure 16.3c). To be effective the area of grounding loops needs to be as small as possible. Multi-point grounding is frequently used within a single package where ground planes on printed circuit boards are used to minimize cross-talk between digital devices.

It is usual to employ single point grounding for packages and subsystems. Within each one multi-point grounding is used, with the reference normally isolated from the package's conducting wall and available on a pin of its electrical connector. All of these reference pins are connected together to a single reference point (frequently at the spacecraft power conditioning unit or central data handling unit), and there connected to the spacecraft body. This connection often has a deliberate impedance to achieve some isolation between the structure and the electronic systems.

The single most common failure of grounding strategies is that the ground impedance may become large at some frequency.

16.3.2 Harness design

Early attention over the design of wiring harnesses is a cheap way of avoiding problems. The objective is to avoid cross-talk, that is signals on one cable inducing interference signals on another.

Frequently separate harnesses are used for power distribution and data subsystems. These should be kept apart and crossings should be at right angles if they are necessary. Otherwise cables should be as short as possible and loop areas small.

The wiring usually uses coaxial or concentric cables for high frequencies and twisted-pair (or triplet) wiring for low. The twisting is to minimize effective loop areas (by making them random). The whole harness is usually shielded and provision should be made for the shield to be grounded, ideally at less than $\frac{1}{20}$ × wavelength intervals, and certainly at both ends (see Figure 16.4).

16.4 SPACECRAFT EMC CONTROL

At the spacecraft level it is important to establish an EMC control programme early. Its nature will naturally vary from project to project, but there are certain general steps. The first stage involves identifying and modelling the coupling between subsystems so as to identify the critical areas. These then will require testing. The most important single aspect is for all the subsystem engineers to be convinced of the need to take the EMC aspects of their work seriously at an early stage.

A frequency utilization plan should be established, showing all the frequencies that may be emitted together with the susceptibilites of all the subsystems. A susceptibility may be that of a receiver (a lower bound to the signals it is expected to receive) or it may that of some circuitry which should not act as a receiver. The obvious use of this frequency plan is to avoid emitters and susceptors at the same frequency (after allowing for the bandwidth of each).

The next stage is to estimate the coupling between all emitters and susceptors for conducted and radiated paths at all relevant frequencies. This may be done analytically or by computer modelling. The task is frequently so large that only the most obvious frequencies and subsystems are considered. The result is the EMI margin where

$$\text{EMI margin} = \frac{\text{susceptibility level at susceptor}}{\text{interference level of susceptor}} \text{(dB)} \qquad (16.4)$$

This EMI margin may be 6 dB for adequate operation of some subsystems, but for safety a margin of 20 dB is more normal. However, the uncertainties in the analysis and the measured levels is frequently 20 dB, which leads to larger margins being required. Thus for many purposes frequency separation or time multiplexing may be necessary.

After testing, the EMI margin of some microwave systems may be less than 6 dB and still yield satisfactory results.

EMC testing has two objectives—to provide data for EMC calculations, and to confirm the design. It is expensive, and the results need to be used with great care. EMC test facilities must have low ambient noise, and usually shielded enclosures are used. These are frequently lined with radio frequency anechoic material (RAM) to absorb radiation emitted by the system being tested. It may also be necessary to provide specialist equipment to exercise the spacecraft in some realistic way while it is being tested. This may be the hardest part of any EMC programme.

Computer modelling in EMC is increasing, but is not as advanced as in thermal or stress analysis. This is mainly because the range of frequencies and the coupling mechanisms make the task very complicated. A model enables a database to be produced and used to assess the effects of modifications, waivers (where a subsystem 'wishes' to be allowed to 'fail' a specification level) and other trade-offs. It can also be used to survey for incompatibilities and to produce specification limits for a system.

16.5   SPECIFICATIONS

There are many EMC specifications and standards. Perhaps the most important for spacecraft engineering are as follows:

- MIL-STD-461—EMC emission and susceptibility requirements for equipment and sub-systems;

- MIL-STD-462—description of how measurements are to be made;

- MIL-STD-463—definitions and systems of units for EMC;

- MIL-STD-1541 (USAF)—EMC requirements for space systems including launch vehicles and ground support equipment;

- MIL-STD-1542 (USAF)—EMC requirements for facilities including grounding, lighting, air conditioning, etc.

- MIL-STD-6051—control of EMC environments and margin estimation for systems.

16.6 FURTHER DEVELOPMENTS

Future developments in EMC engineering for spacecraft will include a trend towards higher frequencies being used. One consequence of this will be that antenna patterns and leakage of microwaves will become more important problems than was previously the case. A further problem will result from the use of high-power transmitters on spacecraft. These will be used, for example, in synthetic aperture radar and may require other subsystems to be tested to have susceptibilities above 100 V/m at frequencies in the 20–40 GHz range.

REFERENCES

[1] Keiser, B.E. (1979) *Principles of Electromagnetic Compatibility*, Artech House, Dedham, USA.
[2] White, D.J.R. and Mardiguian, M. (1985) *EMI Control Methodology and Procedures* (4th edn), ICT/DWCI Holland, Hoogmade, Holland.

17 PRODUCT ASSURANCE

T. Meaker

European Space Agency (ESA)

17.1 INTRODUCTION

The Product Assurance (PA) system ensures that **failure, hazard, and degradation** aspects of the design and software are properly considered in spacecraft engineering.

In order to achieve system 'effectiveness', or 'availability', it is necessary to understand the dependency of design and hardware on PA, and vice versa. The most elegant design in the world is useless if it cannot be: (1) analyzed for its weakness, (2) manufactured and maintained, and (3) tested.

Fundamentally, the **objective** of Product Assurance is to ensure that the consequences of hazards and failures do not affect life, the space vehicle mission, nor the space vehicle itself.

An essential element in all PA disciplines is the *feedback* cycle, in which the results of design, analysis, manufacturing, development, test and operation are immediately fed back to all 'involved' areas. There is a continuous interchange and updating of all data, and a *measure* of its significance and status. This presents a significant organizational and management challenge.

Disciplines which collectively constitute product assurance are reliability, quality, safety, configuration control, parts (electronic, etc.), and materials and processes evaluation.

The **aims** of PA are as follows:

- to protect human life, the investment, the environmemt, public and private property, and the mission;
- to establish and implement parametric derating criteria to ensure that no electrical, mechanical or chemical overstressing occurs during the mission;
- to define the probability that the system will perform successfully according to the specifications;
- to identify and control critical elements within the system such that the success of the mission is not jeopardized;

Spacecraft Systems Engineering. Edited by P.W. Fortescue and J.P.W. Stark
© 1991 John Wiley & Sons Ltd

- to verify that all elements of design, hardware, manufacturing, and test are of an adequate standard, and are consistent and correlatable during all phases of the programme;

- to ensure that all failures, hazards and nonconformances are identified, their total effects understood, and adequate rectification and retest validation carried out.

A general policy of prevention rather than cure is applied.

It will be apparent that PA covers all aspects of spacecraft design, build, test and operation at all contractual levels and, technically, from transistor junctions to the overall space segment. PA is thus an integrated and necessary part of spacecraft engineering. It cannot be emphasized too strongly that an essential ingredient is the application of efficient management techniques to ensure proper and timely identification of status and problems, and resource allocation for their rectification.

17.1.1 History

The extensive implementation of quality control requirements in the space business can be traced back to the failures of the American Redstone, Jupiter and Vanguard rockets in the 1950s. Their problems emphasized the need to take a scientific–engineering approach to reliability; a launcher typically contains over 100 000 component parts! Hence, *reliability engineering* was seriously introduced into the application arena.

In January 1967 a fire in the oxygen environment of the Apollo spacecraft caused the tragic death of three astronauts. This resulted in a complete re-assessment of the role of PA and of safety in particular. NASA subsequently formulated a policy which resulted in the centralization of most PA functions and responsibilities. This was a major landmark since it resulted in a coordinated approach, detailed requirements, and the recognition of PA as an accountable function in spacecraft engineering. The accountability of engineering and design disciplines was also established.

In November, 1980 three fuses in the attitude control system of the $235 million Solar Max satellite blew, resulting in a lack of fine pointing capability for four of the six on-board telescopes, and effectively mission failure after only nine months in orbit. The fuses failed because they were incorrectly sized following a circuit modification. The importance of *Configuration and Change Control* was demonstrated and its vital role in spacecraft engineering underlined.

Now that the space segment is being utilized, the *availability* of functioning equipment and the protection and sustaining of Man (and his investments) are top priority requirements. With the future habitation, industrialization and further exploration of space, PA-related aspects such as human engineering, maintainability, replenishment, and availability/reliability simulation are already being specified.

Over the past decades, with the tremendous increases in the complexity of technology, mission performance and lifetimes, the financial commitments and the impact of the loss of men, or transmitted data, a need has developed to provide an *a priori* confidence that the spacecraft will successfully complete its mission. This is also a prerequisite of the insurance companies. Against this background, which will continue into the future, PA has developed from fundamentally an 'inspection' function to a sophisticated and respected total engineering science.

17.1.2 Qualification

Prior to launch all elements of a spacecraft, including the launch vehicle, must be *Qualified*.

Qualification means that, by test and analysis, the design has been demonstrated to contain adequate margins and factors of safety such that performance, manufacturing and interactive variabilities will not cause excursions outside the flight operational envelope of any individual spacecraft made to that design. In fact, it is then considered that only random failures will occur. Qualification is applicable at part, equipment, and system levels.

Qualification test involves the application of stresses to demonstrate that indigenous failure mechanisms will not degrade the mission. The 'types' of stress relate to those that the spacecraft would experience from lift-off to end-of-life. The stress levels are higher than those that are expected during operation life in order to demonstrate the margins mentioned above; generally they are 1.5 times the expected vibration levels, and $\pm 10°$C on the predicted orbit temperatures for thermal testing. Test durations are either directly related to the operational environment or representative of it, e.g. solar simulation testing includes one or two equinox, solstice, and eclipse test cases. Equipments are usually only energized if they would normally be energized during the corresponding part of the mission. For example, an equipment that is switched on for the first time in geostationary orbit is not energized during vibration testing. In the author's view this is a short-sighted policy since intermittent faults, possibly initiated only during vibration testing *and* the actual launch phase, would not be detected during testing but might cause serious problems during the actual mission.

The two main methods of achieving spacecraft qualification are termed the *prototype* and *protoflight* approaches. Parts qualification is addressed on page 407.

In the *prototype* approach, which is applicable to entirely new designs and missions, dedicated fully instrumented qualification hardware is manufactured and exposed to the full qualification test programme both at the equipment and integrated system levels. After such testing the hardware is considered unfit for flight and in many cases some of it is subjected to destructive analysis to evaluate the effects of the test stresses fully.

For this approach the cost and schedule impact of *one complete additional spacecraft* and test programme must be considered. For a series of satellites, e.g. Intelsats or European ECSs, the costs can be amortized over the series and the risk versus cost trade-off is more attractive.

The *protoflight* approach was introduced after some experience had been gained from a number of space projects. It was argued that many pieces of hardware were being reused with relatively insignificant modifications from previously qualified missions, and that the application of the *prototype* philosophy was both inappropriate and unnecessarily expensive. The requirement for the application of the protoflight approach is that a qualification 'heritage' from similar space missions is demonstrated for each equipment proposed for the new project. This process often forms a major point of altercation between a customer and potential contractors due to different claims being made for 'qualification similarity'. With this approach the qualification test programme is applied to equipments and the fully integrated spacecraft but *half the full test durations* are used at each level. After this the spacecraft is considered fit for a full operational mission, requiring no refurbishment. Instrumentation in the spacecraft is much less than in the *prototype* approach and no additional internal hardware investigation is normally carried out.

17.1.3 Acceptance

Once the design has been qualified then hardware will be manufactured and integrated to form flight spacecraft. At the equipment and integrated spacecraft levels *acceptance* testing is applied to ensure that no manufacturing/workmanship errors have been introduced which could deleteriously affect the mission. In general random vibration and thermal vacuum testing are applied; vibration stress levels equal to predicted orbit levels, and $\pm 5°C$ on the orbit predictions are used. Significant repair/re-work during testing *can* require repeat testing and hence the importance of rigorous qualifications testing and controlled manufacturing to avoid such large additional expenses and schedule impacts.

17.2 FAILURES

A spacecraft must be designed to fulfil its particular mission; it is hoped that it will not be underdesigned nor overdesigned. Underdesign will lead to failure to complete the mission whilst overdesign will result in a reduction in the payload due to non-optimal utilization of mass, power, etc., elsewhere.

Designers in general tend to consider results based on the successful functioning of technology. The PA engineer on the other hand evaluates the consequences of the presence and reaction of failures, hazards and degradations. The *quantitative* 'PA orientated' requirements of a design are stipulated in terms of reliability, mission life, and permissible single point failures; availability and maintainability are now also used. These are 'system' level requirements and are manifest at the circuit design level as derating rules, failure mode effects, and criticality analyses, etc. The overall objective is that during testing, prior to launch, and operation during the orbit mission, the spacecraft shall not experience any 'infant mortality or wear-out' effects and that all other 'failure tendencies' shall be controlled by the design such that the system performance is successful.

The consideration of the failure potential, or mission (technical) risk, of a design and the subsequent hardware and software is an essential part of spacecraft engineering. In order to ensure that the correct design has been established it is necessary to understand all the possible failure mechanisms that may degrade or terminate the performance or life.

Failure mechanisms are the fundamental events, electrochemical or otherwise, which represent an unacceptable change from a previously defined and stable condition. Failure mechanisms are always present and usually become more significant as time increases; it is therefore imperative that their 'end-of-orbit-life' effects are understood and contained within the design envelope. Particulate radiation can seriously affect technology and man; its effects can be instantaneous or time dependent.

In order to facilitate the design process and enable the resolution of otherwise impossibly complex reliability equations, failure mechanisms are divided into two classes; those that occur randomly in time and those that are time dependent. In practice, the former group contains all failure mechansims for which time dependency has not been dis-covered. As missions become more demanding it sometimes happens that a failure mode changes its classification due to a higher stress environment. An example of this is the recent banning of wire filament fuses on some long-life (7–10 year) satellite missions due to gradual glass seal leakage causing the fuse to operate in an unpredictable fashion.

Failure mechanisms and activation energies for electronic parts are determined during the qualification programmes; life testing is carried out to verify that the operational missions will not be deleteriously affected. Life testing is typically of an 'accelerated' nature whereby the component operating temperature, for example, is increased and each test hour is then equivalent to a number of orbit operating hours according to the Arrhenius relationship (see Appendix 1). It is important to realize that each failure mechanism has its own activation energy and test programmes must be designed accordingly.

Failure modes refer to the way in which failures are observed, as opposed to the mechanisms which cause them. In electronics the main failure modes are short, open, and drift circuit conditions. Mechanical modes are usually associated with wear, crack propagation, spalling etc. and are therefore time dependent in nature. Electrochemical failure mechanisms, involving interaction and breakdown effects, also tend to be time dependent.

Table 17.1 Failure modes and mechanism examples.

Type	Failure mode	Failure mechanism
Resistor—	Drift	Faulty thermal treatment. Wire corrosion
wirewound	Open	due to contamination.
Thermistor	Open	Fracture of glass bead types during vibration and temp. cycling. Bond failure of element to mount.
	Drift (calibration change)	Lead fracture. Contamination through poorly manufactured or damaged coating.
Inductors	Drift	Cracked ferrite core during vibration temperature cycling (cause: usually incorrect assembly)
	Open	Movement of core clamps
	Short	Poor locking; wire breakage; shorted turns (workmanship). Dielectric breakdown in high voltage and RF transformers (workmanship)
Capacitor—Tantalum, solid	Short + leakage	Dielectric breakdown
Diode	Open +	Cracked glass
General purpose (glass case)	Drift	Die surface contamination
Transistor— microwave	Drift + open	Unequal current heating in emitter fingers (local heating and metal transportation)
Integrated circuits	Open	Contamination of plating solutions
	Seal faults	Corrosion at lead base; Poor integrity of die/header bond

Table 17.1 contains a listing of some well-known failure mechanisms and modes for selected technologies.

17.2.1 Failure Mode Effects and Criticality Analysis (FMECA)

By analysing the effect of all failure modes on equipment, subsystem and system it is possible to identify, quantitatively, the probability of single point failures and thus assess the criticality of the design.

The FMECA is a 'bottom-up' analysis and procedures exist for carrying it out. It is complemented by the fault tree analysis which concentrates on failures at system level and shows their possible causes by a 'top-down' approach to part or equipment level.

The importance of proper FMECA cannot be over-emphasized. It should be commenced at the conceptual design stage, updated for every design and hardware review, and used for validation of all ground testing and orbit operation. Its role in testing is frequently omitted, often with serious consequences. An example of this is the testing of a particular infra-red Earth sensor. The FMECA indicated the existence of a failure mechanism at the upper operating temperature which would only be detectable if the infra-red sensor was stimulated by an infra-red source at that temperature. Unfortunately the FMECA was not used during the compilation of the test procedure, the sensor was not stimulated during thermal testing, and a fault occurred—in orbit! A $100 million project was thus unnecessarily jeopardized.

Figure 17.1 shows an example of a fault tree.

17.2.2 Failure rates: confidence levels

The failure rate (the instantaneous failure rate is often referred to as hazard rate) is a measure of the rate at which failures occur; whether it is constant, increasing or decreasing with respect to mission time is clearly of critical importance. The two general curves in Figure 17.3 are typical of the failure rate versus time characteristics for 'standard' electronic and mechanical parts.

Failure rates are derived from test and operational data, and since this represents a sample of the total population it is necessary to quote a confidence level to indicate how closely the sample data statistically correlates to the population. In practice, failure rates are usually quoted at a confidence level of 60%. If the sample relates to an exponential distribution then the chi-square statistic can be used to derive confidence levels. They can also be derived from Weibull plots but special statistical tables must be used which give the probability that a defined number of weak parts remain in the batch.

Failure rates are used in reliability calculations in order to evaluate the probability that an electronic circuit, for example, will successfully function (according to specification) for the required time. The failure rate is usually subdivided according to the relative likelihood of occurrence of the applicable failure modes. Hence in the design process the effect of the short circuit failure mode of a transistor on circuit, subsystem and spacecraft performance can be established. A general source of failure rates is the American RADC document: Mil-Hdbk.-217.

A selection of failure rates is shown in Table 17.2.

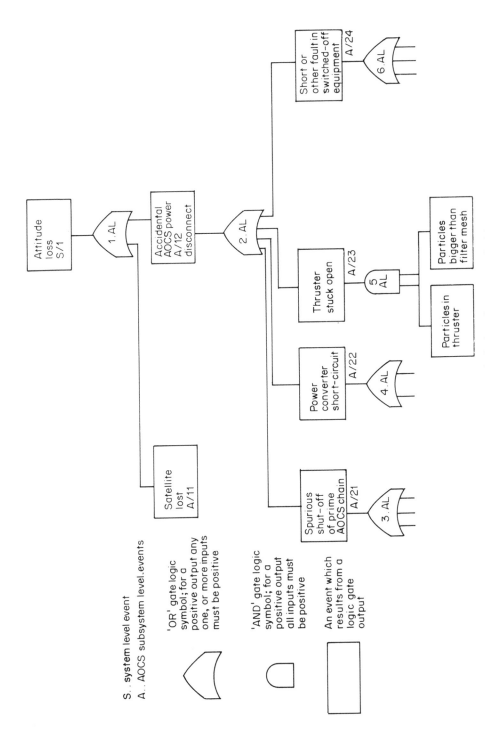

Figure 17.1   Fault-tree analysis (contingency analysis)

Table 17.2 Typical 'failure rate and % occurrence of related failure mode' data for transistors (unit: FITS = failures per 10^9 hours)

Type	Failure rate (FITS)	Failure mode % of occurrence			Notes	
		Short	Open	Drift	Power (W)	Frequency (GHz)
Linear Si	0.8	20	15	65	<1	—
Logic switching Si	0.4	60	30	10	<1	—
Power Si	1.2	20	15	65	1–5	—
Power Si	1.6	20	15	65	5–20	—
Power Si	2.0	20	15	65	20–50	—
RF Si	8.0	40	10	50	<1	<0.2
Microwave bipolar	15	40	10	50	<5	0.2–1.0
Microwave bipolar	100	40	10	50	5–10	0.2–1.0
Microwave bipolar	600	40	10	50	10–30	0.2–1.0
Microwave bipolar	15	40	10	50	<0.5	1–4
Microwave bipolar	200	40	10	50	0.5–5	1–4
Microwave bipolar	1000	40	10	50	5–10	1–4
Microwave bipolar	7000	40	10	50	10–30	1–4
FET linear	1.5	20	15	65		
FET Switching	0.7	20	15	65		
FET RF	5.0	40	10	50		
FET GaAs low noise	50.0	20	15	65		
FET GaAs Driver	360.0	20	15	65		

17.3 RELIABILITY

The reliability of a spacecraft is defined as the probability with which it will successfully complete the specified mission performance for the required mission time. Reliability is not a 'stand-alone' discipline but is dependent on all others.

For a typical telecommunications type spacecraft, such as ECS-1, with a mission life of seven years, a reliability of about 0.8 is currently achievable. This means that there is about an 80% chance that the full specification performance will be achieved for the mission time; or conversely, a 20% chance that it will not. The latter represents the risk that is taken, by the customer for example, and will be manifest in the operational status as *down-time*.

For a 'service' type spacecraft such as a TV communications satellite, down-time, or 'unavailability', constitutes loss of revenue and hence the cost benefits of design improvements to increase reliability can be optimized against their impact on revenue return. This is not a trivial process in practice and can involve a complex Monte Carlo simulation using the spacecraft reliability model to determine the expected availability from failures occurring randomly in time.

Reliability calculations have two main applications:

1. trade-off analysis, usually with other criteria such as cost and mass, to compare, for example, one subsystem with another;

2. assessment of risk, availability, maintenance strategies, etc., using absolute reliability figures.

Reliability is calculated using failure rates, hence the accuracy of the calculations depends on the accuracy and realism of our knowledge of failure mechanisms and modes. For most established electronic parts, failure rates are well known but the same cannot be said for mechanical, electromechanical, and electrochemical parts, or personnel. In modern applications where computers, and their 'embedded' software, are often integrated into the system, the reliability of the software must also be considered. Once again, at this point in time, failure data for software is very sparse and although a number of 'reliability' models exist it will be some years before there is an established industrial standard. Hence, the accuracy of the overall *system* reliability analysis is *dominated* by the lack of precise knowledge in the 'non-electric' and 'software' and man–machine interface areas. Notwithstanding these limitations, which must be carefully considered by the spacecraft engineer, reliability analyses cover significant portions of the spacecraft design and represent, together with safety hazard analyses, practically the only method of assessing technical risk.

Software reliability is a relatively new discipline and is generally defined as a function of indigenous faults within the software, use of the software, and inputs to the software. As far as the user is concerned, software faults usually occur randomly in time; the failure causes are clearly systematic. Typical errors relate to programme interactions, computational value, syntax errors, logic errors and performance (storage, accuracy, etc.). Figure 17.2 shows a software life-cycle management scheme, which indicates the early introduction of PA and the various test phases.

17.3.1 Analysis

As mentioned previously, it is the objective of the 'qualification' analysis/test programmes to verify that the spacecraft design limits and controls the effects from the 'identified' failure mechanisms such that the mission will be successful—with a certain probability. Unfortunately, all variables cannot be analyzed or tested; occasionally manufacturing and materials exhibit high variability and so failures still occur. The latter are considered to occur randomly in time—by 'chance'. Thus, since we are considering only random variables, probability theory can be applied and relatively simple mathematics can be used to evaluate reliability.

It is very important that the spacecraft engineer realizes that, in general, reliability calculations involve only random failures, and constant failure rates. Hence the following functions are *essential* to ensure that all *non-random* failure mechanisms are removed from, or controlled by, the flight design and hardware:

1. a rigorous *qualification* programme;

2. expert and detailed *investigation of all problems* to determine their cause, and rectify accordingly;

3. rigid *configuration control* to ensure proper design and hardware modification and analysis.

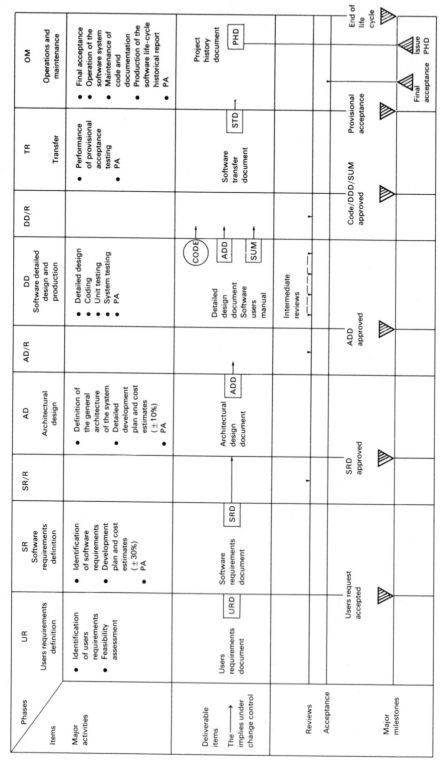

Figure 17.2 Software life-cycle management scheme

17.3.2 Probability distributions

Reliability is defined as the probability that a product will operate successfully for a given time. The relative frequency with which particular events occur is indicated by a *probability distribution*; this is used to define the relative frequency of 'the arrival times at an unsatisfactory state' of the product. For example, the failure rate may be constant, increasing, decreasing, or a combination, depending on the failure mechanisms in operation, i.e. the distribution of arrival times at the failure condition.

Four probability distributions will now be briefly discussed. Binomial and Weibull distributions are applicable in Figure 17.3(a) (regions 1 and 2) and in Figure 17.3(b) (regions 1, 2, and 3).

The binomial distribution relates to a sequence of mutually independent events and takes the form $(p + q)^n$ where p is the probability of a random event, $q = 1 - p$, and n is the number of trials. It is applicable when an event has only two possible outcomes and is thus used for single-shot evaluations such as solid apogee boost motor firings. It is also used to compute the reliability of complex systems and the sensitivity of the overall reliability to various degrees of redundancy.

The Gaussian (Normal) distribution can be used when the failure rate is increasing as a function of time. The main applications in Reliability are as follows:

1. the relationship of a variable operating characteristic to a set of limits;
2. wear-out life of parts;
3. failures due to extended periods of operation.

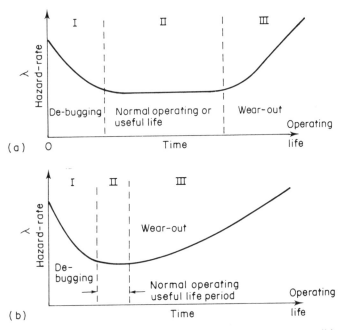

Figure 17.3 Failure rates versus time: (a) electronic parts; (b) mechanical parts

The distribution is described by two parameters; mean and standard deviation.

The *Weibull distribution* is described by three parameters, relating to scale, shape and location. It is extensively used in the analysis of experimental data—a process which is greatly facilitated due to the ready availability of Weibull graph paper that enables the Weibull parameters to be deduced from the plots.

This distribution is used in the evaluation of parts 'burn-in' data, fatigue, and other wear-out data.

The *(negative) exponential distribution*, which is a special case of the Weibull distribution, is used extensively when the failure rate λ is constant and failures occur randomly in time, e.g. electronics. It is described by one parameter, failure rate, and is applicable to region 2 in Figure 17.3(a).

The reliability function takes the form

$$R(t) = e^{-\lambda t}$$

and the failure density function is

$$f(t) = -\frac{dR(t)}{dt} = \lambda e^{-\lambda t}$$

The reliability of elements within a system can thus be computed and the system reliability calculated by considering the *functional way(s)* by which the elements are connected. Reliability block diagrams or models can be constructed on a functional basis which incorporates redundant elements and how they are to be brought into operation.

17.3.3 Reliability enhancement techniques

Redundancy of functional elements is used to *counteract* the effects of failures on the mission: the two main types are *standby* and *active*.

In *standby redundancy*, the redundant element is switched into operation when a failure occurs in the prime unit; it is often not energized prior to operation, and in this 'dormant' state it is assumed to have a failure rate equal to one tenth of its energized rate. A special case which is used increasingly in telecommunication-type satellites is *ring redundancy*. This requires a large number of three or four port switches which enable any transmitter to be switched to any channel.

Active redundancy involves all elements being operational for the entire mission, in such a way that the total load or stress is shared, and thus reduced in proportion to the number of redundant elements. *Voting* redundancy is a special case in which N elements are required to ensure the required performance, but M elements are available, where $M > N$.

Depending on the mission requirements, any mixture of types of redundancy may be used. The impact of different redundancy configurations can be established by carrying out a *sensitivity* analysis. This usually involves the use of the binomial distribution to evaluate the individual probabilities of partially failed configurations; the total number of states is simply the sum of the binomial coefficients. A comparison of the above 'individual' probabilities with the criteria for mission success permits selection of the optimal redundancy design.

With the prospects of multi-spacecraft operation to achieve a particular mission, such as cluster satellites and inter-satellite links, the concept of redundancy being provided by physically detached hardware must be considered. In this scenario maintenance aspects, i.e. repairability and replacement, must also be modelled. The assessment of risk is thus becoming more difficult, and more necessary, and resort must be made to simulation to overcome the complexities and assumptions of analytical techniques.

A summary of redundancy models is given in Figure 17.4.

Radiation shielding is used to enhance reliability.

The possibly catastrophic effects of solar flare activity on man are beyond the scope of this book. However, its effect on electronics, solar cells and external thermal control paints and materials is considered. The 'shielding effect' of the spacecraft structure, equipment walls, printed circuit boards, etc. is evaluated by carrying out a *sector analysis* of the entire

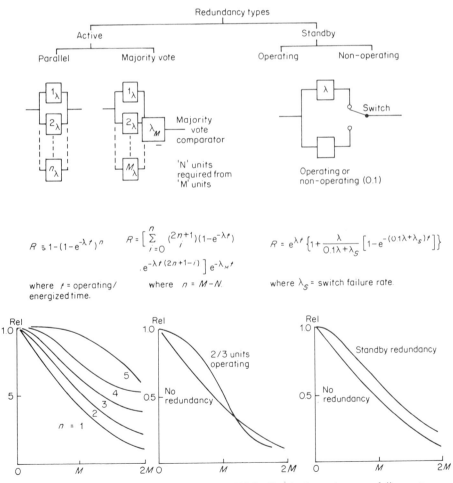

Figure 17.4 Redundancy models. M = MTBF; all units have the same failure rate

spacecraft. In this process the reduction of the incident radiation by the on-board materials is assessed at the location of the sensitive part; this reduction in energy can be three orders of magnitude. Hence careful layout may preclude the need for extra shielding and provide more mass for payload.

Depending on the spacecraft orbit it will be exposed to varying doses of, most significantly, electron and proton radiation. The major effects are bulk atomic displacement and surface ionization, with the main long-term consequences being caused by atomic displacement. The actual failure mechanisms concerned are decreased carrier lifetime, mobility and concentration, etc. Cosmic rays are also problematic in that they can cause changes of state in small-geometry technology such as microprocessors and large memories. These changes, termed single-event upsets (SEU), are manifest as temporary bit errors or permanent latch-up, and are the result of high local ionization caused by individual high-energy heavy ions. The effects can be extremely serious since, with autonomous on-board control, incorrect commands can be given. Shielding is ineffective with this problem; the only method of protection is careful system design (e.g. frequent post-mortem checks) and part selection. On manned space vehicles certain technologies can be prohibited for safety critical functions.

The two main methods of avoiding electron and proton radiation problems are the use of *radiation hard* parts, and metal *shields*. Dose–depth curves exist which indicate the 'stopping power' of various thicknesses of, for example, aluminium, for various increasing strengths of incident radiation. This information, together with the knowledge of the radiation sensitivity of the part measured during the qualification programme, enables a safe and reliable design to be made. A short note of warning is necessary. When charged particles scatter off each other as in the case of electrons impacting shielding, for instance, then a secondary emission of X-rays is produced which also has a damaging effect. This phenomenon is termed *Bremsstrahlung* (German for 'braking radiation') and represents the lowest level of radiation which can be achieved by shielding. In general, Bremsstahlung is more significant for light, high-energy particles and high-density shield materials.

Appendix 2 contains radiation sensitivities for some part types, and a note on radiation units.

De-rating, in order to reduce the stress levels in Parts, is used as a mean of improving reliability.

Parts manufacturers issue *data sheets* which specify the performance of the particular part; this is termed the *rated* performance and represents the maximum stresses to which the part shall be subjected in order to provide a nominal operational life. The life and reliability aspects are often not specified or warranted and hence the evolution of the *Hi-Rel* part which attempts to define such characteristics by very careful material selection, process control and test, and the identification of all failure mechanisms.

To ensure, to the maximum extent possible, that parts will not fail when used in safety critical functions or during long, non-maintainable space missions they are utilized in a *de-rated* condition. The actual amount of de-rating necessary depends on the mission requirements, and economic considerations. However, many procurement authorities specify firm de-rating requirements and stipulate that their specified failure rates are valid only if the de-rating rules, or better, are applied. The failure rate can be reduced if the stress level is below the required de-rating level but in most cases the cost and time to justify such marginal gains is prohibitive. Typical de-rating data is shown in Appendix 3.

17.3.4   Reliability test/demonstration

Reliability testing/demonstration is rarely carried out nowadays in space programmes due to the high cost and long durations involved. It is, however, effectively carried out at 'part' level by considering the cumulative test hours accrued by each part type during life-test programmes. Thus in the American ER-MIL (established reliability) system, passive parts, e.g. resistors, can be procured against a *specified failure rate*. The usual objective of a reliability test is to measure the failure rate, or mean time between failures (MTBF), of a piece of hardware such that the test results are statistically significant. Such a test has to be extremely carefully designed and implemented and all failures have to be investigated to primary cause, i.e. failure mechanism. The test should be continued until at least one failure occurs, and 'truncated' test methods have been developed, such as the AGREE and MIL test methods to minimize test times. Replacement and/or repair of failed elements must be decided and all interfaces and assumptions carefully considered in the statistical calculations to decide the test result. It must be remembered that a success or failure may determine whether a contractor is paid and hence all test criteria, and all possible test outcomes, must be agreed before the test commences.

As an example of the duration of, and investment necessary for, a reliability test the following points are made. Consider a spacecraft equipment that is claimed to have an MTBF of one million hours (equivalent to a failure rate of one thousand fits) with a confidence level of 60%. In order to demonstrate this claim, the contractor would have to test the equipment for 916 thousand hours* with no failures or just over 2 million hours* with one failure. The cost of retaining manpower and test facilities to carry out such a test programme can well be imagined but if the impact of the equipment failure is sufficiently large, or relates to manned safety, then the test may be necessary. The spacecraft engineer should be aware that such testing is not trivial and other methods of risk evaluation and containment, such as the use of Hi-Rel parts and analytical attempts to keep the design point above the stress/strength envelope, are usually used.

17.4   THE USE OF QUALIFIED PARTS

When a Qualification Authority, e.g. European Space Agency, Goddard Space Flight Centre, exists, then it usually issues and maintains a Qualified Parts List (QPL). In general the qualification criteria are similar for the various bodies.

The spacecraft engineer is constrained to select parts from a Preferred Parts List (PPL), which is compiled from QPLs. The list of parts that are actually needed to design the particular spacecraft is termed the Declared Parts List (DPL).

The problems really begin when parts are required which are not on the PPL. It is then necessary to evaluate, or qualify, the 'performance', for the particular application, and the 'technology' for its failure mechanisms. Outputs from the qualification programme enable the screening test criteria (infant mortality) and the technology wear-out (life) point

---

*Based on: $MTBF = 2n/\chi^2(a; 2r + 2)$
   where: n = test hours;
                    r = number of failures;
                    a = confidence level of the χ^2 distribution.
           χ^2 = chi-square statistic.

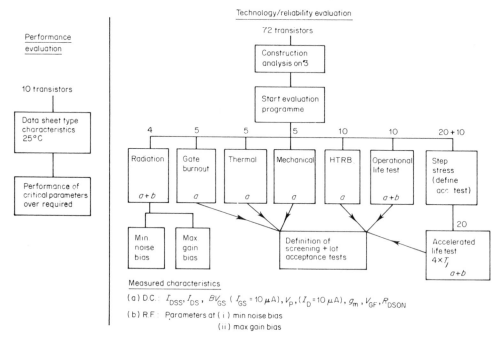

Figure 17.5 Example of a detailed evaluation programme for GaAS FETS (82 transistors)

to be defined (see Figure 17.3). The reliability (risk) calculations are thus dependent on using properly qualified and screened parts.

A typical evaluation/qualificaton programme is given in Figure 17.5. Such a programme often proceeds in parallel with the main project, thus introducing additional risk, cost and schedule constraints.

After (qualified) parts are manufactured they are *all* subjected to 'screening' or 'burn-in' tests to verify that they contain no indigenous or 'infant-mortality' defects (see Figure 17.3). It is possible to integrate the screening tests within the evaluation/qualification programme and thus use some of the parts directly for flight equipment manufacture.

17.5 MATERIALS AND PROCESSES

Probably the most underrated aspect of spacecraft engineering is that related to the performance of materials and processes.

The prime properties under consideration are strength, corrosion resistance, electrochemical interaction, outgassing, compatibility with other materials, toxicity, odour, flammability and offgassing. The latter four properties are specific to manned missions and relate to safety aspects.

One of the major problems associated with using materials in a vacuum is 'outgassing'. The criterion for accepting a material is that its outgassing shall not exceed '1% (... of its ...) total mass loss (TML) and 0.1% collected volatile condensible materials (CVCM)'; the latter refers to the material emitted from a sample at $+125°C$ and collected

Table 17.3 List of prohibited and non-preferred materials

1. All adhesives must be 100% solid
2. Polyvinychloride backing tapes
3. Cellulose, paper, fabric, etc.
4. Varnishes and coatings which rely on solvent evaporation for hardening.
5. Canada balsam; organic glasses in high-precision equipment
6. Direct space exposure of most oils and greases
7. Graphite—is an abrasive in vacuum
8. Cadmium, zinc (whisker growth)
9. Paints should be avoided where possible
10. Polyvinychloride and acetate; cellulose and acetate, plastic films
11. Potting should be avoided where possible
12. Polyester laminates
13. Polysulphide rubbers; rubbers containing plasticizers; chlorinated rubbers.
14. Polyvinyl chloride (PVC) thermoplastic, polyvinyl acetate butyrate, many polyamides.

on a surface at $+25°C$ over a 24-hour period. This criterion is only an indicator of acceptability.

A large data store exists which contains a wide spectrum of information on materials and processes. The job of the materials engineer is to 'fit' materials and processes to the required task within the environment concerned (see Chapter 2). A clear understanding of failure mechanisms and modes is required and 'evaluation' programmes have to be designed and implemented. Also, standards have to be developed, such as for soldering, welding, crimping, material plating, outgassing tests, surface property (absorptance/reflectance/emittance) measurements.

Tables 17.3 and 17.4 contain a listing of non-preferred materials, and some information on materials 'stress/performance' characteristics.

17.6 SAFETY

The overall objective of the safety programme is to ensure that accidents are prevented and all 'hazards', or threats, to the man, the system and the mission are identified and controlled. Safety requirements apply to all programme phases and embrace ground and flight hardware, software and documentation. They also endeavour to protect man from 'man-induced' hazards.

For *unmanned spacecraft* safety is largely a matter of industrial protection and involves such aspects as handling toxic fluids, protection against high-voltage and r.f. supplies, transportation and storage of flight equipment etc.

In the case of *manned spacecraft safety is a severe design requirement and compliance must be demonstrated prior to launch*. Hazards are classified as '**catastrophic**, *critical*, or marginal' depending on their consequences (**loss of life**, *spacecraft loss, injury*, damage, etc.). For example, functions that could result in critical or catastrophic hazards must be controlled by two or three independent inhibits (barriers), respectively. Premature firing of a liquid propellant propulsion system would probably constitute a catastrophic hazard; the designer therefore has to incorporate three independent propellant flow-control

Table 17.4   Stress/performance characteristics of materials

Parameter / Material	Main categories	Vacuum	Particulate radiation	Ultraviolet radiation	Temperature — High	Temperature — Low	Thermal cycling
Adhesives	Epoxies, phenolic, polymethanes, silicones, cyano-acrylates	Outgassing	Outgassing increased	Optical adhesives darken. Outgassing increased.	Causes degradation 300°C max. for polyimide epoxy 170°C max.	Hardening embrittlement	Unmatched expansion coefficients cause failures
Glasses	Silicates, sapphires, fluorides, poly-styrene, acrylic, etc.	Contamination danger only	Most harmful (10^3 Rad.)	Harmful	Thermal shock main problem	Thermal shock main problem	
Lubricants	Hydrocarbons, silicones, esters, MoS_2, WSe_2, Pb	Evaporation, dry-off creep contamination	Metal screening usually effective		Accelerated evaporation etc.		No major problem
Plastic films	Polyolefins, polyesters, fluorinated plastics, etc.	Film stiffeners; contamination	Deformation embrittlement discolouration		Degradation	Embrittlement	Damaging to metalized films, metal film detachment
Potting compounds	Epoxies, silicones, polymethanes	Contamination corona	Minor problem	Minor problem	Chemical degradation	Shrink rigid increase; internal stresses increase.	Cracking debonding
Reinforced and thermosetting resins	Epoxies, phenolics, melamise, polyesters	Outgassing corona contamination	No problem	No problem			can be a problem due to anisotropy of rein-forced plastics
Rubbers	Polybutadiene, poly-chloroprene, acrylics, nitrile etc.	Outgassing from additives and de-polyurization of base polymer. Contamination	Will harden or soften	Will harden or soften	Decomposition	Hardening, stiffening crazing, crushing	See high/low temperature
Thermoplastics	Polyamides, acetal, polyolefins, acrylics, etc.	Degradation due to outgassing of stabilizing additives. Con-tamination	Discolouration, outgassing, hardening		Softens	Hardening, embrittlement	
Paints	Epoxy, silicone etc. binders, ZWO, TLO_2, Al, C for white/black pigment	High outgassing	Absorptance severely affected; embrittlement		Degradation, paint flaking	Minor problem	Degradation of paint flexibility

devices, in series, in the propulsion system. There is an increase of mass, instrumentation, cost, and testing complexity; and a consequent decrease in mass available for carrying (revenue earning) payload. This is a major disadvantage of launching spacecraft on manned orbiters, but it may be offset by other considerations.

Safety and reliability are complementary. Safety relates to the protection of man and investment; reliability relates to success of the mission. The spacecraft engineer should appreciate that spacecraft must be designed to provide adequate safety and reliability, with strict controls on the build and testing. It is almost impossible to modify a spacecraft, once it is built, to meet the safety requirements. As an example an Intelsat satellite was refused permission to fly on the Space Shuttle because it did not comply with the safety requirements; design rectification would have been too expensive so an expendable launcher (unmanned) was finally used.

17.7 CONFIGURATION CONTROL

Everything on a space project, from specifications and batteries to change requests, is represented by a number. These numbers are written on pieces of paper, which are also allocated numbers, and at any particular time the design or build can be defined by a listing of the appropriate numbers. This listing is termed a 'Baseline', and in theory anybody could build a spacecraft from the paperwork identified by the Final Design Baseline.

A typical communications-type satellite contains over three hundred equipments, is designed, manufactured and tested by thirty or more different companies, and is subjected to over five hundred different design, hardware, test and delivery reviews. The specification tree probably contains over a hundred specifications and 'other' documentation includes four to five times that number. During the project's life a total of over two thousand changes will have been raised and processed, each affecting some or all of the above documents.

The problem is one of 'defining, synchronizing, and controlling' all the documents and their updates such that all Baselines represent a design or build standard that is up to date and consistent, both at its internal and external interfaces. The solution is provided by the configuration control system which requires 'accountability' between all documents and 'reconciliation' between all baselines. Implementation is via 'change-control' and 'interface-management' systems.

As an example of the necessity of such a system to the spacecraft engineer, it would be very serious if a long and expensive analysis of, for instance, the functioning of the spacecraft's on-board processor was carried out using drawings and performance documents that were out of date!

17.8 BUILD AND VERIFICATION

17.8.1 Build

Typically, the flight-standard design is approved ('frozen') at the Critical Design Review and permission given to commence the manufacture of flight hardware. The 'Build

Baseline' is thus initiated. As the build progresses, problems are encountered which require expert judgement by spacecraft engineers to ensure that the fully built spacecraft will function according to the original design specification.

Many of the design analyses can be invalidated during the manufacturing phase unless proper quality controls and standards are used. For example, failure rates could become incorrect due to contamination or bad soldering introducing rogue degradation failure mechanisms; RF losses in a transponder could be unacceptably increased due to incorrect assembly of connectors and cables. In order to represent the design exactly in terms of hardware, every facet of the manufacturing and integration activity is planned, monitored, and approved after satisfactory completion, by the quality engineers.

17.8.2 Verification

At the critical design review a verification matrix is approved which identifies how every design and mission requirement as stated in the specifications, will be verified. The verification is normally by analysis and/or test.

Prior to the implementation of any verification aspect, the build baseline is compared with the design baseline; all differences must be satisfactorily explained (or 'reconciled' to use the jargon) before the test can proceed. The test programme is a verification activity and *when all verification activities have been successfully completed the spacecraft is qualified for flight*.

Thus the long and exacting spacecraft engineering process is finalized with the assurance of a fully verified spacecraft (product) on the launch pad.

17.9 ACKNOWLEDGEMENT

The author wishes to acknowledge the permission of Dr A. Holmes-Siedle to use material from his 'Radiation' publications.

Appendix 1 THE ARRHENIUS EQUATION

The *Arrhenius equation* was originally formulated by J. J. Hood and demonstrated by Svante Arrhenius as being applicable to most kinds of reactions. It describes the effect of reaction temperature on the rate of a chemical reaction.

When applied in 'parts' engineering the reaction rate formula used for the chemical and physical processes causing strength degradation is

$$R = A \exp(-E_a/KT) \tag{1}$$

The temperature–lifetime relationship is given by

$$t = C \exp(E_a/KT) \tag{2}$$

where: A = a constant
C = a temperature-independent constant
E_a = activation, energy (see Table 17.5)

Table 17.5 Some typical activation energies

Part type	E_a (eV)
1. Silicon semiconductors	
• surface charge build-up (bipolar)	1.0
• surface charge build-up (MOS)	1.2
• electro-migration	0.6/1.2
• corrosion	0.3/0.6
2. Integrated circuits: (MOSFETS)	
• threshold voltage shift	1.2
3. GaAs microwave transistors	
• contact metal migration	2.3/2.7
4. Carbon comp. resistors	0.6

$$K = \text{Boltzmann's constant}$$
$$R = \text{reaction rate}$$
$$t = \text{time}$$
$$T = \text{absolute temperature}$$

For a particular activation energy (failure mechanism), equation (2) reduces to

$$\log(t_1/t_2) = T_2/T_1 \tag{3}$$

This enables accelerated test times to be calculated. For example, a screening test of 240 h at a junction temperature of 200°C is equivalent to 4368 h at 100°C, the temperature normally used during orbit operation. Hence the screening test is roughly equivalent to six months in orbit.

Appendix 2   PARTICULATE RADIATION DATA

The data given is approximate. The changing of one process step, or even part of it, can have a significant effect on the radiation tolerance of the technology.

1   Radiation-sensitive characteristics of some part types

Integrated circuits

(a) Threshold voltage shifts in the negative direction resulting in an absolute increase for P-MOS and a decrease for N-MOS.

(b) Supply current: rapid increase for CMOS as soon as the threshold voltage of the N-MOS reaches zero volts.

(c) Propagation delay: increases due to increased channel resistance of the P-MOS.

The following ionization dose sensitivities apply, the range referring to the values at which the 'most' and 'least' sensitive type shows significant degradation:

N-MOS   800–10^4 rad(Si)

C-MOS   5000–10^7 rad(Si)

Microwave parts: PIN diodes

Increase in surface leakage current. Fundamentally bulk damage sensitivity (mesa construction) and hence some effect (decrease) on lifetime. No significant problem for seven-year geostationary orbit—mainly ionization radiation.

Microwave transistors

D.c. current gain and collector–base breakdown voltage decrease; collector–emitter breakdown voltage increases. Transistor gain (S21e) is affected.

   Degradation can be expected at 5×10^5 rad(Si) but each part type should be individually considered.

Operational amplifiers

Main degradations are loss of gain and alteration of offsets.

A/D and D/A converters

Calibration drifts as a function of irradiation.

2   Radiation units

Ionizing dose

This is the energy per unit mass element deposited by a beam of particles in an array of atoms of a given material. The receptor material must be specified. Silicon is used as the reference material for spacecraft and the dose is thus quoted as 'rad(Si)'.

$$1 \text{ rad} = 10^2 \text{ erg/g}$$

$$1 \text{ gray} = 10^4 \text{ erg/g} = 1 \text{ joule/kg}$$

Conversion

In order to convert radiation flux (number of particles passing through a unit area in unit time) to ionizing dose, a rule of thumb is

(a) 1 rad $= 3 \times 10^7$ electrons/cm^2.

(b) 1 rad $= 4 \times 10^6$ protons/cm^2.

These apply to energy levels of 2 MeV for electrons and 30 MeV for protons.

3   Radiation equivalence

Qualitatively, the effects of all types of ionizing radiation are the same. Providing that the same energy arrives at a particular location the degradations etc. which develop as the result of increasing flux will be the same for all radiation. Hence, in the laboratory or test house, we can use X-rays to simulate electrons, and so on.

Appendix 3   DE-RATING

De-rating is the reduction of electrical and thermal stress of a component in order to increase its useful lifetime. The de-rating factor is usually given as a percentage of the maximum value of the parameter considered, these maxima being defined in the procurement specifications.

1   Linear integrated circuits ($T_{a\,max} < 85°C$)

Supply voltage	of maximum specified rated value	80%
Input voltage	of maximum specified rated value	70%
Output current	of maximum specified rated value	80%

2   Fuses

Fuses shall be avoided wherever possible, but when needed cermet fuses are perferred.
The largest fuse rating compatible with the source current shall be used.
Absolute maximum allowed ratings:
  65% of the nominal specified current
  100% of the maximum specified voltage after blow.

Wire link fuses should only be envisaged for short-time application and on special request.

3 Transistors

De-rating factor of the minimum specified reverse voltage:	65%
De-rating factor of the maximum specified current	75%
Power de-rating factor of the maximum specified value:	60%
Maximum junction temperature:	115°C

Worst-case analysis: Drift to be considered

Leakage currents:	5 times the specified maximum.
Breakdown voltages:	5% decrease of the specified minimum.
Forward voltages:	10% increase of the specified maximum.
Gain:	25% decrease of the specified minmum.
	25% decrease of the specified maximum.
Saturation voltages	
(Vcesat, Vbesat):	10% increase of the specified maximum.

Currents and voltages shall be maintained at levels compatible with the safe operating area.

For transistor operating *without* heat sink:

$$P_{\text{de-rated}} = 0.6 \frac{(115 - T_a)}{90} P_{\text{max}} \qquad T_a > 25°C$$

$$P_{\text{de-rated}} = 0.6 P_{\text{max}} \qquad\qquad T_a \leq 25°C$$

For transistor operating *with* heat sink:

$$P_{\text{de-rated}} = 0.6 P_{\text{max}} \qquad \text{if } T_c \leq 115°C - 0.6\,(175 - T_m)°C*$$

$$P_{\text{de-rated}} = \frac{115° - T_c}{175 - T_m} P_{\text{max}} \qquad \text{if } T_c > 115° - 0.6\,(175 - T_m)°C*$$

where
P_{max} = maximum rated power at 25°C
T_a = ambient temperature
T_c = case temperature
T_m = max. specified case temperature for max. rated power

---

* Maximum junction temperature is fixed to 175°C for the calculation of the derated power. For lower specified maximum junction temperatures additional de-rating shall be foreseen.

4 Relays

Contact voltage	No de-rating
Coil voltage	No de-rating
Number of operations:[1]	50% of qualification number
Contact current:[2]	Lower than 75% of specified maximum rated current (resistive load)
Contact current transients:[3]	Shall be limited such that $I^2t \leq 6(I_{max})^2\ 10^{-5}$ (A^2 s)
Degradation to be considered for the worst-case analysis:	An increase of four times the limit specified for contact resistance shall be used for more than 1000 switching operations[4]

[1] If the number of operations is below 10 cycles, the specified maximum rated current can be used.
[2] It is not recommended to use a relay with a contact current below 20% of the maximum rated current, except for low-level application (≤ 1 mA)
[3] In the case where the pulse duration is low or equal to 10 μs, the transient current shall be limited to $4I_{max}$.
[4] For a number of switching operations of less than 1000 an increase of twice the contact resistance will be used.

INDEX